# THE LOGNORMALITY PRINCIPLE
## AND ITS APPLICATIONS IN
## E-SECURITY, E-LEARNING AND E-HEALTH

## SERIES IN MACHINE PERCEPTION AND ARTIFICIAL INTELLIGENCE*

ISSN: 1793-0839



---

This book series addresses all aspects of machine perception and artificial intelligence. Of particular interest are the areas of pattern recognition, image processing, computer vision, natural language understanding, speech processing, neural computing, machine learning, hardware architectures, software tools, and others. The series includes publications of various types, for example, textbooks, monographs, edited volumes, conference and workshop proceedings, PhD theses with significant impact, and special issues of the International Journal of Pattern Recognition and Artificial Intelligence.

*Published*

Vol. 88: *The Lognormality Principle and its Applications in e-Security, e-Learning and e-Health*
by Réjean Plamondon, Angelo Marcelli and Miguel Angel Ferrer

Vol. 87: *Fuzzy Systems to Quantum Mechanics*
by Hong-Xing Li

Vol. 86: *Graph-Based Keyword Spotting*
edited by Michael Stauffer, Andreas Fischerand Kaspar Riesen

Vol. 85: *Ensemble Learning: Pattern Classification Using Ensemble Methods (Second Edition)*
by L. Rokach

Vol. 84: *Hybrid Metaheuristics: Research and Applications*
by Siddhartha Bhattacharyya

Vol. 83: *Data Mining in Time Series and Streaming Databases*
edited by Mark Last, Horst Bunke and Abraham Kandel

Vol. 82: *Document Analysis and Text Recognition: Benchmarking State-of-the-Art Systems*
edited by Volker Märgner, Umapada Pal and Apostolos Antonacopoulos

Vol. 81: *Data Mining with Decision Trees: Theory and Applications (Second Edition)*
by L. Rokach and O. Maimon

Vol. 80: *Multimodal Interactive Handwritten Text Transcription*
by V. Romero, A. H. Toselli and E. Vidal

Vol. 79: *Document Analysis and Recognition with Wavelet and Fractal Theories*
by Y. Y. Tang

Vol. 78: *Integration of Swarm Intelligence and Artificial Neural Network*
edited by S. Dehuri, S. Ghosh and S.-B. Cho

*The complete list of the published volumes in the series can be found at
https://www.worldscientific.com/series/smpai

Series in Machine Perception and Artificial Intelligence – Vol. 88

# THE LOGNORMALITY PRINCIPLE
## AND ITS APPLICATIONS IN E-SECURITY, E-LEARNING AND E-HEALTH

Editors

Réjean Plamondon
Polytechnique Montréal, Canada

Angelo Marcelli
Università di Salerno, Italy

Miguel Ángel Ferrer
Universidad de Las Palmas de Gran Canaria, Spain

World Scientific

NEW JERSEY • LONDON • SINGAPORE • BEIJING • SHANGHAI • HONG KONG • TAIPEI • CHENNAI • TOKYO

*Published by*

World Scientific Publishing Co. Pte. Ltd.
5 Toh Tuck Link, Singapore 596224
*USA office:* 27 Warren Street, Suite 401-402, Hackensack, NJ 07601
*UK office:* 57 Shelton Street, Covent Garden, London WC2H 9HE

**British Library Cataloguing-in-Publication Data**
A catalogue record for this book is available from the British Library.

**Series in Machine Perception and Artificial Intelligence — Vol. 88**
**THE LOGNORMALITY PRINCIPLE AND ITS APPLICATIONS IN E-SECURITY, E-LEARNING AND E-HEALTH**



ISBN 978-981-122-682-3 (hardcover)
ISBN 978-981-122-683-0 (ebook for institutions)
ISBN 978-981-122-684-7 (ebook for individuals)

For any available supplementary material, please visit
https://www.worldscientific.com/worldscibooks/10.1142/12006#t=suppl

Défi

Donnez-moi
une feuille
et un crayon,
je soulèverai
le temps!

Challenge

Give me
a page
and a pen,
I will lift
time!

https://doi.org/10.1142/9789811226830_fmatter

# Contents

https://doi.org/10.1142/9789811226830_fmatter

# Foreword

*George Nagy*

What a relief! A book with fresh ideas about explaining ourselves to ourselves, with very little about Artificial Intelligence. Explaining how we move may be easier than explaining how we think, but no less important. Much of the time we do both. We move our heads, eyes, tongue and jaw, limbs, and our wrists and fingers. Voluntary motion is required for hunting and gathering, for observing our environment (real or virtual), and for communication — speech, gestures, writing and typing. Motion is the essence of all athletics and sports. Some of us may also constantly wiggle our toes, but you must look elsewhere for explanation.

I will attempt three tasks in this Foreword. First, I will sketch the elements of Réjean Plamondon's Kinematic Theory for the uninitiated, in the hope that a whiff will encourage them to peruse more scholarly reports here and elsewhere. Second, I will briefly recount the circumstances of my long acquaintance with two of the editors. Third, I will tell you, for whatever it is worth, what aspects of the research presented here I expect to flourish or languish. That is the plan. So back to brain and brawn.

The Central Nervous System processes sensory stimuli and engenders both thought and movement. It generates command sequences of voltage impulses called *action potentials*. Nerve impulses propagate relatively slowly — roughly at car speed, depending on the axonal diameter — by means of electro-chemical depolarization of nerve cell membranes. The commands hop from neuron to neuron through synapses. A *synapse* connects the axon of the pre-synaptic cell to a dendrite or the soma of the post-synaptic cell.

The motor control commands are conveyed to the muscles via conduits of nerve fibers. The upper neurons form an orderly arrangement in the spinal cord. They connect through interneurons to the lower neurons that innervate the muscles via neuro-muscular synapses. Central commands originate in the motor cortex, while local commands are initiated closer to their destination. Each command triggers the contraction of some of the hundreds of skeletal muscles that enable targeted movements.

The trajectories of movements generated by normal healthy individuals are smooth and energy efficient. The corresponding velocity profiles display consistent asymmetric bell-shaped patterns. Over the past century, many devices were developed to record the position of various body parts as a function of time. The currently preferred recording platform for fine human motion is the digitizing tablet (now marketed primarily for human-computer interaction). Tablets with styluses allow sampling and recording motion constrained to a two-dimensional plane.

Millions of signatures and lines of text, real and synthetic, in many languages and scripts, have been saved in annotated databases and made available for experimental research. There are also databases of mouse, trackball and game controller manipulation, eye movement and, increasingly, 3D motion captured by wearable accelerometers or multiple cameras. Some motion, like that of the tongue, remains elusive.

Copernicus tracked the heavenly bodies, Newton calculated the acceleration of falling apples, and Napoleon had good approximations to the trajectory of cannon balls. Models of motion are an integral part of autopilots, self-driving cars, space exploration, torpedoes and guided missiles. We have been far more successful in analyzing and predicting vehicle and projectile motion than live movements. Nevertheless, the chapters that follow indicate that we are making good progress in quantifying the kinematics of human motion.

*What is kinematics?* I must avoid confusing it with kinetics, the other subfield of dynamics. *Kinetics* relates motion to force. *Kinematics* bears only on the geometry of motion, ignoring the cause of movement (force). Its fundamental equations relate position, velocity and acceleration (and higher derivatives of displacement, like *jerk* and *snap*), to time. *Kinesiology* embraces the entire study of the mechanics of body movements. This book focuses on kinesiology and kinematics. The Kinematic Theory (formerly "a" kinematic theory) supports the claim that the LogNormality Principle governs almost all voluntary neuro-muscular activity.

*What is Lognormal?* Here is my take. It is the functional form $e^{-\ln t^2}$, with appropriate shape, scaling and delaying constants, of the impulse response of an elaborate model of many connected linear subsystems. The output of any one of these subsystems, *Y(t)*, is the convolution of its impulse response *H(t)* with its input signal *X(t)*. The convolution operator, in turn, is the integral (or sum) over *T* of the product of *X(T)* and *H(t-T)*. The puzzling minus sign aligns the impulse response with a past input whose effect is reflected in the current output. Let us next consider the output of the lumped subsystems.

The key to the Lognormality Principle is avoiding precise specification of the impulse response *H(t)* of each muscular subsystem by postulating an infinite number of elementary subsystems in series. Each subsystem feeds into the next one in response to the same command prompt. In this configuration, the Central Limit Theorem (CLT) assures a Gaussian output in the time domain. Simulations suggest that combining seven responses is close enough to infinity for a good approximation to the desired bell-shaped output. Simulation does, of course, require specifying the subsystems even if the final output is largely independent of these specifications. The referenced Simulink program used second-order damped exponentials with empirically determined parameters.

The complete lognormal model is derived from several plausible assumptions. The addition of a "proportionate" time delay takes into account the coupling between the different subsystems and converts the response of the resulting model from a Normal to a lognormal impulse response that starts at a specified time, rises rapidly, and decays with an infinite tail. The derivation of this expression, published in 2003 as Part IV of a *Biological Cybernetics* series, is not for the faint of heart.

The next step is to set the difference of two (*agonistic* and *antagonistic*) lognormal responses as a *delta-lognormal* expression. This formula characterizes a primitive stroke in one dimension. The agonistic command expression drives a real or virtual pen to trace the line segment. The antagonistic command stops or reverses it. With appropriate parameters a delta-lognormal can provide an excellent fit to the velocity profile of a line segment with possible reversals (called "glitches) at either end. Concatenating lognormal expressions allows modeling almost any trace.

The *sigma-lognormal* formula extends the 1-D model to arcs in two or three dimensions. Each arc is constrained to specified beginning and end angles. No antagonistic action is required to generate chained strokes, therefore six parameters suffice to define each primitive stroke in two dimensions, eight in three dimensions.

To determine whether recorded samples of on-line handwriting fit the lognormal model, to generate synthetic samples of signatures or text, and to discover deviations from normal patterns due to neuromuscular disorders, it is necessary to estimate the parameters from recorded traces. Our understanding of human physiology is not yet deep enough to derive them from fundamental physiological observations.

The first parameter estimation methods were based on the algebraic relationship of the lognormal parameters to easily observable features of the velocity profiles like extrema and inflection points. These features are affected by various types of environmental and instrumentation noise that must be distinguished from perturbed command signals. Eventually optimization procedures were developed to exploit all the time-stamped coordinate data. Segmentation and stroke fitting of a single line of handwriting may require hundreds of parameters.

Strictly 1-D models are used mainly for diagnosing health issues, but the sigma-lognormal model remains part of many tools for handwriting, signature and drawing analysis. The 2-D models work remarkably well for both writing and signatures. Optimal segmentation and fitting of both trajectory and velocity profile, as opposed to either, remains a trade-off. 3-D modeling is catching up, after being long hampered by the unavailability of accurate sensors. Experiments are reported on activity recognition, biometric identification and air-writing.

*Who are the Editors?* The foreworder is obligated to reveal factoids about the protagonists that are still missing from Wikipedia.

*Réjean Plamondon.* I first became aware of Réjean's work in the early nineties when he launched, with Isabelle Guyon, Lambert Schomaker, Mark Liberman and Stanley Janet, Unipen to collect samples of on-line handwriting generated under controlled conditions. The resulting database filled a crying need in the research community and was widely used. A visit to his Scribens Lab introduced me to the topic of this book. Thereafter I made a point of reading his articles and catching his lectures. However, I eventually had to give up keeping up with his prodigious productivity and now resort to active learning. He gives me ample opportunity to question his assumptions on leisurely walks at opportune conferences. The discussion is, of course, more lively when we disagree. We share our love of Montréal and the hinterlands of La Belle Province.

Réjean's most enduring and endearing contribution is undoubtedly his poetry. I hope that it is also the most widely read. Having learned French before English, I enjoy his bold word plays, like in the titles "Portraits Rebelles d'un Cervonaute" (1998), or "Flânerinages" (2017). Besides neologisms, one poem has an impenetrable four-line integral equation, another geometric allusions, and still others are laid out like parts of crosswords. I like it best when he turns lyrical, as in Foule de l'île. Hearing him recite his emanations adds another dimension. I must remember to ask him whether he indites his poems with a pen or pencil, types them, or dictates them. In any case I anticipate browsing other books of his poetry (four at last count), and perhaps also his police novel.

A radical departure from both muscular lognormal and sparsely punctuated verse was a 200-page treatise, Patterns in Physics – Toward a Unifying Theory (2012). Unification here embraces general relativity and quantum mechanics. What Réjean (seconded by his physicist son) proposed is a theory that predicts, according to various mappings and projections in high-dimensional information spaces, the values of all the physical constants (instead of tweaking existing theory to accommodate these constants' observed values). This quasi-Bayesian theory is supported by the analogy of a tensor product with a joint probability with multivariate components, the Principle of Interdependence, and the Principle of Asymptotic Convergence. Like Lognormality, it draws on the Central Limit Theorem to approximate finite sums. Selling a brand new universal theory to the established physics community is no easy task. The jury deliberating its fate is still out.

Réjean somehow also found time for assorted executive, administrative and service activities. He was president or director of several international and Canadian professional societies, organized and conducted conferences and workshops, edited journals and established international collaborations. After serving as department chair, he was appointed Director General (i.e., President) of the École Polytechnique (now Polytechnique Montréal). During his four-year tenure he erected a 32,500 square meter building. The Pavillon Lassonde, nestled on the steep slopes of Mount Royal — the huge park designed by Frederick Olmstead — is truly beautiful, inside and out. It won several awards for appearance, functionality and record energy efficiency.

To save the blushes of the recipient, I reluctantly forbear listing the many honors, awards and prizes in my Plamondonia.

*Angelo Marcelli.* The venerable Professor Luigi Cordella, dean of Italian pattern recognition, introduced his prize student to me at the Paris ICPR, just one year before he earned his doctorate. And now Angelo is a proud grandfather! I am most grateful for his outstanding hospitality on my visits to Naples, Caserta and Salerno, all of which he considers his home towns — along with Venice. I owe him uncountable espressos as well as other sustenance.

After completing his PhD, Angelo joined the CRIAI research center in Portici, and then the Federico II University (founded in 1224!) of Naples. A few years later he moved to the faculty in Salerno named after cybernetics and neural networks pioneer Eduardo R. Caianiello. The University of Salerno has been growing like topsy ever since (correlation is *not* causation). He is a dedicated and talented teacher and an inventive researcher, but he was not able to keep his administrative talents under wraps and was therefore quickly promoted to Graduate school chair and is now Dean of the Faculty.

In recent years Angelo has been the moving spirit behind the biennial Graphonomics workshops. The first, in 1982 at Nijmegen, was devoted to Motor Aspects of Handwriting. He inherited Professor Cordella's devotion to handwriting and devised several successful approaches with evolutionary and genetic programming. Some of his algorithms are in commercial use. He made significant contributions to classifier theory. When his early passion for physiology and medicine revived, he began integrating related themes into his research on handwriting recognition. I am very glad that I can consider him a family friend.

*Miguel Ángel Ferrer Ballester.* I have not met MAFB. Having read some of his excellent papers, I wish I had! I look forward to some opportunity to visit him on Gran Canaria. When I googled him just now, I found that he has many other interests besides biometrics and the topics in this book. If you are curious, you will have no trouble tracing his contributions beyond this tome.

*Why is this book important?* First and foremost, it is appropriate to have a comprehensive and authoritative source of information about processes that engage all of us daily. It behooves us to know something about how we move just as we ought to know at least a little about what we eat, drink and breathe, and what is under the hood and the furnace cover. The thoroughly-refereed chapters of this book, with their lists of references (that I wish had been merged), admirably fill this need.

Have you ever watched a pre-schooler or an octogenarian try to thread a needle? Anything that helps to speed up maturation and slow down aging is welcome. Several chapters examine the related changes that are readily observable in drawing and handwriting and can now be quantified using lognormality. Well-designed feedback based on the observed changes could improve computer-aided learning for the young. Calligraphy has a low profile in most Western countries, but China has recently mandated instruction in calligraphy for all primary school pupils. Feedback based on monitoring hand and body motion could also assist the old. Another application being investigated is customized gesture-based human-computer interface design. A more remote possibility is that statistical data from large collections of selected populations will inform research on developmental and geriatric physiology and its applications to bioengineering and medicine.

Helping normal children, adults and elderlies is fine, but it seems to me that the most important applications of lognormality will be the detection and diagnosis of *abnormal* conditions. The research reported here heralds non-invasive, inexpensive and simple diagnostic tools based on kinematics. Importantly, their results can be instantly calculated on most any microprocessor (unlike, for example, the results of blood tests or Pap smears). The following chapters list many abnormal neuro-muscular conditions that affect the mechanisms underlying handwriting and drawing. The list includes both trauma-based conditions (accidents, athletic injuries) and naturally arising afflictions.

The success of these tools at the triage level does not depend on the physiological validity of its foundations. However, diagnostic use will require further confirmation of its physiological correlates. The worthy role of electromyography in pursuing this objective is included here only by reference. We must, however, keep in mind the difficulty of acquiring instrumentation and conducting appropriate experiments on live patients in engineering departments.

I am less sanguine about the commercial future of on-line handwriting recognition. Handwriting is disappearing from interpersonal communications. Nowadays it is almost exclusively used for notes to oneself — like a shopping or to-do list or an as a reminder of an elusive name or number. The rising alternative is voice or keyboard entry on mobile devices (that also provide cogent suggestions for shopping, entertainment, and most everything else). Many organizations have downshifted data entry via web forms to their clients. Even with perfect handwriting recognition, few individuals will favor entering text by writing instead of typing or speaking. The use of signatures for authentication follows a similar trend. Many point-of-sale and portable credit card machines still require a token signature. Nevertheless the primacy of signatures over other biometric tools and cryptography is fading. Even the conservative Professional Engineering associations shun analog seals and signatures. "Sign in" is an anachronism in its verbatim sense.

This mildly pessimistic outlook may be colored by my personal, local and North-American perspective. In Europe and the Far East governments are funding diverse initiatives to revive and improve handwriting instruction and to develop reliable movement-based personal identifiers. Most tablet-based products — large and small — include some handwriting and gesture recognition app. *Disclosure:* I was almost held back in Grade 1 because of my illegible scrawl. Much later, during three summer stints as a draftsman, I learned to print properly, but I still envy folks with a distinguished and legible script. And in spite of my arsenal of CAD tools, I much prefer drawing diagrams by hand.

Manu-script may or may not thrive, and security and privacy are essential components of the social contract, but for me health comes first. The prospective benefits of applications of Lognormality to the detection of neuro-physical disorders are manifest. There are plenty of initiatives reported here to advance them. Some of my facts and predictions may be off, but I am convinced that my enthusiasm for this compendium will be widely shared.

George Nagy, PhD 1962 Cornell (on neural networks)
Professor Emeritus, Rensselaer Polytechnic Institute

Troy, NY, USA
February, 2020

https://doi.org/10.1142/9789811226830_fmatter

# Preface

*Réjean Plamondon, Angelo Marcelli and Miguel Á. Ferrer*

On May 17, 2018, in the context of the First International Conference on Pattern Recognition and Artificial Intelligence held in Montréal, Canada, under the auspices of Professor Ching Y Suen from Concordia University, the first **International Workshop on the Lognormality Principle and its Applications** was organized. At the end of the symposium, it was decided that a selection of the best proceedings papers would be published under the editorship of Professor Réjean Plamondon, chief organizer of the workshop, with the help of Professors Angelo Marcelli and Miguel Ferrer-Ballester, two long-time collaborators. A few months later an agreement was made with WORLD SCIENTIFIC PUBLISHING to act as the publisher. In total 15 chapters, expanding and detailing the extended abstracts published in the ICPRAI conference Proceedings were finally accepted for publication, after a blind review by at least two independent reviewers.

The book starts with a chapter presenting a historical overview of the gradual developments, implementations and applications of the Kinematic Theory of human movements made over Professor Plamondon's career at Laboratoire Scribens, Polytechnique Montréal. The goal is to show the different pathways that have led to the statement of the Lognormality Pinciple, which is the corner stone of this book. Various projects achieved at Laboratoire Scribens as well as those realized with students and researchers in the context of numerous national and international collaborations over this period are reported and put into perspective. After describing the Kinematic Theory, a focus is made on the development of its Delta- and Sigma-Lognormal models and their associated parameter extraction algorithms. Then the specific applications of the theory are summarized under the headings of e-Security, e-Recognition, e-Learning and e-Health in the visionary perspective of developing Personal Digital Bodyguards. The paper then converges on the statement of the Lognormality Principle and ends by exploring of some emerging fields of applications.

The next two chapters deals with e-Security problems, and particularly automatic signature verification.

Chapter 2 proposes two efficient methods to modify the fluency of dynamic signatures. The main idea is to modify the number of velocity minima or virtual target points and reconstruct the signature with the new virtual target points. If the number of virtual target points is reduced, the fluency is improved, and vice versa. The modification of the virtual target points is accomplished initially by linking the samples of an on-line signature by 8-connected Bresenham's lines to obtain the interpolated trajectory. Then, two methods are proposed for estimating the most perceptually important points from the corners in the trajectory. The first one is a resampling-wise method adopting a lognormal synthetic velocity profile for resampling the 8-connected trajectory. The second one is a recovering-wise method that estimates the virtual target points from the perceptually important points in the trajectory, links them by circular trajectories, and eventually reconstructs the dynamic trajectory. Both methods can be used to generate synthetic skilled specimens. Their efficiency is tested with automatic verifiers using the method for generating synthetic skilled signatures from static 8-connected trajectories with increasing fluidity.

Chapter 3 focuses on modeling the complexity of biomechanical tasks through the usage of the Sigma-Lognormal model for modeling and generating synthetic handwritten signatures in order to improve the performance of automatic verification systems. The paper reports the experimental work for the usage of the Sigma-Lognormal model to predict the complexity of biomechanical tasks on two case studies: 1) on-line signature recognition, in order to generate user-based complexity groups and develop specific verification systems for each of them, and 2) detection of age groups (children from adults) using touch-screen patterns.

The three following chapters introduce the application of the Lognormality principle in the domain of e-Learning involving children with normal development, ADHD and concussion.

Chapter 4 deals with the relevance of the Sigma-Lognormal model as a tool to identify reliable objective and simple parameters to assist teachers in monitoring children's schooling graphomotor progress involved in handwriting along their entire scholarship. Focusing on the identification of the most suitable variables for representing the kinematic and procedural efficiency of children's procedures for cursive isolated words handwriting as taught to French pupils in elementary school, the paper shows the capacity of the Sigma-Lognormal model to reproduce children's cursive writing movements since the early stage of their scholarship at primary school. Furthermore, the potential interest of a new offline variable, the deltaPenlift, that could be estimated by teachers is underlined. The deltaPenlift and the relevance of the signal to noise ratio (SNR), provided by the online Sigma-Lognormal modeling are used to characterize pupils' level of graphomotor expertise and their capacity to respect the cursive handwriting procedure taught at elementary school.

Chapter 5 presents an exploratory study that examines whether the Sigma-Lognormal model can discriminate between the handwriting strokes produced by children with and without Attention Deficit Hyperactivity Disorder (ADHD). Twelve children with ADHD and 12 controls aged 8–11 years were asked to produce handwriting strokes on a digitizing tablet, and the strokes were then analyzed using the Sigma-Lognormal model. The results show that strokes made by children with ADHD reflected poorer motor control, action planning and execution than strokes made by controls. Based on these findings, the authors trained different machine learning models to classify the subjects according to the discriminatory parameters used as features. Although the sample size and data are modest, the results indicate that the Sigma-Lognormal model may be a useful tool in the assessment of ADHD. Among other things, it is shown that using a Support Vector Machine (SVM), the maximum compromise between the specificity and the sensitivity of the classification varies between 80% and 85%.

Chapter 6 reports a pilot study investigating whether and to which extent the Sigma-Lognormal model can detect differences between handwriting strokes produced by youth with mild traumatic brain injuries (mTBI). In the study, thirty-two participants, aged between 6 and 17 years, who sustained mTBI were asked to produce handwriting strokes on a digitizing tablet during two sessions, 1 and 3 months after their injury, respectively. Strokes were reconstructed and analyzed using the Sigma-Lognormal model. Consistent with initial study expectations, the majority of children in the sample showed reduced fine motor control between 1-month and 3-month post-injury, as reflected by changes in the Sigma-Lognormal parameters. Moreover, some parameters show significant differences between the two sets of strokes produced in the two sessions, with the quality of strokes improving over time for ten participants, decreasing for seventeen participants and showing no changes in the remaining four cases. These results are eventually compared to the results of the Post-Concussion Symptom Inventory questionnaire. Altogether, these findings point out the potential benefits of using the Sigma-Lognormal model to assess and monitor objectively the integrity of neuromotor systems following pediatric mTBI.

The next group of three chapters deals with applications for e-Health, and particularly Parkinson's disease, and introduces a new tool for multimodal data acquisition.

Chapter 7 focuses on Parkinson's disease as manifested in handwriting and voice and proposes one single lognormal parameter to evaluate the influence of the illness on both voice and handwriting, as the symptoms affecting these two behaviors have a common origin. The velocity profile obtained to characterize voice between the first and second formant is computed by a spatio-temporal approximation. In handwriting, the velocity profile is obtained from the sampled positions of the pen on a digitizer tablet. The comparison between the parameters of the lognormal model of the two profiles points out similarities between the voice and the handwriting of Parkinson's patients, that are reflected by a single, common parameter: the averaged value of the mean stroke logresponse time sequential difference, which is lower for PD patients than for healthy subjects.

Chapter 8 reports the first experiment testing the possibility of analyzing inertial measurement units (IMU) signals with the Kinematic Theory through its signature. Considering that, when assessing mobility performance in humans, the specific way by which a task is performed is often referred to as its movement signature, the authors hypothesize that if one can determine a normal signature for a task, deviations from that reference signature can potentially be used to identify and quantify mobility deficits. The chapter presents a conceptual framework for assessing turn mobility deficits based on movement signatures derived from the Sigma-Lognormal model, using orientation data captured at multiple segments with eighteen inertial measurement units (IMUs). The concept tested on a small sample of older adults and participants with early Parkinson's disease, suggests that even at early stages of the disease, the complexity and fluidity of the turn signature are impacted, and that some of the parameters of the main lognormal may have the potential of capturing the effect of medication on turning deficits.

Chapter 9 investigates the discriminating potential of the Sigma-Lognormal modeling for an automatized Parkinson's disease diagnosis based on handwriting. The experiments were performed on a freely available dataset PaHaW, which includes 37 PD patients and 38 age and gender-matched healthy control (HC) subjects, and aimed at evaluating the impact on the classification performance of Sigma-Lognormal-based features (alone or in combination with more classic measures). The experimental results confirm that in-air movements carry on non-redundant information with the on-surface movement as well as that some of the most relevant discriminant features are Sigma-Lognormal-based. They also showed that for discriminating automatically Parkinson's handwriting from healthy controls using the Sigma-Lognormal model the top results are obtained by adopting a Support Vector Machines (SVMs) with radial basis function (RBF) kernel.

The next two chapters introduce the applications to the design of Human-Computer Interface (HCI) and art.

Chapter 10 deals with the interest of using the Kinematic Theory to design Human Machine Interface (HCI). Indeed, gesture recognizers usually require a large number of examples to achieve good accuracy which normally requires a series of time-consuming and expensive experiments, e.g. preparing the lab, recruiting participants, collecting and labeling data, and often reporting reviewing boards. In this context, the Kinematic Theory is exploited to easily bootstrap gesture data generation. The synthesized data, in turn, may enable further applications of interest. In this chapter, the foundation of synthetic stroke gestures generation is surveyed i.e. the synthesis of data sequences comprising 2D points and associated timestamps, derived from electronic pens and touch screens. Synthesized gestures are shown not only to perform equally similar to gestures generated by human users, but also they "look and feel" the same. One interesting feature of the synthesized gestures is that they can be used to estimate production time, which is a fundamental measure of performance in Human-Computer Interaction. Ultimately, this work benefits researchers and designers who wish to create gesture-driven prototypes or use the synthesized data to build more sophisticated applications.

Chapter 11 demonstrates how to use the Sigma-Lognormal model to reproduce computationally calligraphic traces, such as found in the art practices of graffiti and various forms of more traditional calligraphy, while mimicking their production process. To this end, a method that allows to reconstruct kinematics solely from the geometric samples of handwritten traces is presented. The kinematics possibly embedded in the data is ignored in order to treat online data and vector patterns with the same procedure. At the heart of this algorithm, a robust method to identify curvilinear shape features based on an analysis of local symmetry axes is developed. These features determine the segmentation of a trace into circular arcs and guide an iterative reconstruction of the input kinematics and geometry in the form of Sigma-Lognormal parameters. This parameterization can be used to generate plausible kinematics for a static input trace, and to exploit parameter variations to generate traces that resemble the ones seen in real instances of human made calligraphy and graffiti.

The next four chapters introduce the most recent tool and algorithmic developments exploiting the lognormality principle to acquire, represent, describe and analyze human movements.

Chapter 12 describes an integrated and updated version of writing movement acquisition system used for many years at Scribens laboratory and referred to as Sign@medic. This new system, called the Lognometer, takes advantage of all the features of the Wacom Cintiq technology to integrate the hardware parts of the Sign@medic system. To this end, the Lognormeter includes the acquisition system, the audio-visual stimulus and the guide sheets in a unique tablet environment. This feature allows users to work directly on the screen of the Cintiq 13HD tablet. The two systems are compared pointing out that the Lognormeter is a more versatile, efficient and affordable tool than Sign@medic. Thanks to an easier integration of new tests, the new system can be used in various domains of pattern recognition and artificial intelligence relying on studies of human movements.

Chapter 13 presents an algorithm for segmenting handwriting movements into strokes, each of which corresponds to a command stored into the motor program, while disregarding those that may depend on local control. Indeed, motor control theories suggest that handwriting movements result from both central and local control, assuming that some strokes of a sequence may not be encoded into the stored motor program but rather generated locally to keep the actual movement as close as possible to the intended one. According to the Kinematic Theory, it is assumed that handwriting is a time superimposition of a sequence of commands issued by the motor system, each command producing a stroke, i.e. a movement with a lognormal velocity profile. Experiments on handwriting samples show that the proposed algorithm detects the same number of central strokes across multiple executions of a handwriting task by the same subject, and this set of strokes also provides a good reconstruction of the movement.

Chapter 14 presents an algorithm based on interval arithmetic to construct envelope models that contain all possible parameter values of a Delta-Lognormal function that can match an observed human movement. Indeed, one important problem with the current parameter extraction algorithm is that they are non-global estimators. The basic idea behind the proposed algorithm is to exploit the isotonicity and monotonicity over an interval to sharply bound the range of the Delta-Lognormal functions within a finite space of its parameters. The new algorithm is described, tested and compared with a previous branch and bound algorithm developed for the same purpose. The simulations show that the proposed algorithm gives better results in terms of speed and accuracy and enables the creation of new tools for the basic study of Delta-Lognormal strokes, as observed in any pointing movements.

Chapter 15 describes a model for representing movements in three dimensions, which naturally extends the Sigma-Lognormal approach to 3D. So far, the Kinematic Theory of rapid human movements has been extensively and exhaustively tested on one-dimensional movements with the Delta-Lognormal model and on two-dimensional movements (and 2.5 D movements) with the Sigma-Lognormal model. The theory has been confirmed in a large number of experimental evaluations, achieving a high reconstruction quality when compared with observed trajectories and providing pertinent features for biomedical applications as well as biometric identification, opening the possibilities to analyze any type of movements under any controlled or uncontrolled conditions. The algorithm is evaluated on two action recognition datasets and an air-writing dataset, demonstrating a high reconstruction quality for modeling rapid 3D movements in all cases. In a final experiment, the different types of original and reconstructed movements are classified using dynamic time warping (DTW).

Finally, since the 2018 International Workshop, many new projects have been launched. We present at the very end Appendixes that briefly report on each of these.

We want to thank the reviewers for their help in making this book, providing useful constructive comments to both the authors and the editors:

Andreas Fischer,
*DIVA Group, University of Fribourg, Switzerland*

Antonio Parziale,
*DIEM Natural Computation Laboratory, University of Salerno, Italy*

Arend W.A. Van Gemmert,
*Louisiana State University, USA*

Aythami Morales,
*BiDA Laboratory, Universidad Autonoma de Madrid, Spain*

Claudio De Stefano,
*Università degli studi di Cassino e del Lazio Meridionale, Italy*

Cristina Carmona,
*Universidad de Las Palmas de Gran Canaria, Spain*

Daniel Berio,
*Computing Dept., Goldsmiths, University of London, UK*

Daniel Martín Albo,
*Universidad Politécnica de Valencia, Spain*

Donato Impedovo,
*Università degli Studi di Bari Aldo Moro, Italy*

Frederic Flo Leymarie,
*Goldsmiths, University of London, Computing Department, UK*

Giovanni Spagnuolo,
*Università degli Studi di Salerno, Italy*

Giuseppe Pirlo,
*Università degli Studi di Bari Aldo Moro, Italy*

Julian Fierrez,
*BiDA Laboratory, Universidad Autonoma de Madrid, Spain*

Karina Lebel,
*Université de Sherbrooke, Canada*

Luis Leiva,
*Aalto University, Finland*

Moises Diaz,
*Universidad del Atlantico Medio, Spain*

Nadir Faci,
*Laboratoire Scribens, Polytechnique Montreal, Canada*

Nicole Dalia Cilia,
*Università degli studi di Cassino e del Lazio Meridionale, Italy*

Nicole Vincent,
*Laboratoire d'Informatique Paris Descartes, France*

Pedro Gómez Vilda,
*Neuromorphic Speech Processing Lab, Universidad Politécnica de Madrid, Spain*

Richard Guest,
*School of Engineering and Digital Arts, University of Kent, UK*

Rosa Senatore,
*DIEM Natural Computation Laboratory, University of Salerno, Italy*

Rubén Tolosana,
*BiDA Laboratory, Universidad Autonoma de Madrid, Spain*

Rubén Vera,
*BiDA Laboratory, Universidad Autonoma de Madrid, Spain*

Simon Pierre Boyogueno Bidias,
*Laboratoire Scribens, Polytechnique Montreal, Canada*

We finally thank Professor Ching Y Suen, the general chairman of ICPRAI 2018 for inviting us to organize the first International Workshop on the Lognormality Principle and its Applications and Professor George Nagy for kindly accepting to write the foreword of this book.

https://doi.org/10.1142/9789811226830_0001

# Chapter 1

# The Lognormality Principle: A Personalized Survey

Réjean Plamondon
*Laboratoire Scribens, Département de génie électrique,*
*Polytechnique Montréal, Montréal, (Québec) Canada, H3C 3A7*
*rejean.plamondon@polymtl.ca*

This introductory chapter presents a survey of the stepwise developments, implementations and applications of the Kinematic Theory of human movements made over the last 25 years, culminating with the statement of the Lognormality principle, which is the corner stone of the present book. The focus is put on the various projects achieved at Laboratoire Scribens as well as those realized with many students and researchers in the context of numerous national and international collaborations. After describing the Kinematic Theory, a focus is made on the development of its Delta- and Sigma-Lognormal models and their associated parameter extraction algorithms. Then the specific applications of the theory are summarized under the headings of e-Security, e-Recognition, e-Learning and e-Health in the perspective of developing Personal Digital Bodyguards. The following section is devoted to the statement of the Lognormality Principle and an exploration of some emerging fields of applications is presented prior to conclusion.

## 1. Introduction

The study of human movements is a multidisciplinary research field investigated by neuroscientists, computer scientists, robot designers, kinesiology experts, physicists, orthopedists, experimental psychologists, movement disorder specialists, education experts, human computer interface designers, biomedical and biomechanical engineers, video game designers, exoskeleton developers, expert document analysts, etc. All these research teams focus on specific aspects of this complex behavior, from the fundamental understanding of human motor control to the design of automatic systems exploiting numerous pattern recognition and artificial intelligence technologies.

Many book chapters and journal papers have been published on these topics in numerous international journals, book series as well as in a huge variety of international conferences. As a young faculty interested in the design of automatic online signature verification systems I have been introduced to the field. After designing a couple of verification systems using instrumented pens [Plamondon *et al.*, 1983; Plamondon and Brault, 1983; Brault and Plamondon, 1983; Baron and Plamondon, 1986, 1989] and conventional digitizers [Plamondon *et al.*, 1992; Plamondon, 1994], it became evident that to make a breakthrough in this specific application field, a better understanding of human motor control was a prerequisite. Why some parts of a signature seemed stable while others were not? [Brault and Plamondon, 1993a]. How can we segment signature signals to properly extract these parts? [Brault and Plamondon, 1993b]. All these questions were pointing at a more fundamental one: What is a handwriting stroke? A simple question, but a complex answer that, at that time, I did not know I would spend a large part of my research career on trying answer it.

Handwriting strokes are at the basis of any well-controlled human movements, a kind of the hydrogen atom upon which a full periodical chart of movements can be built. I thought that understanding its generation would open up many new pathways to provide new solutions to open problems in the various disciplines listed above.

After several trials and errors in attempting to model strokes [Plamondon, 1989, 1992], it became evident that an ecological approach [Kelso, 1995] where the basic properties of strokes would be emergent features of a complex system, was a key solution. In the next few years, I developed what I called naïvely at that time, the Kinematic Theory of rapid human movements. Over the years, this paradigm has grown and culminated recently in what is now referred to as the Lognormality Principle, which is the conducting thread of the present book.

This introductory chapter reports and summarizes the various theoretical, fundamental and applied studies dealing with the theory and its basic principle. The survey is almost exclusively dedicated to this topic and mostly refers to the various projects where I have been personally involved. It is by no means a general survey of the field and the very few external references that are cited here and there as typical examples of a given topic are pointed out when these are necessary to understand the broader context in which the research projects were realized. In a sense, this chapter is more like an annotated list of my publications on the lognormal modeling of human movements, a subset of my curriculum vitae.

Many models have been proposed to approach the problem of studying human movement production in general and handwriting in particular: models relying on neural networks [Bullock and Grossberg, 1988; Schomaker, 1991; Gangadhar *et al.*, 2007; Kalveram, 1998], equilibrium point models [Feldman, 1966; Feldman and Latash, 2005; Bizzi *et al.*, 1978, 1992], behavioral models [Schmidt and Lee, 1999; Thomassen *et al.*, 1983; van Galen and Teulings, 1983], coupled oscillator models [Hollerbach, 1981; Kelso, 1995; Zanone *et al.*, 2005], kinematic models [Plamondon, 1995a, b; Plamondon and Djioua, 2006], Push pull models [Johnson *et al.*, 2012], motor synergetic models [Nielson and Nielson, 2010], non-Euclidean [Biess, 2013] and Riemanian [Biess *et al.*, 2011; Nielson *et al.*, 2015] models as well as models exploiting minimization principles [Wada and Kawato, 1995; Engelbrecht, 2001], minimization of the acceleration [Neilson, 1993; Neilson and Neilson, 2005], of the energy [Nelson, 1983], of the time [Tanaka *et al.*, 2006; Enderle and Wolfe, 1987; Hermes and LaSalle, 1969], of the jerk [Hogan, 1984; Flash and Hogan, 1985], of the snap [Edelman and Flash, 1987], of the torque changes [Uno *et al.*, 1989] and of the sensory-motor noise [Harris and Wolpert, 1998], sensory-motor optimal control [Todorov, 2004)]. Finally, many models exploit the properties of various mathematical functions to reproduce human movements: exponentials [Plamondon and Lamarche, 1986], second order systems [Denier van der Gon and Thuring, 1965; Dooijes, 1983], Gaussians [Leclerc *et al.*, 1992], beta functions [Alimi, 2003], splines [Morasso *et al.*, 1983] and trigonometric functions [Maarse, 1987].

In this perspective, I will not make any overview of nor comparison with these models; this is beyond the scope of the present book. There is still a lot of open problems in this field [Karniel, 2011] and the reader interested in taking a broader view to position the modeling approach in a larger perspective is invited to consult various text books addressing human motor control [Shadmehr *et al.*, 2005; Rosenbaum, 2009; Ivancevic and Ivancevic, 2010; Schmidt and Lee, 2011; Latash, 2012; and Kandel *et al.*, 2013]. Consequently, I will thus focus on the strengths of the Kinematic Theory pointing out what it makes stand out in the context of the lognormality principle.

Section 2 presents the key elements of the Kinematic Theory, which is the emergence of the lognormal impulse response of a neuromuscular system. Follows two sections, one dedicated to the Delta-Lognormal model and the second, to the Sigma-Lognormal model. The Delta-Lognormal Sec. 3 focuses on simple pointing movement studies while Sec. 4 describes the Sigma-Lognormal model and reports its numerous applications grouped under four sub-headings: e-Security, e-Recognition, e-Learning and e-Health. Section 5 introduces the lognormality principle while Sec. 6 deals with some emerging new fields of applications. The paper concludes with brief suggestions of few open fields of applications.

## 2. The Kinematic Theory

In the eighties and early nineties, numerous studies were dealing with the control of rapid aimed movements. A database of common observations had been built over the last century, which raised many questions in terms of motor control. The main peak of the velocity profiles was described as asymmetric bell-shaped. There was a debate about the presence or absence of corrective sub-movements in pointing tasks [v.g. Meyer *et al.*, 1988 for a survey]. Regular discussions were concerning which signal (position, time, velocity, acceleration, jerk, torque, etc.) was the control variable used by the Central Nervous System (CNS) in a given task [v.g. Stein, 1982, for a discussion]. Regarding acceleration, two different types of patterns were observed depending on the experimental conditions, isometric or isotonic [v.g. Gielen, 1985; Ghez and Gordon, 1987]. Similarly, different speed-accuracy trade-offs were observed depending on the type of spatial or temporal constraints the subjects had to respect in a given protocol [Fitts, 1954; Meyer *et al.*, 1990 for a survey, as well as Plamondon and Alimi, 1997].

After some exploratory studies [Plamondon *et al.*, 1990, 1991; Plamondon and Clément, 1991; Schomaker and Plamondon, 1990] and the development of some intermediate models to take into account these questions and observations [Lamarche and Plamondon, 1984; Plamondon and Lamarche, 1986; Plamondon, 1989a, 1989b, 1990, 1991, 1992a, 1993a, 1993b, 1995a, 1995b; Plamondon and Yergeau, 1990; Leclerc *et al.*, 1992], the Kinematic Theory of rapid human movements was finally developed and formalized [Plamondon, 1995c, 1995d, 1998; and Plamondon *et al.*, 2003]. The starting point was the question raised by the commonly accepted asymmetric bell-shaped velocity profile observed in numerous experimental conditions. This suggested that the velocity might be linked to the control variable exploited by the CNS to generate movements and that its basic profile was the manifestation of an underlying emerging phenomenon. Considering the velocity of a trajectory as linked to the control variable was making sense since this is the sole vector tangent to any given trajectory, allowing on-line pattern learning by looking at its execution or off-line pattern learning by examining its trace. Figure 1 summarizes the whole modelling approach: considering the central nervous system and the upper limb as a linar black box.

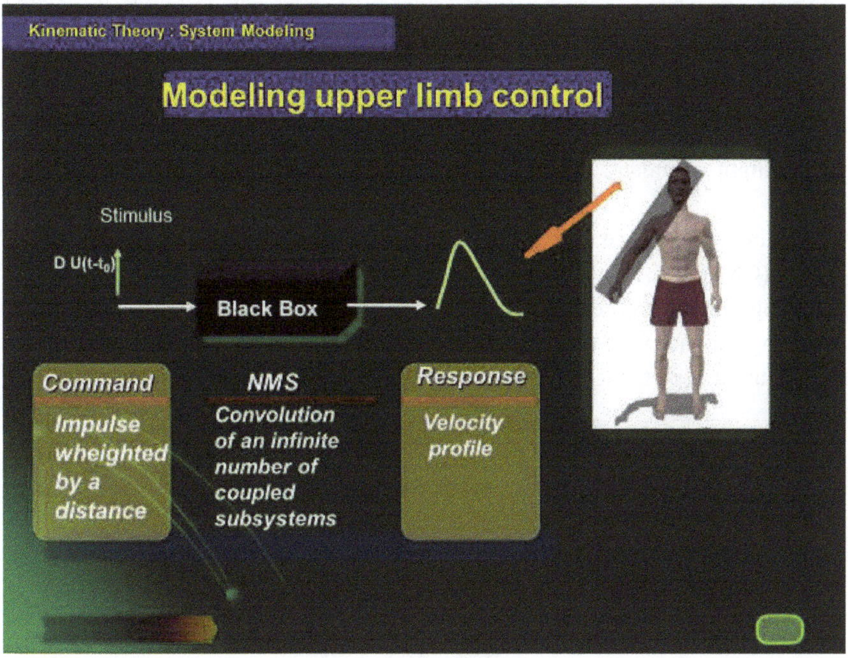


Fig. 1. Modeling the impulse response of the central nervous system and the upper arm as a black box.

The conducting thread to model this emergence phenomenon was the Central Limit Theorem [Plamondon *et al.*, 2014b] as applied to describe networks made up of an infinite number of coupled linear subsystems, this coupling being expressed in terms of the proportional cumulative time delays of the command propagation as hypothetically measured at various points along a given neuromuscular network. In other words, under these two hypotheses:

$$n \to \infty \tag{1}$$

$$T_j = \left(1+\varepsilon_j\right)T_{j-1} \tag{2}$$

the Central Limit Theorem predicted that the impulse response of a neuromuscular system would be a lognormal and the resulting velocity output of such a system, once stimulated with a command D at a given time $t_0$, would be described by:

$$\left|\vec{v}(t)\right| = D\Lambda(t;t_0,\mu,\sigma^2) \tag{3}$$

and the impulse response by:

$$\Lambda(t;t_0,\mu,\sigma^2)=\frac{1}{\sigma\sqrt{2\pi}(t-t_0)}\exp\left(\frac{-[\ln(t-t_0)-\mu]^2}{2\sigma^2}\right) \tag{4}$$

where the set of parameters describing a lognormal velocity pulse refers to:
*D*: the input command, which is the intended distance to be covered with the pulse;
$t_0$: the time occurrence of the command, as instantiated in the central nervous system (CNS);
$\mu$: the logtime delay (the time delay on a logarithmic time scale);
σ: the logresponse time (the response time on a logarithmic time scale).

Figure 2 depicts the key concepts of the whole modelling approach.

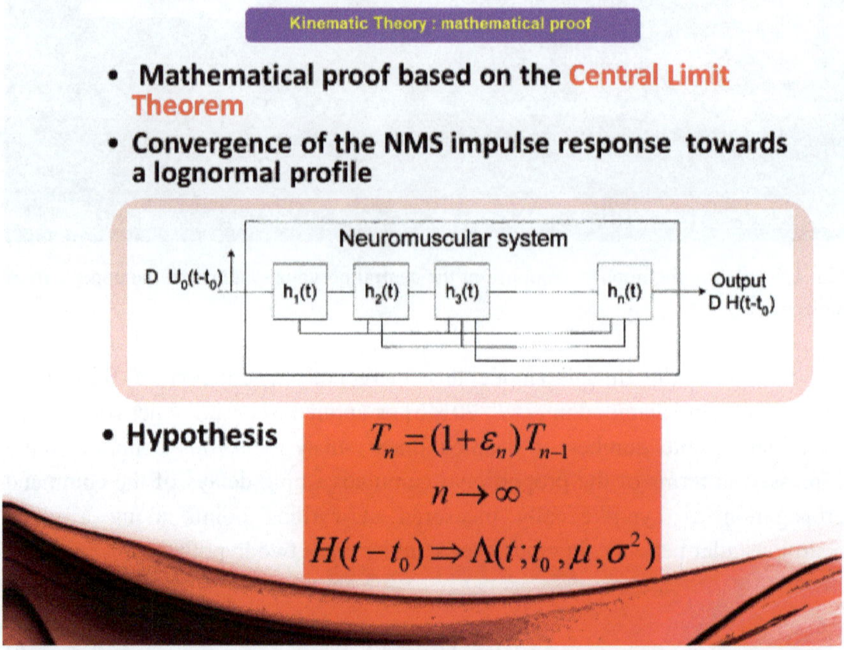


Fig. 2. Conditions for a lognormal convergence as predicted by the central limit theorem.

In other words, the corner stone of the Kinematic Theory is the existence of a specific movement primitive that encompasses all the fundamental properties of the velocity profiles of rapid pointing movements as consistently reported in studies dealing with the major psychophysical phenomena involving human motor control. This primitive is assumed to emerge at early age through gesture babblings and reflects the impulse response of a neuromotor system made up of a very large number of coupled neural and muscular subsystems. This converging behavior is predicted by the Central Limit Theorem and results in a lognormal impulse response. This convergence phenomenon was also artificially generated using a Simulink model [Djioua and Plamondon, 2004] to highlight the importance of the timing condition defined in equation (1.2).

From a fundamental point of view, the theory makes two predictions: first, as expressed in equation (1.2), there must be a proportional time coupling between the numerous subsystems constituting a neuromuscular system and second, there is a volition time $t_0$ reflecting the above-threshold activity of a cell population in the neuromotor cortex providing the commands to the system. These two predictions have been experimentally supported by electromyographic (EMG) and electroencephalographic (EEG) studies. Indeed, on the one hand, delays between bursts in surface EMG signals of six muscles of the upper limb activated during a rapid hand movement were highly correlated, supporting the proportionality of the cumulative time delays [Djioua and Plamondon, 2003; Djioua *et al.*, 2011; Plamondon *et al.*, 2013a]. On the other hand, the model-based kinematic analysis of rapid handwritten strokes and their corresponding EEG signals confirmed the prediction of the occurrence of a motor command evoked response potential (ERP) at $t_0$. The average map of these ERPs was stronger in electrodes close to the contralateral motor area [O'Reilly *et al.*, 2013]. Finally, it has been mathematically demonstrated that the Kinematic Theory is the ultimate minimization theory and the process of trajectory selection could thus be described as the process of recruiting a sufficient number of neuromuscular subsystems to approximate the lognormal profile [Djioua and Plamondon, 2009c, 2010]. Moreover, the theory has been used for analyzing various types of signals from different 1D, 2D and 3D data acquisition devices: inertial sensors, digitizers, tablets, cameras and even whiteboard [Fischer *et al.*, 2014] and mices [O'Reilly and Plamondon, 2011a].

## 3. The Delta-Lognormal Model

A first interesting property of (1.5) is its intrinsic bell-shaped asymmetry, with the rising phase steeper than the descent [Plamondon *et al.*, 1993]. Moreover, when neuromuscular systems are used to produce a straight pointing movement, two of these neuromuscular networks form a synergy: an agonist network pushing toward the target and an antagonist one pulling in the opposite direction. The resulting velocity profile is then described by what has been defined as a Delta-Lognormal equation:

$$|\vec{v}(t)| = D_1\Lambda(t;t_{01},\mu_1,\sigma_1^2) - D_2\Lambda(t;t_{02},\mu_2,\sigma_2^2) \quad (5)$$

where the subscripts 1 and 2 refer to the agonist and antagonist neuromuscular system respectively.

Numerous studies have been conducted over the last decades to point out that the basic properties of Delta-Lognormal pointing movements were in agreement with the set of common observations regularly reported in the field. This settled a unifying framework to integrate this prior knowledge into a consistent model that also provides simple answers to some debated questions. The following list summarizes the main studies and their conclusion:

1-The lognormal family of curves is the best fitting function to reproduce velocity profiles [Plamondon and Maarse, 1989; Plamondon *et al.*, 1993; Feng *et al.*, 2002]. The other analytical models, as described in the context of minimum principles, correspond, with more or less accuracy, to an approximation of the Delta-Lognormal equation [Plamondon and Djioua, 2010]. Moreover, another support to the lognormal convergence was brought using a Simulink Matlab tool to model an artificial neuromuscular system made up of a large number of coupled linear subsystems. It was demonstrated that the agonist and antagonist synergy could be simulated and that the Delta-Lognormal model was emerging from the set-up, its outputs mimicking human behaviors [Djioua and Plamondon, 2004].

2-From a psychophysical coherence point of view, the Delta-Llognormal model is providing a basic definition for a movement primitive [Woch and Plamondon, 2001, 2003, 2004], which includes bidirectional primitives [Woch and Plamondon, 2010; Woch *et al.*, 2011], offering a clear operational criterion to distinguish a primitive from a corrective sub-movement. The resulting theory also explains under the same umbrella the various speed-accuracy trade-offs observed in rapid aimed movements [Alimi and Plamondon, 1995, 1996; Plamondon and Alimi, 1998; O'Reilly and Plamondon, 2011b, 2013] and the parameters extracted from these movements were psychologically coherent with the predictions of various experimental conditions [Alimi and Plamondon, 1993a, 1993b, 1994a].

3-Regarding the parameter extraction to reproduce a given pointing movement, a few algorithms have been developed. The first approach, INFLEX, was based on a geometrical estimation of the inflexion points of a velocity curve [Guerfali and Plamondon, 1994; Plamondon and Guerfali, 1995a], the second, INITRI, focused on the initial raising phase of the velocity [Plamondon *et al.*, 2007] and the third, XZERO, exploited the analytical relationships that exist between the four parameters of a lognormal and three specific time indices: its maximum and two inflexion points [Djioua and Plamondon, 2009a]. In this latter study, these three methods were then combined into the IIX algorithm to pick up the optimal solution among the three outputs provided. The performance of this combined system was then compared with an evolutionary approach [Djioua *et al.*, 2005, 2007; Marcelli *et al.*, 2006], where it was shown that in free-noise conditions, the deterministic approach led to better results than the evolutionary one, while under extremely noisy conditions, the evolutionary approach seems to be less sensitive to noise, but it was still less successful than the deterministic search. Even under these conclusive results, there were no warrants that the best solution was globally optimal. To explore this question, two new branch and bound algorithms were developed for the Delta-Lognormal model [O'Reilly and Plamondon, 2012a; Boyoguéno-Bibias *et al.*, 2018]. In the first algorithm, the search space is divided into subspaces defined by box constraints. These subspaces are explored, minimizing an objective cost function, covering the whole space and stopping when the global minimum is found. The second one is based on interval arithmetic which offers the advantage of providing a range of confidence for the global optimal solution. A chapter of this book is dedicated to this new approach. These optimal performances were obtained at the expense of increasing the computation time.

## 4. The Sigma-Lognormal Model

According to the Kinematic Theory, the production of complex movements is reached through the exploitation of its lognormal primitive, using it as a building block, by time superimposing and, summing up lognormal vectors [Plamondon and Djioua, 2006]. From a mathematical point of view, this can be expressed by a vector summation that will be called here after the Sigma-Lognormal equation:

$$\vec{v}(t)=\sum_{i=1}^{N}\vec{v}_i\left(t;t_{0i},\mu_i,\sigma_i^2\right)=\sum_{i=1}^{N}D_i\begin{bmatrix}\cos\left(\theta_i(t)\right)\\ \sin\left(\theta_i(t)\right)\end{bmatrix}\Lambda_i\left(t;t_{0i},\mu_i,\sigma_i^2\right);N\geq 2 \tag{6}$$

where:

$$\theta_i(t)=\theta_{si}+\frac{(\theta_{ei}-\theta_{si})}{2}\left\{1+erf\left[\frac{\ln(t-t_{0i})-\mu_i}{\sigma_i\sqrt{2}}\right]\right\} \tag{7}$$

and where $t_{0i}$, $D_i$, $\mu_i$ and $\sigma_i$ keep on their previous meaning for each $i^{th}$ stroke and $\theta_{si}$, $\theta_{ei}$ stand respectively for the starting and ending angular direction of each discontinuous primitive represented in a motor program. In this context, the production of a complex movement starts with the instantiation of an action plan made up of a sequence of discontinuous virtual targets linked together by circular arcs defining their length and their starting and ending angular directions, as if the corresponding movements were made around a virtual pivot. Discontinuous lognormal strokes are thus sequentially produced and a continuous velocity profile emerges from the time superimposition of these vectors.

The development of this 2D model has been made in three steps. A first version [Plamondon and Guerfali, 1998a] assumed an independent controller for the tangential and the angular velocities. Each stroke was described by a Delta-Lognormal velocity profile, requiring nine parameters per stroke. This latter model was successful in reproducing many language-independent patterns collected with a digitizer [Plamondon, 1994; Plamondon *et al.*, 1997, 1998]. A second group showed that the angular velocity did not have to be controlled independently, the vector superimposition of strokes was sufficient to reproduce the kinematics of a given handwriting pattern [Guerfali and Plamondon, 1995b, 1995c; Guerfali and Plamondon, 1997]. Several psychophysical phenomena related to two-dimensional movements were explained and analyzed in the context of this model: the speed/accuracy trade-offs, spatial scaling, the isochrony principle, the two thirds power law [Plamondon and Guerfali, 1997, 1998b], effector independence [Plamondon and Alimi, 1997; Plamondon and Guerfali, 1998c]. The next simplification [Plamondon and Djioua, 2006] came from the idea that complex movement did not require antagonist components except for stopping at the end of a lognormal stroke sequence. The whole concept of antagonist system being useless in the absence of any real stopping targets and privileged directions. The partial velocity slowdowns were taken into account by the change of direction provoked by the forthcoming next stroke. The vector summation of single lognormal strokes was thus sufficient to generate various types of handwritings, signatures and graphs, using only six parameters per strokes, instead of nine.

A few robust algorithms have been developed to perform the Sigma-Lognormal parameter extraction. The first generations were dealing with the preliminary models [Guerfali and Plamondon, 1994] but the real breakthrough came with the development of the ScriptStudio platform which could extract the sequence of parameters from any complex trajectories [O'Reilly and Plamondon, 2008, 2009a]. Further extensions were then developed to track specific problems like oscillatory movements [Bou-Hernandez *et al.*, 2015] often referred to as omega-lognormal extractor [O'Reilly and Plamondon, 2010], prototype based stereotypical handwriting tasks [O'Reilly *et al.*, 2014], best first search implementation [Martin Albo *et al.*, 2015]. All these algorithms aimed at optimizing the reconstruction of velocity profiles. Recently a new algorithm, called iDeLog, that focused on the reverse engineering of both the velocity and trajectory was developed [Ferrer *et al.*, 2020]. iDeLog consists of two steps, the first one derives an initial action plan defined by a set of virtual points and angles from the trajectory and a sequence of lognormal curves from the velocity. In the second step, the virtual target points of the action plan are iteratively moved to improve the matching between the observed and the reconstructed trajectory and velocity. It has been shown that iDeLog is better than ScriptStudio to reconstruct the trajectory of continuous, long and complex movements like handwritten signatures. However, iDeLog provided poorer results than ScripStudio in reconstructing velocity, because of the trade-off between trajectory and velocity reconstruction in iDeLog.

One key study dealing with the Sigma-Lognormal model [Plamondon and Djioua, 2005; Djioua and Plamondon, 2007, 2009b] that has been at the basis of many projects dealing with the synthetic generation of handwriting specimens described in the next sections was dealing with the understanding of the movement variability introduced by the neuromuscular system. Using a reference action plan, a wide range of handwriting distortions were artificially generated by acting on the Sigma-Lognormal parameters to understand scale changes and rotational deformations and to rank the impact of each parameter in increasing order, $t_0$ being the most sensitive one, followed by the central parameter, D, $\theta_s$, $\theta_e$ and the peripheral parameters $\mu$ and $\sigma$ being the less sensitive. Overall this study provided practical cues for the development of huge databases of letters, words, signatures or gestures needed to train and test on-line handwriting classifiers and recognizers [Djioua *et al.*, 2006; Djioua and Plamondon, 2008a]. At about the same time, the Sigma-Lognormal model was also used to propose a new methodology to improve EMG signal processing [Djioua and Plamondon, 2006, 2008b]. A survey paper, [Plamondon *et al.*, 2009] in French, summarized the different milestones reached at that time.

Finally, all the test protocols that have been used over the years to validate the theory and explore its numerous fields of applications have been recently integrated into a Wacom Cintiq 13HD environment [Faci *et al.*, 2018]. The whole set up is described in a chapter of this book.

Over the last decade, the Sigma-Lognormal model has been used in numerous applications contributing to the development of the Personal Digital Bodyguard vision [Plamondon *et al.*, 2018]. These can be grouped under four headings:

### 4.1 *e-Security*

Three main problems have been tracked in e-Security so far: signature verification, writer identification and CAPTCHA generation. The Kinematic Theory has been first used in the context of signature verification. Preliminary prototypes were dealing with the theory concepts [Plamondon *et al.*, 1991; Plamondon, 1993c, 1994, 1995e] and the best one was tested in the context of a large-scale public experiment involving more than 80,000 attempts to defeat the system with forgeries. Only two failures were reported. [Plamondon *et al.*, 1998, 1999]. The most recent applications have been using the Sigma-Lognormal features directly to design a system [Fischer and Plamondon, 2017] or to improve skilled forgery detection [Gomez-Barrero *et al.*, 2015] to evaluate the quality of signatures [Galbally *et al.*, 2010, 2011]. It has also been used for data augmentation that is, generating synthetic signatures that preserve the biometric traits of a signer [Galbally *et al.*, 2012a, 2012b; Plamondon *et al.*, 2014a]. This culminates recently with the design of an automatic system that could work from a single reference specimen [Diaz *et al.*, 2015, 2016]. Moreover, [Pirlo *et al.*, 2014] used the Sigma-Lognormal model to analyze and compare the stability of signatures. A lognormality-oriented survey of the online signature verification state-of the art was also published in the Handbook of Document Image Processing and Recognition by [Plamondon *et al.*, 2014c] and more recently a second [Diaz *et al.*, 2019], based on 270 papers, provided benchmarks to point out the new research directions in the field, while [Faundez-Zanui *et al.*, 2020] pointed out the importance of considering security, health, and metadata from a joint perspective.

Since the compressed representation of an online handwritten pattern can be used to reconstruct an original signal, applying some perturbations on these parameters can result in generating different distortions of this initial pattern. This concept was applied by [Ramahia *et al.*, 2013, 2014, 2015] to design an automatic online CAPTCHA generation system to protect and control access to web sites. Since the present performances of automatic online handwriting recognition are worse than the performances of offline printed text recognizers, it was demonstrated that within a proper range of parameter variations, it was possible to design a robust system where the human performances greatly exceeded the best online handwriting recognizers. The first prototype was based on words while the second generation was based on characters, which offer the advantage of using non-words to make the task more complicated. These results were also improved when noise was added and ideas inspired by accents in handwriting were applied to the problem [Ramahia *et al.*, 2015].

### 4.2 *e-Recognition*

Since the Sigma-Lognormal model primarily focuses on velocity profile, most of its applications have been dealing with online pre-processing and segmentation techniques as well as data augmentation [Plamondon, 1992b; Plamondon *et al.*, 1999], although it provided some cues to track the segmentation of offline signals and the recovery of motor temporal information [Privitera and Plamondon, 1995a; Plamondon and Privitera, 1999] as well as setting the table for the design of neural networks dedicated to handwriting generation and learning [Privitera and Plamondon, 1995; Plamondon and Privitera, 1995b]. Indeed, processing on-line handwriting signals with the Kinematic Theory provide automatic segmentation of the signal into a superimposed sequence of lognormal strokes each stroke being described by the previously defined six parameters [Plamondon and Guerfali, 1996].

The lognormal representation of individual handwritten patterns has also been used to augment existing databases of cursive words and command gestures for the design of better human computer interfaces. In the case of word databases, [Battacharya *et al.*, 2017] used the Sigma-Lognormal model to augment a database of Bangla words. Estimating the acceptable range of parameter variations to produce duplicated specimens from real online arbitrary script acquired from various digitizers, validation through human perception experiments, computer evaluations with a HMM recognizer and statistical hypothesis testing confirmed the structural similarities, including the naturalness of such synthetic prototypes and their usefulness in designing and testing language independent automatic handwriting recognizers. Similarly, [Martin-Albo *et al.*, 2014] used a similar approach to train a user-specific text recognizer, improving by 28% the relative accuracy of the best adaptation baseline.

In the case of gestures, three teams have been investigating the interest of using a similar methodology for database augmentation. [Almaksour *et al.*, 2011] used synthetic gestures in an incremental learning scenario to enhance the prediction capacity of a gesture command classifier in applications dedicated to online interactive form filling with electronic ink on tablets. Experimental results show that the Sigma-Lognormal deformed samples help to improve significantly the classification performance as well as boosting the learning on the fly process. Moreover the evolving and customizable classifier better resists to the introduction of new classes of commands, reaching a better stability using a few real samples. [Plamondon *et al.*, 2014].

[Reznova *et al.*, 2015, 2017] also exploited synthetic data for user-friendly on-line handwritten gesture command recognition. Emphasis was made on the problem of forgetting unused classes by applying a random buffer and Elastic Memory Learning (EML). Since, from the user perspective, it is better to let her or him define his own gesture commands, it was shown that the use of synthetic samples derived from their Sigma-Lognormal representation by adding some noise on the $\mu$ and $\sigma$ parameters can compensate for the lack of data at the initialization. These studies also focused on finding the proper amount of synthetic data needed for each command sample to be learned as well as on the most appropriate time to stop the synthetic generation for learning purposes.

In a series of papers, [Leiva *et al.*, 2016a, 2016b, 2017] have been working on the automatic generation of commands for the design of optimal human computer interfaces (HCI) using the Sigma-Lognormal model. They have made their tool public through an interactive website [Leiva *et al.*, 2016a] referred to as "Gestures à Go Go" (G3), a web service plus an accompanying web application for bootstrapping stroke gesture samples based on the Kinematic Theory. The user has only to provide a gesture example once, and G3 generates from tens to thousands of synthetic humanlike samples which can be used for testing and designing recognizers to be integrated into a HCI. In the follow-up paper [Leiva *et al.*, 2017], it was shown through various perception experiments that the synthesized stroke commands look and feel like human gestures and hold the same statistical characteristics. In the second paper of 2016 [Leiva *et al.*, 2016b], mouse cursor movements were analyzed with the Sigma-Lognormal model to decompose these into strokes, to recover the gesture action plan and discriminate between intentional and unintentional movements based on the observation that when the user's intent is clear the pointing movements are faster and better reconstructed. [Leiva *et al.*, 2020] pursued along this pathway and proposed a new high-performing technique, named Keytime, to make accurate predictions of users' stroke gesture production times, very close to the performance of real users. The mid-term goal here was to provide researchers with unprecedented levels of accuracy and sophistication to characterize users' a priori time performance in order to improve gesture input-based user interface designs. A chapter of this book extends the works described in [Leiva *et al.*, 2018b].

Moreover, a first study aiming at characterizing the bimanual and unimanual neuromotor coordination under the Kinematic Theory paradigm was recently reported [Pan *et al.*, 2019]. Findings from the Sigma-Lognormal model suggested that the dominant and non-dominant hand contribute differently to the manual conditions, which may lead to new heuristics for implementing gesture commands systems, for example to interact with both hands on large digitizers, for video game playing or rehabilitation training.

### 4.3 *e-Learning*

The acquisition of skilled handwriting is a lengthy and complex process. The Sigma-Lognormal model can be of great help to characterize and monitor the evolution of the learning progress as well as for the design of learning tools. Several studies have been conducted in this perspective, first to design interactive tools to help children learn handwriting as well as using the Kinematic Theory to characterize the children fine neuromotricity and monitor its evolution with age and years of practice.

In the early nineties, a first learning tool was developed combining a transparent digitizer with in house electronic controller [Plamondon and Carrières, 1993; Carrières and Plamondon, 1994]. Unable to follow the fast-growing market for on-screen writing digitizing tablets, we decided to focus on the study of the learning process itself using the first generation of commercial handheld digitizing tablets. Formal letter models were developed [Djeziri *et al.*, 1999] and a second generation of interactive learning tool, referred to as Scriptôt, was then implemented [Audet *et al.*, 2000; Djeziri *et al.*, 2002].

In a series of papers, [Duval *et al.*, 2013, 2015; Rémi *et al.*, 2015, 2018a] demonstrated that the Sigma-Lognormal model, initially developed to study well mastered fluent handwriting could be used to reconstruct handwriting from young kids, as young as three years old, and that the extracted parameters could be used to evaluate the fine neuromotor control of children, providing a valuable tool for studying their graphomotor control learning strategies and following their evolution as they become better practitioners. Overall these convincing results have led to the launch of the @MaGma project, defined with the Academic direction of the Guadeloupian Region to collect in a secured way and to analyze a large amount of various contextualized online and offline handwritten data produced in real scholar situations by numerous pupils from kindergarten up to middle school. The first report was dealing with the online acquisition of various levels of difficulty of scribbles produced by middle school pupils using various tablet devices [Rémi *et al.*, 2018b]. A chapter of this book is dedicated to this topic.

Recently a comparative study of French and Québec Kindergarten (5 years old) and first year of primary school (6 years old) children pointed out that the Kinematic theory could be used to characterize and compare the neuromotor abilities of these kids as well as providing objective comparison criteria between two different handwriting learning methods: cursive vs block letter writing, the former seeming more efficient in developing fluency. A short abstract summarizes this project in the Appendix section.

### 4.4 *e-Health*

The idea of analyzing online handwriting to detect and evaluate aging and health problems is not new but it is becoming a driving force of the Sigma-Lognormal applications. Over the last ten years, several studies have demonstrated the interest of using the Kinematic Theory for biomedical applications. Here is a list of these studies.

The first projects [Woch and Plamondon 2007, 2010; Woch *et al.*, 2011] dealt with healthy young and aged subjects producing bidirectional Delta-Lognormal strokes. A substantial increase of neuromuscular response delays and a decrease of the command amplitudes with age was reported similarly on the agonist and antagonist systems. Moreover, the age has a proportional effect on the various time characteristics of the movements and also induced correlations between the agonist and the antagonist neuromuscular systems imputable to the combination of the Delta-Lognormal model properties and the task constraints. Regarding aging, a preliminary study on how aging could affect signatures over a short time period was also investigated [Gomez-Barrero *et al.*, 2013]. The second series of experiments [O'Reilly *et al.*, 2009c; O'Reilly and Plamondon, 2011b, 2012b; O'Reilly *et al.*, 2014)] a set of nine neuromuscular tests has been designed to infer the probability of presence of brain stroke risk factors using logistic regression and linear discriminant analysis with leave-one-out cross-validation on the lognormal parameters extracted from the different tasks. The clinical potential value of movement information for stroke prevention was assessed with AUC curves ranging from 0.6 to 0.9, depending on the neuromuscular test and the risk factor investigated. A book chapter [O'Reilly and Plamondon, 2009c] also showed how the whole methodology could be applied in rehabilitation of patients by designing neuromotor tests to monitor the recovery of upper limb movement control after a brain stroke, here again exploiting the lognormality principle to evaluate the recovery progresses. The same methodology could be applied to design neuromuscular disorder diagnostic systems [O'Reilly and Plamondon, 2007, 2012c]. [Impedovo *et al.*, 2013] proposed a similar approach to study Alzheimer's disease.

Two studies dealt with child health problems. The first [Laniel *et al.*, 2018, 2019] analyzed young kids with Attention Deficit Disorder with or without Hyperactivity (ADHD). The production of pen strokes by 12 children with ADHD as compared to a control group of 12 normal kids was analyzed with the Sigma-Lognormal model. Longer $t_0$ and smaller D demonstrated poorer motor planning in ADHD kids. Similarly, a poorer execution was observed (larger NbLog) as well as a greater variability in motor control (smaller SNR/nbLog). The parameters extracted from the pen strokes were significantly correlated with performance on other motor and handwriting measures, providing preliminary evidence that this test protocol could be a useful tool for rapidly detecting motor skill problems associated with ADHD. A chapter of this book is analyzing these data from a pattern recognition and machine learning perspective.

The second investigation [Faci *et al.*, 2018] analyzed handwriting strokes made by children and adolescents who sustained mild traumatic brain injuries, one and three months after the accident. Thirty-two participants, aged between 6 and 17 years, were asked to produce handwriting strokes on a digitizing tablet and these strokes were analyzed using the Sigma-Lognormal model. The majority of children in the sample showed reduced fine motor control between 1-month and 3-month post-injury, as reflected by significant differences for at least one parameter when comparing their result at 1- and 3-month post-injury. Subjects were objectively classified into four groups: Better, stable, worse, no decision. The results were also compared with the answer to the Post-Concussion Symptom Inventory questionnaire to check for concordance with the parent subjective answers. A concordance was observed between the two tests for only nine participants, which suggests that further research will be needed to examine associations between neurocognitive measures and the Sigma-Lognormal parameters with regards to recovery. A chapter of this book reports further analyses.

Some preliminary studies with Parkinson's patients [van Gemmert *et al.*, 2013], have also shown the interest of using the Kinematic Theory to characterize this illness. This paper was the first attempt to use the Sigma-Lognormal model to investigate the influence of Parkinson's disease on previously captured handwriting movements executed with a requirement on the target sizes. Here again it was demonstrated that the Kinematic Theory could be useful for such studies and it was pointed out that PD patients used significantly more lognormal primitives than age-matched controls. This paved the way to some new projects on this topic.

[Nadeau *et al.*, 2018a] investigated if aerobic exercise training could help improve upper limb functions in sedentary people with Parkinson's disease and healthy adults. Participants were training on a stationary bicycle for 12 weeks and asked after each training session to draw straight lines on a digitizer. The Delta-Lognormal model was used to characterize and quantify the different motor control commands involved in the task. Significant decreases of the antagonist commands ($D_2$), as well as of the antagonist logresponse time ($\sigma_2$) were observed, highlighting a significant positive impact of the bike training on these subjects as measured from their improved capacity to draw lines in a more efficient way.

Finally, [Unguean *et al.*, 2018] investigated stroke gestures articulated on touchscreens by people with motor impairments, showing again that the Kinematic Theory can successfully model their strokes and reflect their motor skills entailed by gesture articulation. Here again it was demonstrated that computer-synthesized gestures could be successfully generated with the same geometric and kinematic characteristics of those produced by people with spinal cord injury and cerebral palsy, a first study on these types of motor impairments.

Another domain of applications that is under investigation deals with eye movements. Indeed, as early as 1995, Plamondon showed that single eye saccades could be described by Delta-Lognormals. In an ongoing project, Vincent *et al.* in [Nadeau *et al.*, 2018b] wanted to check if the model could be used to reconstruct sequences of saccades as currently used in some Alzheimer disease tests. Preliminary results have shown that for long sequences of saccades, the Sigma-Lognormal model is more appropriate since in the majority of the cases the movements are not straight saccades. An abstract of the Appendix reports these findings.

## 5. The Lognormality Principle

All these successful projects, theoretical, physiological and experimental, as well as the numerous applications that have been designed so far on the basis of the Kinematic Theory and its family of lognormal models has led me to postulate the lognormality principle [Plamondon *et al.*, 2013b] as the fundamental premise to position and evaluate any direct or indirect motor control investigation. This principle can be stated as follows:

*The emergent lognormality of the neuromuscular impulse response of a given human motor system, which is the result of an asymptotic convergence described by the Central Limit Theorem, is a basic global feature reflecting the behavior of individuals who are in perfect control of their movements.*

The production of complex movements is accomplished through the exploitation of this principle by time superimposing and, summing up lognormal vectors, with the goal of minimizing their number in a given task, to produce efficient and fluent gestures and optimize the energy required to generate these.

As a corollary, if researchers focus on the basic mathematical convergence toward lognormality, motor control learning in young children can be interpreted as a migration toward lognormality. Then, for the greater part of their lives, healthy human adults take advantage of their lognormality to control their movements efficiently. Finally, as aging and health issues intensify, a progressive departure from lognormality is occurring.

## 6. Emerging Fields and New Projects

So far lognormal patterns have been used and observed in numerous neuromuscular system movements: fingers, hand, wrist, arm, eye, head, trunk, as well as in speech, making this model quite universal and complementing the exhaustive list of lognormal phenomena observed in neuroscience. Although we are working here with deterministic impulse responses, as compared to the random variable statistical distribution analysis generally reported in these fields [Burzsaki and Mizuseki, 2014], both interpretations are intrinsically equivalent and lead asymptotically to identical results. Indeed, the fact that both approaches rely on the Central Limit Theorem and that the mathematical convergence predicted by this theorem reflects a basic property of the convolution operator when it is applied sequentially to a large number of positive functions, a physical system, made up of a large number of subsystems, can be asymptotically described, using the Central Limit Theorem, by a multidimensional Gaussian function, the individual constituting subsystems being either of probabilistic or deterministic nature. Under the coupling hypothesis described in equation (1.2), a deterministic lognormal convergence is predicted.

Finally, recent projects analyzing or exploiting lognormality have been launched to characterize various populations of subjects and applied in various domains. Among the recent applications, [Laurent *et al.*, 2019, 2020] have applied Kinematic Theory to see if there is a possibility to detect shoulder fatigue. Results show that some kinematic parameters extracted from the strokes produced by a given subject reflected these motor control changes. With the distinction between the agonist and antagonist components of each stroke, it was possible to point out which subsystem in a given subject was more altered by the fatigue and, thanks to the Sigma-Lognormal parameter extraction, it was also possible to see if fatigue impacts the central or peripheral systems or both. To the best of our knowledge, no tools taken separately can discriminate a peripheral from a central fatigue to measure human muscle fatigue, which opens up to a new series of potential applications.

Another unexpected field of applications has been proposed by [Berio and Fol Leymarie, 2015]: the use of the Kinematic Theory to design interactively curves that closely resemble those observed in graffiti art. These ideas were further developed over the last three years in [Berio *et al.*, 2017, 2018a, 2018b] focusing on the stylization and beautification of calligraphic trajectories via curves that reflect a smooth, fluent and rapidly executed movement as well as on the generation of multiple synthetic specimens from a single tag example. In a recent development, [Berio *et al.*, 2019] proposed a data-driven path stylization system implemented with a Recurrent Mixture Density Network that extracts and learns dynamic and visual qualities from a small number of human handwriting gestures.

The theory has been mathematically extended to 3D movements as early as 2001 [Leduc and Plamondon, 2001] but efficient algorithms to automatically process these higher dimension curves are still under development. Recently a 2.5D has been proposed in [Chidami *et al.*, 2018] and a full 3D parameter extractor [Schindler *et al.*, 2018] have been developed on these grounds. The 2.5D approach is based on the assumption that every 3D short and fast hand stroke is a limb restricted 2D movement on a 3D coordinates plane. The least error plane is thus approximated with a Principal Component Analysis and the best extracted projection plan is then analyzed with the Delta-Lognormal model. In the 3D algorithm, the Sigma-Lognormal model is fully extended mathematically to 3D and the extraction of the corresponding eight parameters is done with a curve velocity optimization process similar to the one used for 2D but incorporating the estimation of two supplementary angles for strokes executed on a sphere instead of a circle. This latter algorithm is extensively described in a chapter of this book.

Moreover, in the context of the growing trend to assess human kinematics with wearable sensors, which enable recording of raw inertial signals (acceleration and angular velocity) from which the orientation of the body parts can be estimated, [Lebel *et al.*, 2017] investigated the use of the Kinematic Theory to characterize the cranio-caudal turn signature as assessed by attitude and heading reference systems. It was demonstrated among other things, that the metrics derived from this approach could, be a promising avenue to develop and track markers measuring early mobility deficits in Parkinson's disease [Lebel *et al.*, 2018a] and even to capture the change in medication state early in the disease [Lebel *et al.*, 2018b]. A chapter of this book is dedicated to this new topic.

Finally, another breakthrough has been made in the last three years, when [Carmona-Duarte *et al.*, 2016a] have extended the domain of applications of the Kinematic Theory to speech processing. The key idea behind that extension is that the velocity is calculated based on the relationship between the speech formants and the tongue-jaw dynamic reference center. The first and second formants of the voice signal are thus used to define a XY reference system from which velocity can be computed. It has been shown that this velocity can be described by a time superimposition of lognormal patterns. Moreover, it has been pointed out that such a methodology could be exploited for neurological disorder assessments by evaluating speech from people with neurodegenerative and non-neurologically related pathologies. [Carmona-Duarte *et al.*, 2016b, 2017a, 2017b]. A chapter of this book extends this work in progress by comparing the speech and handwriting methodologies for tracking Parkinson's disease [Carmona-Duarte *et al.*, 2018]. Moreover, a recent paper [Carmona-Duarte *et al.*, 2019] suggested that the new methodology could also be useful to improve the speech formant extraction or to evaluate the quality of a speech synthesizer. Carmona-Duarte also proposed a tailored version of the Kinematic Theory to design a neuromotor inspired handwriting synthesizer [Carmona-Duarte *et al.*, 2017c]. Finally, based on previous works on handwriting learning [Senatore and Marcelli, 2012; Marcelli *et al.*, 2013, 2017] linked the lognormality principle to the central and local controllers and proposed an algorithm for segmenting handwriting movements into strokes, each of which corresponds to a command issued by the central nervous system, while disregarding those that may depend on local control. A chapter of this book is also dedicated to this representation problem.

## 7. Conclusion

The fourteen following chapters of this book describe in more details some of the previous applications and, in my mind, this is the tip of an iceberg. In the forthcoming years, I foresee numerous new developments where the Kinematic Theory and its lognormal models will be exploited under the umbrella of the lognormality principle [Lefebvre *et al.*, 2015] and the Personal Digital body guard vision [Plamondon *et al.*, 2018]. For example, combined with FMRI techniques it could provide a better understanding of the healthy and non-healthy motor control. It is also expected that 3D applications will boom once the 3D parameter extractors have been robustly tested. I foresee potential applications for the design of exoskeletons, anthropomorphic robots, human-like avatars and in the vast domain of the evaluation and monitoring of athlete and astronaut performances, to name a few.

## References

Alimi, A., Plamondon, R. (1993a). Performance Analysis of Handwritten Strokes Generation Models, *Proc. 3rd International Workshop on Frontiers in Handwriting Recognition*, Buffalo, USA, pp. 272–283.

Alimi, A., Plamondon, R. (1993b). Parameter Analysis of Handwriting Strokes Generation Models, *Proc. 6th International Conference on Handwriting Drawing*, Paris, France, pp. 4–6.

Alimi, A., Plamondon, R. (1994). Analysis of the Parameter Dependence of Handwriting Generation Models on Movements Characteristics, *Advances in Handwriting and Drawing: A Multidisciplinary Approach*, C. Faure, G. Lorette, A. Vinter, P. Keuss (Eds.), Europia, Paris, pp. 363–378.

Alimi, A., Plamondon, R. (1995). Speed/Accuracy Tradeoffs Formulations: Toward a Robust Law, *Proc. 7th Biennial Conference of the International Graphonomics Society*, London, Ontario, Canada, pp. 17–18.

Alimi, A., Plamondon, R. (1996). A Comparative Study of Speed/Accuracy Tradeoffs formulations: The case of Spatially Constrained Movements where both Distance and Spatial Precision Are Specified, in M. Simner, G. Leedham, A.J.T.W.M. Thomassen (Eds.), *Handwriting and Drawing Research: Basic and Applied Issues*, IOS Press, Amsterdam, pp. 127–142.

Alimi, M.A. (2003). Beta neuro-fuzzy systems, In: Duch, W., Rutkowska, D. (Eds.), *TASK Quarterly J.*, Special Issue on Neural Networks, vol. 7(1), pp. 23–41.

Almaksour, A., Anquetil, E., Plamondon, R., O'Reilly, C. (2011). Synthetic Handwritten Gesture Generaiton Using Sigma-Logormal Model for Evolving Handwriting Classifier, *Proc. 15th Biennial Conf. of the International Graphonmics Society* (IGS 2011), Cancun, Mexico, pp. 98–101.

Audet, M., Robert, J.M., Plamondon, R. (2000). Scriptôt: un système d'aide à l'apprentissage de l'écriture manuscrite qui utilise une tablette et un crayon électronique, *Comptes Rendus Congrès Ergo-IHM 2000*, Biarritz, France, pp. 16–24.

Baron, R., Plamondon, R. (1986). Conditionnement du signal d'un crayon accélérométrique pour la vérification des signatures, *Comptes rendus de la 3e Conférence Canadienne sur l'Informatique Industrielle*, Montréal, Canada, p. 66.1 66.5.

Baron, R., Plamondon, R. (1989). Acceleration Measurement with an Instrumented Pen for Signature Verification and Handwriting Analysis, *IEEE Transactions on Instrumentation and Measurement*, vol. 38, no. 6, pp. 1132–1138.

Bhattacharya, U., Plamondon, R., Chowdhury, S. D., Goyal, P., Parui, S.K. (2017). A Sigma-Lognormal model-based approach to generating large synthetic online handwriting sample databases, *International Journal on Document Analysis and Recognition*, vol. 20, no. 3, pp. 155–171.

Berio, D., Fol Leymarie, F. (2015). *Computational Models for the Analysis and Synthesis of Graffiti Tag Strokes*, in Computational Aesthetics in Graphics, Visualization, and Imaging, EXPRESSIVE 2015, Paul Rosin Ed., pp. 35–47.

Berio, D., Akten, M., Fol Leymarie, F., Grierson, M., Plamondon, R. (2017). Calligraphic Stylisation Learning with a Physiologically Plausible Model of Movement and Recurrent Neural Networks, *Proc. of 4th Int. Conf. on Movement Computing* (MOCO), 6 pages.

Berio, D., Fol Leymarie, F., Plamondon, R. (2018a). *Expressive Curve Editing with the Sigma Lognormal Model*, in EG2018 conference, Diamanti, O., Vaxman, A., (Eds.), Eurographics Ass., Publ. 6 pages.

Berio, D., Fol Leymarie, F., Plamondon, R. (2018b). Kinematic Reconstruction of Calligraphic Traces from Shape Features, *Proc. Int. Conf. on Pattern Recognition and Artificial Intelligence: Workshop on the Lognormality Principle and its Applications*, CENPARMI, Montréal, Canada, pp. 762–767.

Berio, D., Fol Leymarie, F., Atken, M., Plamondon, R. (2019). Calligraphic Stylisation Learning with a Physiologically Plausible Model of Movement and Recurrent Neural Networks, *21st Human-like Computing Machine Intelligence Workshop*, 30th June to 3rd July 2019, Cumberland lodge, Windsor Great Park, UK, 8 pages.

Biess, A., Flash, T. and Liebermann, D. G. (2011). Riemannian geometric approach to human arm dynamics, movement optimization, and invariance, *Physical Review*, E, 83, 031927.

Biess, A. (2013). Shaping of arm configuration space by prescription of non-Euclidean metrics with application to human motor contro, *Physical Review*, E, 87, 012729.

Bizzi, E., Dev, P., Morasso, P., Polit, A. (1978). Effect of Load Disturbances During Centrally Initiated Movements, *J. Neurophysiol.* 41, 542–556.

Bizzi,E., Hogan, N., Mussa-Ivaldi, F.A., Giszter, S. (1992). Does the nervous system use equilibrium-point control to guide single and multiple joint movements? *Behav. Brain Sci.* 15, 603–613.

Bou-Hernandez, A., Fischer, A., Plamondon, R. (2015). Omega-Lognormal Analysis of Oscillatory Movements as a Function of Brain Stroke Risk Factors, *Proc. 17th Biennial Conference of the International Graphonomics Society*, Pointe-à-Pître, Guadeloupe, pp. 59–62.

Boyoguéno-Bibias, S.P., David, J-P., Savaria, Y., Plamondon, R. (2018). On the Use of Interval Arithmetic to Bound Delta-Lognormal Rapid Human Movement Models, *Proc. Int. Conf. on Pattern Recognition and Artificial Intelligence: Workshop on the Lognormality Principle and its Applications*, CENPARMI, Montréal, Canada, pp. 738–742.

Brault, J.J., Plamondon, R., Robillard, P.N. (1983). Handwriting Modelling and Simulation for the Development of an Instrumented Pen, *Proc. Int. Symp. on Modelling and Simulation*, Bermuda, pp. 3–7.

Brault, J.J., Plamondon, R. (1993a). Segmenting Handwritten Signatures at their Perceptually Important Points, *IEEE Transactions on Pattern Analysis and Machine Intelligence*, vol. 15, no. 9, pp. 953–957.

Brault, J.J., Plamondon, R. (1993b). A Complexity Measure of Handwritten Curves: Modelling Dynamic Signature Forgery, *IEEE Transactions on System, Man and Cybernetics*, vol. 23, no. 2, pp. 400–413.

Bullock, D., Grossberg, S. (1988). *The VITE model: a neural command circuit for generating arm and articulator trajectories*. In: Kelso, J.A.S., Mandell, A.J., Shlesinger, M.F. (Eds.), *Dynamic Patterns in Complex Systems*. World Scientific Publishers, Singapore, pp. 305–326.

Buzsaki, G., Mizuseki, K. (2014). The Log-Dynamic Brain: How Skewed Distributions affect Network Operations, *Nature Reviews*, Neuroscience, April, pp. 264–278.

Carrières, P., Plamondon, R. (1994). An Interactive Handwriting Teaching Aid, *Advances in Handwriting and Drawing: A Multidisciplinary Approach*, C. Faure, G. Lorette, A. Vinter, P. Keuss (Eds.), Europia, Paris, pp. 207–239.

Chidami, S., Archambault-Caron, M., Plamondon R. (2018). The Delta-Lognormal model in 2.5D, *Proc. Int. Conf. on Pattern Recognition and Artificial Intelligence: Workshop on the Lognormality Principle and its Applications*, CENPARMI, Montréal, Canada, pp. 743–747.

Carmona-Duarte, C., Alonso, J.B., Diaz, M., Ferrer, M.A., Gómez-Vilda, P., Plamondon, R. (2016a). Kinematic modelling of dipthong articulation, *Recent Advances in Nonlinear Speech Processing. Smart Innovation, Systems and Technologies*, vol. 48, E. A. *et al.* (Eds.) Switzerland: Springer, Cham.

Carmona-Duarte, C., Plamondon, R., Gómez-Vilda, P., Ferrer, M.A., Alonso, J.B., Londral, A.R.M. (2016b). Application of the lognormal model to the vocal tract movement to detect neurological diseases in voice, *Innovation in Medicine and Healthcare 2016. Smart Innovation, Systems and Technologies*, vol. 60, J. L. Chen Y.W., Tanaka S., Howlett R. (Eds.) Switzerland: Springer, pp. 25–35.

Carmona-Duarte, C., Góme-Vilda, P., Ferrer, M.A., Plamondon, R., Londral, A. (2017a). Study of several parameters for the detection of amyotrophic lateral sclerosis from articulatory movement, *Loquens*, vol. 4, January issue, pp. 1–5.

Carmona-Duarte, C., Gomez-Vilda, P., Plamondon, R., Rodellar-Biarge, V., Palacio-Alonso, D., Ferrer, M.A., Alonso-Hernandez, J.B. (2017b). Phonatory and Articulatory Correlates in Kinematic Neuromotor Degeneration, *Converging Clinical and Engineering Research on Neurorehabilitation* Ii vol. 1 and vol. 2, 2017, vol. 15, pp. 203–208.

Carmona-Duarte, C., Ferrer, M.A., Parziale, A., Marcelli, A. (2017c). Temporal Evolution of Synthetic Handwriting, *Pattern Recognition*, 68, pp. 233–244.

Carmona-Duarte, C., Ferrer, M., Gómez-Vilda, P., Van Gemmert, A.W.A., Plamondon, R. (2018). A common framework to evaluate Parkinson's disease in voice and handwriting, *Proc. Int. Conf. on Pattern Recognition and Artificial Intelligence: Workshop on the Lognormality Principle and its Applications*, CENPARMI, Montréal, Canada, pp. 795–799.

Carmona-Duarte, C., Ferrer, M.A., Gómez-Vilda, P., Plamondon, R., Gómez-Rodellar, A. (2020). Kinematic Modelling for Speech Articulation, *Cognitive Computation Journal*, 12 pages.

Denier Van Der Gon, J.J., Thuring, J.P.H. (1965). The guiding of human movements. *Kybernetik* 14, 145–148.

Diaz, M., Fischer, A., Plamondon, R., Ferrer, M.A. (2015). Towards an Automatic On-Line Signature Verifier Using Only One Reference per Signer, *Proc. 13th IAPR International Conference on Document Analysis and Recognition* (ICDAR 2015), pp. 631–635.

Diaz, M., Fisher, A., Ferrer, M.A., Plamondon, R. (2016). Dynamic Signature Verification System Based on One Real Signature, *IEEE Trans. Cybernetic*, 48 (1), 228–239.

Diaz, M., Ferrer, M. A., Impedovo, D., Malik, M.I., Pirlo, G., Plamondon, R. (2019). A perspective analysis of handwritten signature technology, *ACM Computing Surveys* 51(6): 39 pages.

Djeziri, S., Plamondon, R., Robert, J.M. (1999). A Letter Models Generator to Assist Handwriting Teaching, *Proc. 9th Biennial Conference of the International Graphonomics Society* (IGS 1999), Singapore, pp. 181–185.

Djeziri, S., Guefali, W., Plamondon, R., Robert, J.M. (2002). Learning Handwriting with Pen Based Systems: Computational Issues, *Pattern Recognition*, Special Issue, vol. 35, no. 5, pp. 1049–1057, 2002.

Djioua, M., Plamondon, R. (2003). Relationship between Proportionality Law and Lognormality of a Coupled System Response, *Proc. 11th Biennial Conference of the International Graphonomics Society*, (IGS 2003), Scottsdale, Arizona, pp. 81–85.

Djioua, M., Plamondon, R. (2004). The Generation of Velocity Profiles with an Artificial Simulator, *International Journal of Pattern Recognition & Artificial Intelligence* (IJPRAI), Special Issue: Connecting Graphonomic Research and Pattern Recognition, vol. 18, no. 7, pp. 1207–1219.

Djioua, M., Plamondon, R., Della Cioppa A., Marcelli, A. (2005). Delta-Lognormal Parameter Estimation by Non-Linear Regression and Evolutionary Algorithm: A Comparative Study, *Proc. 12th Biennial Conference of the International Graphonomics Society* (IGS 2005), Salerno, Italy, pp. 44–48.

Djioua, M., O'Reilly, C., Plamondon, R. (2006). An Interactive Trajectory Synthesizer to Study Outlier Patterns in Handwriting Recognition and Signature Verification, *Proc. 18th International Conference on Pattern Recognition* (ICPR '06), Hong Kong, vol. 1, pp. 1124–127.

Djioua M., Plamondon R. (2007). Analysis and Synthesis of Handwriting Variability using the Sigma-Lognormal Model, *Proc. 13th Biennial Conference of the International Graphonomics Society*, (IGS 2007), Melbourne, Australia. pp. 19–22,

Djioua, M., Plamondon, R., Della Cioppa, A., Marcelli, A. (2007). Deterministic and Evolutionary Extraction of Delta-lognormal Parameters: Perfomance Comparaison, *International Journal of Pattern Recognition and Artificial Intelligence* (IJPRAI), vol. 21, no. 1, pp. 21–41.

Djioua, M., Plamondon, R. (2008a). An Interactive System for the Automatic Generation of Huge Handwriting Databases from a few Specimens, *Proc. International Conference on Pattern Recognition* (ICPR 2008), Tampa, Florida, 4 pages.

Djioua, M., Plamondon, R. (2008b). A New Methodology to Improve Myoelectric Signal Processing Using Handwriting, *Proc. 11th International Conference on Frontiers in Handwriting Recognition* (ICFHR 2008), Montréal, pp. 112–118.

Djioua, M., Plamondon, R. (2009a). A New Algorithm and System for the Characterization of Handwriting Strokes with Delta-Lognormal Parameter, *IEEE Transactions on Pattern Analysis and Machine Intelligence*, vol. 31, no. 11 pp. 2060–2072.

Djioua, M., Plamondon, R. (2009b). Studying the Variability of Handwriting Patterns using the Kinematic Theory, *Human Movement Science*, vol. 28, no. 5, pp. 588–601.

Djioua, M., Plamondon, R. (2009c). A Comparison of Some Analytical Models Describing the Velocity Profiles of Rapid Movements, *Proc. 14th Biennial Conf. of the International Graphonomics Society* (IGS 2009), Dijon, France, pp. 221–225.

Djioua, M., Plamondon, R. (2010). The Limit Profile of a Rapid Movement Velocity, *Human Movement Science*, 29 (1), pp. 48–61.

Djioua, M., Plamondon, R., Mathieu, P.A. (2011). Experimental Confirmation of the Proportional Effect Hypothesis of the Kinematic Theory of Rapid Movements, *Proc. 15th Biennial Conference of the International Graphomics Society* (IGS 2011) Cancun, Mexico, pp. 90–93.

Dooijes, E.H. (1983). Analysis of handwriting movements, *Acta Psychol.* 54, 99–114.

Duval, T., Rémi, C., Plamondon, R., O'Reilly, C. (2013). On the Use of the Sigma-Lognormal Model to Study Children Handwriting, *Proc. 16th Biennial Conf. of the International Graphonomics Society*, Nara, Japan, pp. 26–29.

Duval T., Rémi, C., Plamondon, R., Vaillant, J., O'Reilly, C. (2015). Combining Sigma-Lognormal modeling and classical features for analyzing graphomotor performances in kindergarten children, *Human Movement Science*, 43 pp. 183–201,

Edelman, S., Flash, T. (1987). A model of handwriting, *Biol. Cybern.* 57, 25–36.

Enderle, J.D., Wolfe, J.W. (1987). Time-optimal control of saccadic eye movements, *IEEE Trans. Biomed. Eng.* 34, 43–55.

Engelbrecht, S.E. (2001). Minimum principles in motor control, *J. Math. Psychol.* 45, 497–542.

Faci, N., Boyogueno Bibias, S.P., Plamondon, R., Bergeron, N. (2018). A New Experimental Set-up To Run Neuromuscular Tests, *Proc. Int. Conf. on Pattern Recognition and Artificial Intelligence: Workshop on the Lognormality Principle and its Applications*, CENPARMI, Montréal, Canada, pp. 753–757.

Faci, N., Désiré, N., Beauchamp, M., Gagnon, I., Plamondon, R., Lognormality in Child with Mild Traumatic Brain Injury: A Preliminary Pilot Study, (2018), *Proc. Int. Conf. on Pattern Recognition and Artificial Intelligence: Workshop on the Lognormality Principle and its Applications*, CENPARMI, Montréal, Canada, pp. 790–794.

Faundez-Zanuy, M., Fierrez, J., Ferrer, M.A., Diaz, M., Tolosana, R., Plamondon R. (2020). Handwriting Biometrics: Applications and Future Trends in e-Security and e-Health. in Press, *Cognitive Computation*, 14 pages, vol. 12, pp: 940–953.

Feldman, A.G. (1966). Functional tuning of the nervous system with control of movement or maintenance of a steady posture. II. Controllable parameters of the muscle, *Biophysics* 11, 565–578.

Feldman, A.G., Latash, M.L. (2005). Testing hypotheses and the advancement of science. recent attempts to falsify the equilibrium point hypothesis, *Exp. Brain Res.* 161 (1), 91–103.

Feng, C., Woch, A., Plamondon, R. (2002). A Comparative Study of Two Velocity Profile for Rapid Stroke Analysis, *Proc. 16th International Conference on Pattern Recognition* (ICPR 2002), Québec, Canada, vol. 4, pp. 52–55.

Ferrer, M.A. Diaz, M., Carmona-Duarte, C., Plamondon, R. (2020). iDeLog: Iterative Dual Spatial and Kinematic 2 Extraction of Sigma-Lognormal Parameters, *IEEE Trans. on Pattern Analysis and Machine Intellligence* (PAMI), 42(1): 114–125.

Fischer, A., Plamondon, R., O'Reilly, C., Savaria, Y. (2014). Neuromuscular Representation and Synthetic Generation of Handwritten Whiteboard Notes, *Proc. 14th International Conference on Frontiers on Handwriting Recognition*, Crete, Greece, pp. 222–227.

Fischer, A., Plamondon, R. (2017). Signature Verification Based on the Kinematic Theory of Rapid Human Movements, *IEEE Transactions on Human-Machine Systems*, vol. 47, no. 2, pp. 169–180.

Fitts, P.M. (1954). The Information Capacity of the Human Motor System in Controlling the Amplitude of Movements, *J. Experimental Psychology*, 47(6), pp. 381–391.

Flash, T., Hogan, N. (1985). The coordination of Arm Movements: An experimentally confirmed mathematical model, *The Journal of Neuroscience* 5, 1688–1703.

Galbally, J., Fierrez, J., Martinez-Diaz, M., Ortega-Garcia, J., Plamondon, R., O'Reilly, C. (2010). Kinematical Analysis of Synthetic Dynamic Signatures Using the Signal-Lognormal Model, *Proc. 12th International Conference on Frontiers in Handwrting Recognition* (ICFHR 2010), Kolkata, India, pp. 113–118.

Galbally, J., Fierrez, J., Martinez-Diaz, M., Ortega-Garcia, J., Plamondon, R. (2011). Quality Analysis of Dynamic Signature Based on the Sigma-Lognormal Model, *Proc. Int. Conf. On Document Analysis and Recognition* (ICDAR2011) Beijing, pp. 633–637.

Galbally, J., Plamondon, R, Fierrez, J., Ortega-Garcia, J. (2012a). Synthetic On-Line Signature Generation Part I: Methodology and Algorithms, *Pattern Recognition* 45 (7): 2610–2621.

Galbally, J., Fierrez, J., Ortega-Garcia, J., Plamondon, R. (2012b). Synthetic On-Line Signature Generation Part II: Experimental Validation, *Pattern Recognition*, 45 (7) 2622–2632.

Gangadhar, G., Joseph, D., Chakravarthy, V.S. (2007). An oscillatory neuromotor model of handwriting generation, *Int. J. Doc. Anal. Recognit.* 10 (2), 69–84.

Ghez C., Gordon J. (1987). Trajectory control in targeted force impulses. I. Role of opposing muscles, *Exp. Brain Res.* 67: 225–240.

Gielen C.C.A.M., Oosten K. van den, Pull ter Gunne F. (1985). Relation between EMG activation patterns and kinematic properties of aimed arm movements, *J. Motor Behav.* 11: 421–442.

Gomez-Barrero, M., Galbally, J., Fierrez, J., Ortega-Garcia, J., Plamondon, R. (2015). Enhanced On-Line Signature Verification Based on Skilled Forgery Detection Using Sigma-Lognormal Features, *Proc. 8th IAPR International Conference on Biometrics* (ICB), pp. 501–506.

Gomez-Barrero, M., Galbally J., Fierrez, J., Ortega-Garcia, J., Plamondon, R. (2013). Variations of Handwritten Signatures with Time: A Sigma-Lognormal Analysis, *Proc. 6th International Conference on Biometrics*, Madrid, Spain, Ps3.1–Ps3.6.

Guerfali, W., Plamondon, R. (1994). Robust Parameter Extraction Techniques for the Delta-Lognormal Model, *Proc. Vision Interface '94*, Banff, Alberta, Canada, pp. 218–225.

Guerfali, W., Plamondon, R. (1995a). Control Strategies for Handwriting Generation, *Proc. 7th Biennial Conference of the International Graphonomics Society*, London, Ontario, Canada, pp. 64–65.

Guerfali, W., Plamondon, R. (1995b). The Delta-Lognormal Theory for the Generation and Modelling of Cursive Characters, *Proc. 3rd International Conference on Document Analysis and Recognition*, Montréal, Canada, pp. 495–498.

Guerfali, W., Plamondon, R. (1995c). Signal Processing for the Parameter Extraction of the Delta-Lognormal Model in C. Archibald and P. Kwok (Ed.), *Research in Computer and Robot Vision,* World Scientific, Singapore, London, New Jersey, Hong Kong, pp. 217–232.

Guerfali, W., Plamondon, R. (1997). Effect of Variability on Letter Generation with the Vectorial Delta-Lognormal Model, *Proc. 1st Bresilian Symposium on Document Image Analysis*, pp. 74–83.

Guerfali, W., Plamondon, R. (1998). A New Method for the Analysis of Simple and Complex Planar Rapid Movements, *Journal of Neuroscience Methods*, vol. 82, no.1, pp. 35–45.

Harris, C.M., Wolpert, D.M. (1998). Signal-dependent noise determines motor planning, *Nature* 394, 780–784.

Hermes, H., LaSalle, J.P. (1969). *Functional analysis and time optimal control*. New York.

Hollerbach, J.M. (1981). An oscillation theory of handwriting, *Biol. Cybern.* 39 (2), 139–156.

Hogan, N. (1984). An organization principle for a class of voluntary movements, *The Journal of Neurosciences* 4, 2745–2754.

Impedovo, D., Pirlo, G., Mangini, F.M., Arbuzzi, D., Rollo, A., Balestrusi, A., Impedovo, S., Sarcinella, L., O'Reilly, C., Plamondon, R. (2015). Writing Generation Model for Health Care Neuromuscular System Investigation, *Computational Intelligence Methods for Bioinformatics and Biostatistics*, revised selected papers, Springer Post proceedings: pp. 1–16.

Ivancevic, V. G. and Ivancevic, T. T. (2010). *Human-like biomechanics. A unified mathematical approach to human biomechanics and humanoid robotics*. Dordrecht: Springer.

Johnson, M.D., Hyngstrom, A.S., Manuel, M., Heckman, C.J. (2012). Push–Pull Control of Motor Output, *The Journal of Neuroscience*, 32(13):4592–4599.

Kandel E.R., Schwartz J.H., Jessell T.M., Siegelbaum S.A., Hudspeth A.J. (Eds.) (2013). *Principles of Neural Science*, 5th ed. New York, NY: McGraw-Hill Professional.

Karniel, A., (2011), Open Questions in Computational Motor Control, *Journal of Integrative Neuroscience*, vol. 10, No. 3 pp. 385–411.

Kalveram, K.T. (1998). A neural oscillator model learning given trajectories, or how an allo-imitation algorithm can be implemented into a motor controller. In: Piek J.P.

(Ed.), *Motor Behavior and Human Skill: A Multidisciplinary Approach*, Human Kinetics, pp. 127–140.

Kelso, S. (1995). *Dynamic patterns: The self-organization of brain and behavior*. MIT Press, Cambridge, MA, USA.

Latash, M.L. (2012). *Fundamentals of Motor Control*, 1st ed. Academic Press, 364p.

Lamarche, F., Plamondon, R. (1984). Segmentation and Feature Extraction of Handwritten Signature Patterns, *Proc. 7th International Conference on Pattern Recognition*, Montréal, Canada, pp. 756–759.

Laniel, P., Faci, N., Plamondon, R., Beauchamp, M.H., Gauthier, B. (2018). Kinematic Analysis of Fast Pen Strokes in Children with ADHD using the Sigma-lognormal Model, *Proc. Int. Conf. on Pattern Recognition and Artificial Intelligence: Workshop on the Lognormality Principle and its Applications*, CENPARMI, Montréal, Canada, pp. 784–789.

Laniel, P., Faci, N., Plamondon, R., Beauchamp, M.H., Gauthier, B. (2019). Kinematic analysis of fast pen strokes in children with ADHD, *Applied Neuropsychology: Child*, 1–16.

Laurent, A., Plamondon, R., Begon, M. (2019). Pre-screening for Central or Peripheral Shoulder Fatigue using the Sigma-Lognormal Model, *Proc. Int Conf. of the International Graphonomics Society*, Cancun, Mexico, 6 pages.

Laurent, A., Plamondon, R., Begon, M. (2020). A proof of concept for pre-screening central and peripheral shoulder fatigue using the Sigma-Lognormal model, *Frontiers in Human Neuroscience* 14 May 2020, Special issue on Graphonomics and Your Brain on Art, Creativity, and Innovation. DOI: 10.3389/fnhum.2020.00171.

Lebel, K., Nguyen, H., Duval, C., Plamondon, R., Boissy, P. (2017). Capturing the Cranio-Caudal Signature of a Turn with Inertial Measurement Systems: Methods, Parameters Robustness and Reliability, *Front Bioengineering Biotechnologies*, vol. 5, pp. 51.

Lebel, K., Duval, C., Nguyen, H., Plamondon, Réjean, Boissy, P. (2018a). Turn cranio-caudal signature assessment from inertial systems for mobility deficit identification in Parkinson's disease patients, *Parkinsonism & Related Disorders* 46 (E24-E25).

Lebel, K, Nguyen, H., Duval, C., Plamondon, R, Boissy, P. (2018b). Cranio-Caudal Kinematic Turn Signature Assessed with Inertial Systems as a Marker of Mobility Deficits in PD, *Frontiers in Neurology*, 29 January 2018 https://doi.org/10.3389/fneur.2018.00022.

Leclerc, F., Plamondon, R. (1990). Validation d'un modèle de la fonction de transfert du bras sur des signatures manuscrites, *Proc. Vision Interface '90*, Halifax, Canada, 105–111.

Leclerc, F., Plamondon, R., Lorette, G. (1992). Des gaussiennes pour la modélisation des signatures et la segmentation des tracés manuscrits, *Traitement du Signal*, vol. 9, no. 4, pp. 347–358.

Leduc, N. Plamondon, R. (2001). A new approach to study human movements: The three-dimensional Delta-Lognormal model, *Proc. 10th Biennial Conference of the*

*International Graphonomics Society* (IGS 2001), University of Nijmegen, the Netherlands, pp. 98–102.

Leiva, L. A., Martin-Albo, D., Plamondon, R. (2016a). Gestures a Go Go: Authoring Synthetic Human-Like Stroke Gestures Using the Kinematic Theory of Rapid Movements, *ACM Transactions on Intelligent Systems and Technology*, 2016, vol. 7, no. 2.

Leiva, L. A., Huang, J., Plamondon, R. (2016b). Strokes of insight: User intent detection and kinematic compression of mouse cursor trails, *Information Processing & Management*, 2016, vol. 52, no. 6, pp. 989–1003.

Leiva, L. A., Martin-Albo, D. (2017). The Kinematic Theory Produces Human-Like Stroke Gestures, *Interacting with computers*, vol. 29, no. 4, pp. 552–565.

Leiva, L.A., Martín-Albo, D., Plamondon, R., Vatavu, R.D. (2018a). KeyTime: Super-Accurate Prediction of Stroke Gesture Production Times, *CHI 2018*, April 21–26, 2018, Montreal, QC, Canada, 12 pages.

Leiva, L., Martin-Albo, D., Vatatu, R.D., Plamondon, R. (2018b). Gesture Synthesis for Human Computer Interaction, *Proc. Int. Conf. on Pattern Recognition and Artificial Intelligence: Workshop on the Lognormality Principle and its Applications*, CENPARMI, Montréal, Canada, pp. 758–761.

Leiva, L.A., Vatavu, R.D., Martín-Albo, D., Plamondon, R. (2020). Omnis Prædictio: Estimating the Full Spectrum of Human Performance with Stroke Gestures, *Int. J. Hum. Comput. Stud.* 142, DOI:10.1016/j.ijhcs.2020.102466.

Lefebvre, O., Riba, P., Fournier, C., Fornes, A., Llados, J., Plamondon R., Gagnon-Marchand, J. (2015). Monitoring Neuromotricity On-line: A Cloud Computing Approach, *Proc. 17th Biennial Conference of the International Graphonomics Society*, Pointe-à-Pitre, Guadeloupe, pp. 63–66.

Maarse, F.J. (1987). *The Study of Handwriting Movement: Peripheral Models and Signal Processing Tehcniques*. Swets & Zertlinger, Lisse, The Netherlands.

Marcelli, A., Della Cioppa, A., Djioua, M., Plamondon, R. (2006). Reverse Handwriting from Electronic Ink to Generation Model Parameter Estimation, A Comparative Study, *Second Italian Workshop on Evolutionary Computation*, Collegio S. Chiara, Siena (Italy) pp. 1–10.

Marcelli, A., Parziale, A., Senatore, R. (2013). Some Observations on Handwriting from a Motor learning Prespective, *Proc. ICDAR_AFHA*, Washington DC, pp. 6–10.

Martin-Albo, D., Plamondon, R., Vidal, E. (2014). Training of On-line Handwriting Text Recognizers with Synthetic Text Generated Using the Kinematic Theory of Rapid Human Movements, *Proc. 14th International Conference on Frontiers in Handwriting Recognition*, Heraklion, pp. 543–548.

Martin-Albo, D., Plamondon, R., Vidal, E. (2015). Improving Sigma-Lognormal Parameter Extraction, *Proc. 13th International Conference on Document Analysis and Recognition* (ICDAR 2015), Nancy, France, pp. 286–290.

Meyer, D.E., Smith, J.E.K., Kornblum, S., Abrams, R.A., Wright, C.E. (1990). Speed-Accuracy Trade-Offs in Aimed movements: Toward a Theory of Rapid Voluntary

Action, In: *Attention and Performance XIII: Motor Representation and Control*. M. Jannerod (Ed.), Erlbaum.

Morasso, P., Mussa Ivaldi, F.A., Ruggiero, C. (1983). How a discontinuous mechanism can produce continuous patterns in trajectory formation and handwriting, *Acta Psychol.* 54, 83–98.

Nadeau A., Lungu O., Boré A., Plamondon R., Duchesne C., Robillard M.-È., Bobeuf F., Lafontaine A.-L., Gheysen F., Bherer L. and Doyon J. (2018a). A 12-Week Cycling Training Regimen Improves Upper Limb Functions in People With Parkinson's Disease, *Front. Hum. Neurosci.* 12:351. doi: 10.3389/fnhum.2018.0035.

Nadeau,A., Lebel, K., Carbo, J., Boré, A., Lungu, O., Doyon, J., Nguyen, H., Duval C., Boissy, P., Vincent, N., Zapula K, Plamondon, R. (2018b). Exploiting the Lognormality Principle: Three Ongoing Projects, *Proc. Int. Conf. on Pattern Recognition and Artificial Intelligence: Workshop on the Lognormality Principle and its Applications*, CENPARMI, Montréal, Canada, pp. 800–805.

Neilson, P.D. (1993). The problem of redundancy in movement control: the adaptive model theory approach, *Psychol. Res.* 55, 99–106.

Neilson, P.D., Neilson, M.D. (2005). An overview of adaptive model theory: Solving the problems of redundancy, resources and nonlinear interactions in human movement control, *J. Neural Eng.* 2 (3), 279–312.

Neilson, P. D. and Neilson, M. D. (2010). On theory of motor synergies, *Human Movement Science*, 29, 655–683.

Nielson, P.D., Nielson, M.D., Bye, R.T. (2015). A Riemannian geometry theory of human movement: The geodesic synergy hypothesis, *Human Movement Science*, 44, pp. 42–72.

Nelson, W.L. (1983). Physical principles for economies of skilled movements, *Biol. Cybern.* 46, 135–147.

O'Reilly, C., Plamondon, R. (2007). A Software Assistant for the Design and Analysis of Neuromuscular Tests, *Biomedical Circuits and Systems Conference* (Biocas 2007), Healthcare Technology Conference Proceedings, Montréal, Canada, 4 pages.

O'Reilly C., Plamondon R. (2008). Automatic Extraction of Sigma-Lognormal Parameters on Signatures, *Proc. 11th International Conference on Frontiers in Handwriting Recognition* (ICFHR 2008), pp. 216–222.

O'Reilly, C., C., Plamondon, R. (2009a). Development of a Sigma-Lognormal Representation for On-Line Signatures, *Pattern Recognition*, Special Issue on Frontiers in Handwriting Recognition, pp. 3324–3337.

O'Reilly, C., Plamondon, R. (2009b). A Lognormal Framework for Human Movement Rehabilitation, Tan Yen Kheng, *Rehabilitation Engineering* available online at: http//sciyo.com/books/show/title/rehabilitation-engineering, INTECH Publishing: 157-172, 2009.

O'Reilly, C., Plamondon, R., Lebrun, L.H., Mathieu, P.A., Clément, B. (2009). Movement Analysis for Stroke Susceptibility Assessment, *Proc. 14th Biennial Conf. of the International Graphonomics Society* (IGS 2009), Dijon, France, pp. 48–51.

O'Reilly, C., Plamondon, R. (2010). Prototype-based Methodology for the Statistical Analysis of Local Features in Stereotypical Handwriting Tasks, *Proc. Int. 20th Conf. Pattern Recognition* (ICPR 2010), Istanbul, Turkey pp. 1864–1867.

O'Reilly, C., Plamondon, R. (2011a). Can computer mice be used as low-cost devices for the acquisition of planar human movement velocity signals? *Behavior Research Methods*, 43(1): 229–238, 2011.

O'Reilly, C., Plamondon, R. (2011b). Impact of the principal stroke risk factors on human movements, *Human Movement Science* 30(4): 792–806, 2011.

O'Reilly, C., Plamondon, R. (2011c). The Coupling of Agonist Commands in Speed/Accuracy Tradeoffs, *Proc. 15th Biennial Conf. of the International Graphonomics Society* (IGS 2011), Cancun, Mexico, pp. 94–97.

O'Reilly, C., Plamondon, R. (2012a). A globally Optimal Estimator for the Delta-Lognormal Modelling of Fast Reaching Movements, *IEEE Transactions on Systems, Man, and Cybernetics, Part B, Cybernetics*, 42(5): 1428–1442.

O'Reilly, C., Plamondon, R. (2012b). Looking for the Brain Stroke Signature, *Proc. 21st International Conference on Pattern Recognition* (ICPR 2013), Tsukuba, Japan, pp. 1811–1814.

O'Reilly, C., Plamondon, R. (2012c). Design of a Neuromuscular Disorders Diagnostic System using their applications, *Proc. ISSPA 2012*, Montréal, Canada, pp. 787–792.

O'Reilly, C., Plamondon, R. (2013). Agonistic and antagonistic Interaction in speed/accuracy trade-off: A Delta-Lognormal perspective, *Human Movement Science*, vol. 32, issue 5: 1040–1055.

O'Reilly, C., Plamondon, R., Landou, M.K., Stemmers, B. (2013). Prediction of the Time Occurence a Motor ERP based on the Kinematic Analysis of Movement with the Delta-Lognormal Model, *European J. Neuroscience*, 27: 173–180.

O'Reilly, C, Plamondon, R., Lebrun, L-H. (2014). Linking Brain Stroke Risk Factors to human Movement Features for the Development of Preventive Tools, *Frontiers in Aging Neuroscience*, doi: 10.3389/fnagi. 2014. 00150.: 1–13, 2014.

Pan, Z., Talwar, S., Plamondon; R., van Gemmert, A. (2019). Characteristics of Bi-Directional Unimanual and Bimanual Drawing Movements: The application of the Delta-Lognormal and Sigma Lognormal Models, *Pattern Recognition Letters*, vol. 121, pp. 97–103.

R. Parisi, A. Parziale, A. Marcelli. (2018). Some observations on lognormality and motor control in handwriting, *Proceedings of the International Conference on Pattern Recognition and Artificial Intelligence Montreal: Workshop on the Lognormality Principle and its Applications*, CENPARMI, Concordia University, pp. 732–737.

Pirlo, G., Impedovo, D., Plamondon, R., O'Reily, C. (2014). Stability Analysis of Online Signatures in the Generation Domain. G. Pirlo, D. Impedovo, M. Fairhurst (Eds.). *Advances in Digital Handwritten Signature Processing, A Human Artefact for e-Society*, Selected papers from the Workshop on "Emerging Aspects in Handwritten Signature Processing", World Scientific ISBN: 978-981-4579-62-9, pp. 1–12.

Plamondon, R., Brault, J.J., Robillard, P.N. (1983). Optimizing the Design of an Accelerometric Pen for Signature Verification, *1983 Conference on Crime Countermeasure and Security*, Lexington, Kentucky, May 1983, pp. 35–40.

Plamondon, R., Brault J.J. (1983). A System for Signature Analysis and Verification Based on an Accelerometric Pen, *Proc. 1983 International Carnahan Conference on Security Technology*, Zurich, Switzerland, October 1983, pp. 157–164.

Plamondon, R., Lamarche, F. (1986). Modelization of Handwriting: a System Approach, in *Graphonomics: Contemporary Research in Handwriting*, H.S.R. Kao, G.P. Van Galen, R. Hoosain (Eds.), Elsevier Science Publishers, North Holland, Amsterdam, New York, Oxford, Tokyo, pp. 169–183.

Plamondon, R., Maarse, F.J. (1987). A Neuron Oriented Representation to Compare Biomechachal Models, *Proc. 3rd International Symposium on Handwriting and Computer Applications*, Montréal, Canada, pp. 2–4.

Plamondon, R. (1987). What Does Differential Geometry Tell Us About Handwriting Generation, *Proc. 3rd International Symposium on Handwriting and Computer Applications*, Montréal, Canada, pp. 11–13.

Plamondon, R. (1989a). A Handwriting Model Based on Differential Geometry, *Computer Recognition and Human Production of Handwriting*, R. Plamondon, C.Y. Suen, M. Simner (Eds.) World Scientific Publish, Singapore, New Jersey, London, Hong Kong, pp. 179–192.

Plamondon, R. (1989b). Handwriting Control, a Functional Model, *Models of Brain Functions*, R. Cotterill (Ed.), Academic Press, Cambridge, New York, Port Chester, Melbourne, Sydney, pp. 563–574.

Plamondon, R., Maarse, F.J. (1989). An Evaluation of Motor Models of Handwriting, *IEEE Transactions on Systems, Man and Cybernetics*, vol. 19, no. 5, pp. 1060–1072.

Plamondon, R., Stelmach, G.E., Teasdale, N. (1990). Motor Program Coding Representation from Handwriting Generator Model: The Production of Lines Responses, *Biological Cybernetics*, vol. 63, pp. 443–451.

Plamondon, R. (1990). On the Origin of Bell-Shaped Velocity Profiles in Rapid-Aimed Movements, *NATO ASI Tutorials in Motor Neuroscience*, Kluwer Academic Publ., Dordrecht, Corsica, France, pp. 283–295.

Plamondon, R., Yergeau, P. (1990). A System for the Analysis and Synthesis of Handwriting, *Proc. International Workshop on Frontiers in Handwriting Recognition*, Montréal, Canada, pp. 167–179.

Plamondon, R., Clément, B. (1991). Dependence of Peripheral and Central Parameters Describing Handwriting Generation on Movement Direction, *Human Movement Science*, Thematic Issue on Handwriting, vol. 10, pp. 193–221.

Plamondon, R., Yu, L., Stelmach, G., Clément, B. (1991). On the Automatic Extraction of Biomechanical Information from Handwriting Signals, *IEEE Transactions on Systems, Man and Cybernetics*, vol. 21, no. 1, pp. 90–101.

Plamondon, R. (1991). What Does Velocity Information Reveals About Handwriting Generation? *Proc. 5th Handwriting Conference*, Tempe, Arizona, USA, pp. 9–10.

Plamondon, R., Yergeau, P., Brault, J.J. (1991). A Multilevel Signature Verification System, *Proc. 2nd International Workshop on Frontiers in Handwriting Recognition*, Bonas, France, pp. 293–301.

Plamondon, R. (1992a). A Theory of Rapid Movements, *Tutorials in Motor Behavior II*, G.E. Stelmach, J. Requin (Eds.), Elsevier Science Publishers, North Holland, Amsterdam, New York, Oxford, Tokyo, pp. 55–69.

Plamondon, R. (1992b). A Model-Based Segmentation Framework for Computer Processing of Handwriting, *Proc. 11th International Conference on Pattern Recognition*, The Hague, The Netherlands, pp. 303–307.

Plamondon, R., Yergeau, P., Brault, J.-J. (1992). A Multi-Level Signature Verification System, *From Pixels to Features III*, S. Impedovo, J.C. Simon (Eds.), Elsevier Publ. Co. Amsterdam, New York, London, Tokyo, pp. 293–301.

Plamondon, R., Alimi, A, Yergeau, P. Leclerc, F. (1993). Modelling Velocity Profiles of Rapid Movements: A Comparative Study, *Biological Cybernetics*, vol. 69, no. 2, pp. 119–128.

Plamondon, R. (1993a). Understanding Stroke Generation: A Global Neuro-Muscular Approach, *Proc. 6th International Conference on Handwriting and Drawing*, Paris, France, pp. 1–3.

Plamondon, R. (1993b). Looking at Handwriting Generation from a Velocity Control Perspective, *Acta Psychologica*, Special Issue on Motor Control of Handwriting, vol. 82, pp. 89–101.

Plamondon, R. (1993c). A Model-Based Dynamic Signature Verification System, *Fundamentals in Handwriting Recognition*, S. Impedovo (Ed.), serie F Computer and Systems Sciences, no. 124, pp. 417–434.

Plamondon, R., Carrières, P. (1993). An Interactive Tool for Handwriting Learning, *Proc. 6th International Conference on Handwriting and Drawing*, Paris, France, pp. 183–185.

Plamondon, R. (1994a). Handwriting Generation: The Delta-Lognormal Theory, *Proc. 4th International Workshop on Frontiers in Handwriting Recognition*, Taipei, Taiwan, pp. 1–10.

Plamondon, R. (1994b). The Design of an On-Line Signature Verification System: From Theory to Practice, *International Journal of Pattern Recognition and Artificial Intelligence*, Special Issue on Signature Verification, vol. 8, no. 3, pp. 795–811.

Plamondon, R. (1995a). The Emergence of Regularity and Invariance in Neuromuscular Systems: The Case of Rapid Human Movement, *Proc. 9th Conference of the International Society for the Study of Time*, St-Adèle, Québec, Canada, pp. 27.

Plamondon, R. (1995b). A Delta-Lognormal Model for Handwriting Generation, *Proc. 7th Biennial Conference of the International Graphonomics Society*, London, Ontario, Canada, pp. 126–127.

Plamondon, R. (1995c). A Kinematic Theory of Rapid Human Movements: Part I: Movement Representation and Generation, *Biological Cybernetics*, vol. 72, no. 4, pp. 295–307, 1995.

Plamondon, R. (1995d). A Kinematic Theory of Rapid Human Movements: Part II: Movement Time and Control, *Biological Cybernetics*, vol. 72, no. 4, pp. 309–320, 1995.

Plamondon, R. (1995e). The Handwritten Signature as a Biometric Identifier: Psychophysical Model and System Design, *Proc. European Conference on Security and Detection*, Brighton, UK, pp. 23–28.

Plamondon, R., Privitera, C.M. (1995). A Neural Model for Generating and Learning a Rapid Movement Sequence, *Biological Cybernetics*, vol. 74, no. 2, p. 117–130.

Plamondon, R., Guerfali, W. (1996). Why Handwriting Segmentation Can Be Misleading?, *Proc. 13th Int. Conf. on Pattern Recognition*, vol. D, Vienna, Austria, pp. 396–400.

Plamondon, R., Alimi, A. (1997). Speed/Accuracy Tradeoffs in Target Directed Movements, *Behavioral and Brain Sciences*, vol. 20, no. 2, pp. 279–349.

Plamondon, R., Guerfali, W. (1997). The Origin of the 2/3 Power Law, *Proc. 8th Biennial Conference of the International Graphonomics Society* (IGS 1997), Genova, Italy, pp. 17–20.

Plamondon, R., Guerfali, W., Li, X. (1997). The Generation of Oriental Characters: New Perspectives for Automatic Handwriting Processing, *Proc. 17th Int. Conf. on Computer Processing of Oriental Languages*, Hong Kong, pp. 113–120.

Plamondon, R. (1998). A Kinematic Theory of Rapid Human Movements: Part III: Kinetic Outcomes, *Biological Cybernetics*, vol. 78, pp. 133–145, 1998.

Plamondon, R., Guerfali, W. (1998a). The Generation of Handwriting with Delta-Lognormal Synergies, *Biological Cybernetics*, vol. 78, pp. 119–132, 1998.

Plamondon, R., Guerfali, W., Li, X. (1998b). The Generation of Oriental Characters: New Perspectives for Automatic Handwriting Processing, *International Journal of Pattern Recognition and Artificial Intelligence* (IJPRAI), vol. 12, no. 1, pp. 31–44.

Plamondon, R., Guerfali, W. (1998c). The 2/3 Power Law: When and Why?, *Acta Psychologica*, vol. 100, pp. 85–96.

Plamondon, R., Guerfali, W., Lalonde, M. (1998). Vérification automatique de signatures: compte rendu d'une expérience pilote grand public, *Comptes Rendus 1er Colloque International Francophone sur l'Écrit et le Document*, Québec, pp. 325–334.

Plamondon, R., Guerfali, W., Lalonde, M. (1999). Automatic Signature Verification: A Report on a Large-Scale Public Experiment, *Proc. 9th Biennial Conference of the International Graphonomics Society* (IGS 1999), Singapore, pp. 9–13.

Plamondon, R., Privitera, C.M. (1999). The Segmentation of Cursive Handwriting: An Approach Based on Off-Line Recovery of the Motor-Temporal Information, *IEEE Transactions on Image Processing*, vol. 8, no. 1, pp. 80–91.

Plamondon, R., Lopresti, D., Schomaker, L.R.B., Srihari, R. (1999). On-Line Handwriting Recognition, *Wiley Encyclopedia of Electrical and Electronics Engineering*, J.G. Webster (Ed.), John Wiley & Sons, N.Y., vol. 15, pp. 123–146.

Plamondon, R., Feng, C., Woch, A. (2003). A Kinematic Theory of Rapid Human Movements: Part IV: A Formal Mathematical Proof and New Insights, *Biological Cybernetics*, vol. 89, pp. 126–138.

Plamondon, R., Djioua, M. (2005). Handwriting Stroke Trajectory Variability in the Context of the Kinematic Theory, *Proc. 12th Biennial Conference of the International Graphonomics Society* (IGS 2005), Salerno, Italy, pp. 250–254.

Plamondon, R., Djioua, M. (2006). A Multi-Level Representation Paradigm for Handwriting Stroke Generation, *Human Movement Science*, vol. 25, no, pp. 586–607.

Plamondon, R., Li, X., Djioua, M. (2007). Extraction of Delta-Lognormal parameters from Handwriting Strokes, *Frontiers of Computer Science in China*, vol. 1, no. 1, pp. 106–113.

Plamondon, R., Djioua, M., O'Reilly, C. (2009). La Théorie Cinématique des Mouvements Humains Rapides: Développements Récents, *Traitement du Signal*, Numéro Spécial: Le Document Écrit, vol. 26, No. 5, pp. 377–394.

Plamondon, R. (2010). Neuromuscular Studies of Handwriting Generation and Representation, *Proc. 12th International Conference on Frontiers in Handwriting Recognition*, Kolkata, India, pp. 261.

Plamondon, R., Djioua, M., Mathieu, P.A. (2013a). The Kinematic Theory of Rapid Human Movements: Experimental Observation of the Proportional Effect Hypothesis, *Human Movement Science*, vol. 32, issue 5, pp. 1026–1039.

Plamondon, R., O'Reilly, C., Rémi, C., Duval, T. (2013b). The lognormal handwriter: learning, performing and declining, *Frontiers in Psychology: Cognitive Science* doi: 10.3389/fpsyg.2013.00945, Special Issue in Cognitive Science, *Writing words: From brain to hand(s)*, Topic Editor(s): Sonia Kandel, Marieke Longcamp, pp. 1–14.

Plamondon, R., O'Reilly, C., Galbally, J., Almaksour, A., Anquetil, E. (2014a). Recent Developments in the Study of Rapid Human Movements with the Kinematic Theory: Applications to Handwriting and Signature Synthesis, *Pattern Recognition Letters*, vol. 35, (1): 225–235.

Plamondon, R., O'Reilly, C., Ouellet-Plamondon, C. (2014b). Strokes against Stroke-Strokes for Strides, *Pattern Recognition*, 44 (3): 929–944.

Plamondon, R., Pirlo, G., Impedovo, D. (2014c). On-line Signature Verification, Doermann D., Tombre K., *Handbook of Document Image Processing and Recognition*, Springer, Chapitre 27, 917–948.

Plamondon, R., O'Reilly, C., Galbally, J., Almaksour, A., Anquetil, E. (2014). Recent Developments in the Study of Rapid Human Movements with the Kinematic Theory: Applications to Handwriting and Signature Synthesis, *Pattern Recognition Letters*, vol. 35, (1): 225–235, 2014.

Plamondon, R., Pirlo, G., Anquetil, É., Rémi, C., Teulings, H-L., Nakagawa M. (2018). Personal digital bodyguards for e-security, e-learning and e-health: A Prospective Survey, *Pattern Recognition*, 81, pp. 633–659.

Privitera, C.M., Plamondon, R. (1995a). A System for Scanning and Segmenting Cursively Handwritten Words into Basic Strokes, *Proc. 3rd International Conference on Document Analysis and Recognition*, Montréal, Canada, pp. 1047–1050.

Privitera, C., Plamondon, R. (1995b). A Self-Organizing Neural Model for Learning and Generating Target-Directed Movements in the Context of a Delta-Lognormal Synergy, *Proc. IEEE International Conference on Neural Networks*, Perth, Australia, vol. 4, pp. 1999–2004.

Ramaiah, C., Plamondon, R., Govindaraju, V. (2013). Handwritten CAPTCHA generation based on the Sigma-Lognormal model. *Proc. 16th Biennial Conf. on the International Graphonomics Society*, Nara Japan, pp. 105–108.

Ramahia, C., Plamondon, R., Govindaraju, V. (2014). A Sigma-Lognormal Model For Handwritten Text CAPTCHA Generation, *Proc. 22nd International Conference on Pattern Recogniton* (ICPR 2014), Stockholm, Sweden, pp. 250–255.

Ramahia, C., Plamondon, R., Govindaraju, V. (2015). A Sigma-Lognormal Model For Character Level CAPTCHA Generation, *Proc. 13th International Conference on Document Analysis and Recognition* (ICDAR 2015), Nancy, France pp. 966–970.

Rémi, C., Vaillant, J., Plamondon, R., Prévost, L., Duval, T. (2015). Exploring for scribbling profiles in kindergarten children, *Proc. 17th Biennial Conference of the International Graphonomics Society*, Pointe-à-Pitre, Guadeloupe, pp. 79–82.

Rémi, C., Vaillant, J., Nagau, J., Plamondon, R. (2018a). Can the Sigma-Lognormal Modelling Help to Monitor Child Graphomotor Skill Progress? *Proc. Int. Conf. on Pattern Recognition and Artificial Intelligence: Workshop on the Lognormality Principle and its Applications*, CENPARMI, Montréal, Canada, pp. 780–783.

Rémi C., Nagau, J., Vaillant, J., Dorville, A., Plamondon, R. (2018b). Multimodal acquisition and analysis of children handwriting for the study of the efficiency of their handwriting movements: the @MaGma challenge, *Proc. 18th Int. Conf. on Frontiers in Handwriting Recognition*, pp. 459–464.

Reznakova, M., Tencer, L., Plamondon, R., Cheriet, M. (2015). The generation of synthetic handwritten data for improving on-line learning, *Proc. 17th Biennial Conference of the International Graphonomics Society*, Pointe-à-Pitre, Guadeloupe, pp. 55–58.

Reznakova, M., Tencer, L., Plamondon, R., Cheriet, M. (2017). Forgetting of unused classes in missing data environment using automatically generated data: Application to on-line handwritten gesture command recognition, *Pattern Recognition*, vol. 72, pp. 355–367.

Rosenbaum, D.A. (2009). *Human Motor Control*, 2nd ed. Academic Press, 528p.

Schindler, R., Bouillon M., Plamondon, R., Fischer, A. (2018). Extending the Sima-Lognormal Model of the Kinematic Theory to Three Dimensions, *Proc. Int. Conf. on Pattern Recognition and Artificial Intelligence: Workshop on the Lognormality Principle and its Applications*, CENPARMI, Montréal, Canada, pp. 748–752.

Shadmehr, R., Wise, S.P. (2005). *The computational neurobiology of reaching and pointing: A foundation for motor learning*. Cambridge, Mass.: MIT Press.

Schmidt, R.A.; Lee, T.D. (2011). *Motor control and learning: A behavioral emphasis.* Champaign, IL: Human Kinetics.

Schomaker, L.R.B., Plamondon, R. (1990). The Relation between Axial Pen Force and Pen point Kinematics in Handwriting, *Biological Cybernetics*, vol. 63, pp. 277–289.

Schomaker, L.R.B. (1991). *Simulation and recognition of handwriting movement: A vertical approach to modeling human motor behavior*, Ph.D. Thesis, Nijmegen University, Netherlands.

Senatore, R., Marcelli, A. (2012). A neural scheme for procedural motor learning of handwriting, *Proc. Frontiers in Handwriting Recognition* (ICFHR2012), pp. 659–664.

Stein, R.B. (1982). What muscle variable(s) does the nervous system control in limb movements? *Behavioral and Brain Science*, 5(4), pp. 535–541.

Tanaka, H., Krakauer, J.W., Qian, N. (2006). An optimization principle for determining movement duration, *J. Neurophysiol.* 95, 3875–3886.

Thomassen, A.J.W.M., Keuss, P.J.G., van Galen, G.P. (1983). Motor aspects of handwriting, *Acta Psychologica* 54 (1–3), 354.

Todorov, E. (2004). Optimality principles in sensorimotor control, *Nature Neuroscience Review*, 7(9), pp. 907–915.

van Galen, G.P., Teulings, H.L.T. (1983). The independent monitoring of form and scale factors in Handwriting, *Motor Aspects of Handwriting*. North Holland, Amsterdam, The Netherlands.

van Gemmert, A., Plamondon, R., O'Reilly, C. (2013). Using the Sigma-Lognormal model to investigate handwriting of individuals with Parkinson's disease, *Proc. 16th Biennial Conf. on the International Graphonomics Society*, Nara Japan, pp. 119–122.

Ungurean, O.-C., Vatavu, R.-D., Leiva, L.A. Plamondon, R. (2018). Gesture Input for Users with Motor Impairments on Touchscreens: Empirical Results based on the Kinematic Theory. *CHI '18 Extended Abstracts*, April 21–26, 2018, Montreal, QC, Canada, 6 pages.

Uno, Y., Kawato, M., Suzuki, R. (1989). Formation, Control of optimal trajectory in human multijoint arm movement, *Biol. Cybern.* 61, 89–101.

Wada, Y., Kawato, M. (1995). A theory for cursive handwriting based on the minimization principle, *Biol. Cybern.* 73 (1), 3–13.

Woch, A., Plamondon, R. (2001). Rapid Movement Analysis with the Delta-Lognormal Model: Towards a Better Understanding of Movement Generation, *Proc. 10th Biennial Conference of the International Graphonomics Society* (IGS 2001), Nijmegen, The Netherlands, pp. 165–169.

Woch, A., Plamondon, R. (2003). The Problem of Movement Primitives in the Context of the Kinematic Theory, *Proc. 11th Biennial Conference of the International Graphonomics Society* (IGS 2003), pp. 67–71.

Woch, A., Plamondon, R. (2004). Using the Framework of the Kinematic Theory for the Definition of a Movement Primitive, *Motor Control*, Special Issue, vol. 8, no. 4, pp. 547–557.

Woch, A., Plamondon, R. (2007). Analysis of Movement Primitives with the model: Insights on the Age Effect, *Proc. 13th Biennial Conference of the International Graphonomics Society* (IGS 2007), Melbourne, Australia, pp. 56–60.

Woch, A., Plamondon R. (2010). Characterization of Bi-directional Movement Primitives and their Agonist-Antagonist Synergy with the Delta-Lognormal Model, *Motor Control*, vol. 14 (1), pp. 1–25.

Woch, A., Plamondon, R., O'Reilly, C. (2011). Kinematic Characteristics of Bidirectional Delta-Lognormal Primitives in Young and Older subjects, *Human Movement Science* 30(1), pp. 1–17.

Zanone, P-G., Athènes, S., Sallagoity, I., Albaret, J-M. (2005). Switching among graphic patterns is governed by coordination dynamics of handwriting, *Proc. 12th Biennial Conference of the International Graphonomics Society*, Salerno, Italy, pp. 255–260.

https://doi.org/10.1142/9789811226830_0002

# Chapter 2

# Improving Handwritten Signatures Fluency via the Lognormality Principle

Moises Diaz*, Miguel A. Ferrer†, Cristina Carmona‡ and Réjean Plamondon§

*Universidad del Atlantico Medio,
Carretera del Fondillo 4, Las Palmas de Gran Canaria, 35017, Spain
Email: moises.diaz@atlanticomedio.es

†Instituto Universitario para el Desarrollo Tecnológico y la Innovación en Comunicaciones, Universidad de Las Palmas de Gran Canaria, Spain
Email: miguelangel.ferrer@ulpgc.es

‡Instituto Universitario para el Desarrollo Tecnológico y la Innovación en Comunicaciones, Universidad de Las Palmas de Gran Canaria, Spain
Email: ccarmona@idetic.eu

§Laboratoire Scribens, Département de génie électrique,
Polytechnique Montréal, Montréal, PQ, Canada
Email: rejean.plamondon@polymtl.ca

This chapter proposes two efficient methods to modify the fluency of dynamic signatures. The main idea is to modify the number of velocity minima or virtual target points and reconstruct the signature with the new virtual target points. If the number of virtual target points is reduced, the fluency is improved, and vice versa. The modification of the virtual target points is accomplished initially by linking the samples of an on-line signature by 8-connected Bresenham's lines to obtain the interpolated trajectory. Then, the most perceptually important points are estimated from the corners in the trajectory. To this end, two methods are proposed. The first method, which we term *resampling-wise*, develops a lognormal synthetic velocity profile used for resampling the static trajectory. The second method, *recovering-wise*, consists in estimating the virtual target points from the perceptually important points in the trajectory, linking them by circular trajectories, and reconstructing the dynamic trajectory. Additionally, both methods can be used to generate synthetic on-line signatures from static trajectories. Both methods' efficiency has been tested in automatic signature verification by increasing skilled forgeries' fluidity with the proposed methods.

## 1. Introduction

Fluent handwriting can be understood as the handwriting produced by a mature person, a behavior which is fully automated after over-learning and deliberate practice. This is particularly true for signatures that are defined, over-learned and practiced over the years, often incorporating design flourishes [Diaz-Cabrera *et al.* (2015)]. The result is a naturally smooth and articulated trajectory, often referred to as ballistic, which optimizes the number of strokes needed to write a particular autograph. In the context of the lognormality principle, the learning objective to produce any fluent signature is to minimize the number of lognormals needed to produce it, minimizing the energy required to generate these. From a practical point of view, the number of strokes, as well as the kinematics of the signing process, can be analyzed through the Sigma-Lognormal model [Plamondon and Djioua (2006)], which decomposes the whole movement into a combined sum of single stroke vectors. Such a decomposition can be achieved by using either the ScriptStudio [O'Reilly and Plamondon (2009)] or iDeLog [Ferrer *et al.* (2020)] software. The former iteratively processes an observed velocity signal from a rapid human movement with a vector sum of lognormals, while iDeLog also uses lognormal functions but it does so to optimize both the velocity and the trajectory at the same time.

The lognormality principle and the Sigma-Lognormal model can thus help in identifying fluid and non-fluid signatures, a non-fluid one being defined as a non-optimal trajectory that uses too many and short strokes. This is often the case for example when someone tries to imitate or forge a signature with little effort or practice. The detection of skilled forged signatures has been demonstrated in the literature. Recent papers [Gomez-Barrero *et al.* (2015); Fischer and Plamondon (2017)] have shown an improvement in the verification of dynamic signatures when Sigma-Lognormal parameters and function-based features are combined. It has been shown that what is generally referred to as skilled forgeries are less fluid signatures with more strokes than what is produced by genuine writers: when the fluidity improves, the number of strokes and their associated virtual target points decrease since the lognormal parameters better approximate a given trajectory.

To gain insights into the underlying processes that produce fluent handwritten signatures and the transition from non-fluent to fluent signatures and vice versa, studying how the fluency of a given piece of the signature can be modified is essential. This is the goal of the present study.

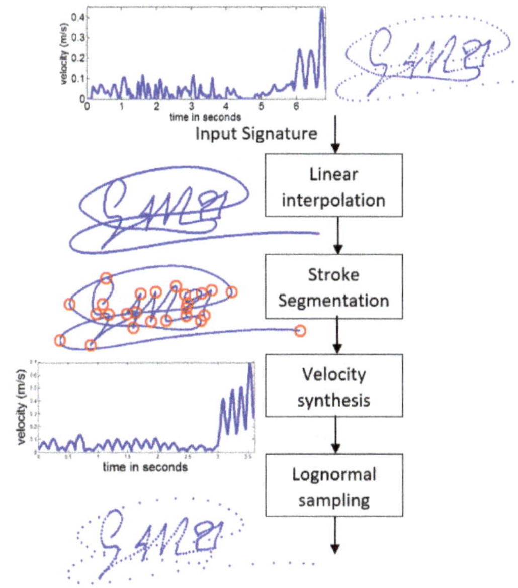


(a) Resampling-wise Method: Modifying the fluency by signature resampling.

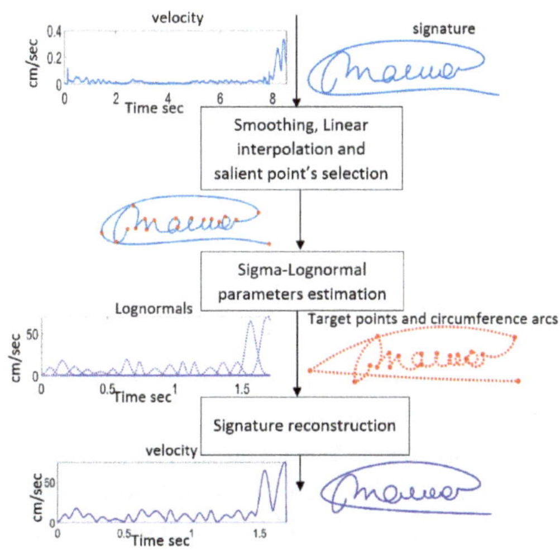


(b) Recovering-wise Method: Modifying the fluency by signature recovering.

Fig. 1. Overview of the two methods to improve handwritten signature fluency.

Indeed, in this chapter, we specifically present two methods to improve the fluency of a given on-line signature automatically. Our target is to improve the fluency of skilled forgeries in order to create more challenging specimens to verify. Both methods are based on linking the samples points of an on-line signature by 8-connected Bresenham's lines to obtain its trajectory. Nevertheless, both methods can work in any ordered sequence of points or dense polylines. Then, the most perceptually important points are estimated from the corners in the trajectory. These points are bound to the stroke limits, which are also known as starting and ending points of the strokes. As such, estimating them is a crucial step in improving handwritten signature fluency of an observed trajectory. Automatically controlling the correct number and position of the perceptually important points leads to handwritten signature patterns with different fluidity levels. Specifically, missing several such points would reflect a deficient signature recovery, whereas detecting too many will result in a non-fluent signature reconstruction. To this aim, we propose a procedure to artificially modify a signature fluency by modifying the number of strokes that describe an observed trajectory.

The next step of our methodology is to assign an occurrence time to each perceptually important point. Once these steps are completed, we propose two methods for the final signature reconstruction. In the case of the *resampling-wise* method, a lognormal function between two perceptually important points is assumed. Such a lognormal is temporally overlapped with its adjacent lognormal functions to obtain a synthetic velocity profile. Consequently, the new on-line fluent signature is obtained by resampling the trajectory of the original specimen with its synthetic velocity profile. In the *recovering-wise* method, the perceptually important points are used to estimate the virtual target points. Here, circular trajectories are drawn between virtual target points. The parameters of each lognormal are worked out similarly to the former method, but between virtual target points. Hence, we reconstruct the dynamic trajectory by sampling each circular trajectory with its lognormal and overlapping them. Figure 1 gives an overview of the two methods, which modify on-line autograph fluency. While the first method maintains the original trajectory, the second modifies both the trajectory and kinematics properties.

The paper is organized as follows: Section 2 describes the procedure to obtain the perceptually important points in signatures. Sections 3 and 4 explain the two proposed methods to improve signature fluency. The efficiency of the methods is experimentally validated in Sec. 5 by improving the fluency of forged signatures. We conclude our paper in Sec. 6 with some future work perspectives in other domains of applications like e-health and e-learning.

## 2. Perceptually Important Points Estimation in Signatures

The perceptually important points are identified in a static trajectory as the points of maximum curvature. These points correspond to the minima of the velocity profile. In the framework of the Sigma-Lognormal model, a perceptually important point has an associated virtual target point [Brault and Plamondon (1993)].

Prior works have suggested thresholding algorithms to detect the perceptually important points [Brault and Plamondon (1993); De Stefano *et al.* (2004)] with promising performances cursive handwriting and signatures. However, we found that more efforts are required when signatures are long, complex, and contain cursive text along with flourishes [Diaz-Cabrera *et al.* (2015)]. They, therefore, produce corners with completely different sharpnesses, which are not correctly detected by a single thresholding operation.

These limitations can be alleviated with the Two-Steps Perceptually Important Points Estimator (TS-PIPE) [Ferrer *et al.* (2017a)]. In the first step, TS-PIPE interpolates on-line trajectories to create continuous trajectories with $M$ samples. Next, the trajectories are scaled up by a factor of 100, smoothed with a sliding window of size 100 and, downscaled to the original size. This procedure reduces the tremor and doubtful handwritten trajectories. The resulting trajectory is denoted as $(x_o(j), y_o(j))\,, \forall j \in 1, \ldots, M$.

In the second stage, the curvature of each point is defined at different scales. It generates a multiscale matrix $A(i,j) = \alpha_{i+2}(j/M)$ of $M-2$ rows and $M$ columns, with $\alpha$ being the angle between two connected segments in the trajectory, $i$ the scale, and $j$ the point in the trajectory.

The third stage produces an approximation of the curvature of the handwriting, denoted as $C(j)$, for each sample $j$. This curvature $C(j)$ is performed as follows: $C(j) = \sum_{i=1}^{M-2} A(i,j)/(M-2)$. The perceptually important points are empirically selected as those peaks whose height/width ratios are greater than $\frac{\max(C)-\min(C)}{45}$ and separated by more than 1/10 the width of the signature, according to heuristic observations in fluid handwritten signatures. We can thus detect the most salient corners in the trajectory.

However, other perceptually important points are not detected with this approach because of the threshold sensitivity in the previous step. Although adjusting the thresholds would be an option to select these points, we noticed that it would be inefficiency in a number of situations where false salience points were selected. Instead, a different strategy to search missing salient point is preferred. Inspired by the implementation of the Sigma-Lognormal model [O'Reilly and Plamondon (2009); Ferrer *et al.* (2020)], we searched for circular curves in the piece of trajectories between two perceptually important points, obtained in the first step. This strategy aims to detect the salient points located generally in the wide arc zones, since their peaks in the curvature $C(l)$ are lower and wider. Thus, if the trajectory is circular enough between two perceptually important points, TS-PIPE does not add a new perceptually important point. Otherwise, if the piece of the trajectory requires several circular arcs to be fitted, a new perceptually important point is detected and stored at the crossing point between the circles in the trajectory. In other words, let $\{i_c\}_{c=1}^{C}$ be the indexes of the $C$ salient points already detected, this step tries to find if a salient point between $\{i_c\}_{c=1}^{C}$ was missed. In Fig. 2, we can see the perceptually important points estimated by the TS-PIPE on two trajectories. The estimation of perceptual important points is crucial in this work, since a bad estimation would result in an accumulation of future errors. For this reason, this procedure has two steps in order to make a less sensitive and more robust result in signatures. We refer the interested reader to [Ferrer *et al.* (2017a)] for a more detailed explanation of the TS-PIPE algorithm.

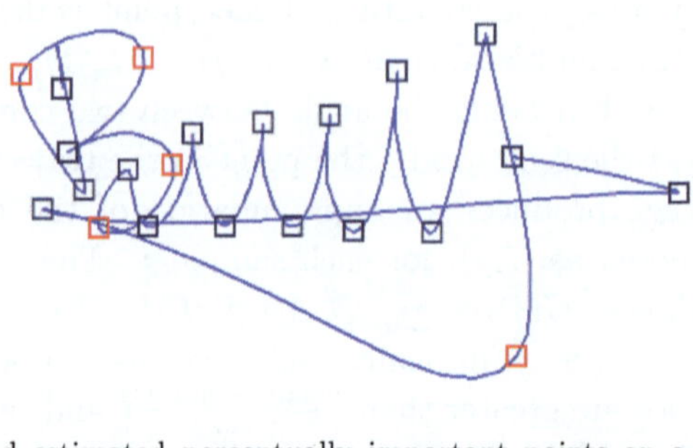

Fig. 2. Highlighted estimated perceptually important points on a signature ballistic trajectory. The squared blue points are first detected by TS-PIPE, and then, the square red points.

## 3. Resampling-wise Method: Modifying the Fluency by Signature Resampling

This method allows modifying the signature fluency when we are faced with a restriction disallowing the trajectory modification. Therefore, the method is focused on resampling the trajectory of the signature with the Sigma-Lognormal model. The result is a synthetic velocity profile whose kinematic properties are similar to those of a fluid signature.

We define synthetic velocity profiles assuming that the perceptually important points are the limits of the strokes and that they correspond to the local minimum points of the velocity signals. As a result, a lognormal is defined for each stroke.

### 3.1. *Lognormal resampling*

As suggested by the Kinematic Theory of rapid human movements, the velocity profile of a signature may be interpreted as a vectorial summation of consecutive lognormals [Ferrer *et al.* (2020)]. A single lognormal speed is described as follows:

$$|v_i(t)| = \frac{D_i}{\sqrt{2\pi}\sigma_i(t - t_{0_i})} \exp\left(-\frac{(\ln(t - t_{0_i}) - \mu_i)^2}{2\sigma_i}\right) \tag{1}$$

where $D_i$ is the amplitude of the strokes, $t_{0_i}$ the occurrence timing of the stroke $i$, $\sigma_i$ the stroke response time and $mu_i$ the stroke time delay. The global movement in its vectorial form holds:

$$\vec{v}(t) = \begin{bmatrix} v_x(t) \\ v_y(t) \end{bmatrix} = \sum_{i=1}^{N} D_i \begin{bmatrix} \cos\phi_i(t) \\ \sin\phi_i(t) \end{bmatrix} D_i \Lambda\left(t; t_{0_i}, \sigma_i^2, \mu_i\right) \tag{2}$$

where $N$ is the number of total lognormal or strokes required to define the signature and $\phi_i(t)$ the angular position defined as:

$$\phi_i(t) = \theta_{s_i} + \frac{\theta_{e_i} - \theta_{s_i}}{2}\left[1 + \operatorname{erf}\left(\frac{\ln(t - t_{0_i}) - \mu_i}{\sigma_i\sqrt{2}}\right)\right] \tag{3}$$

with $(\theta_{s_i}, \theta_{e_i})$ being the starting and ending angles of the arc linking the two virtual target points of the stroke $i^{th}$. Finally, the reconstructed trajectory is worked out as:

$$\vec{s}(t) = \begin{bmatrix} x_r(t) \\ y_r(t) \end{bmatrix} = \sum_{i=1}^{N} \frac{D_i}{\theta_{e_i} - \theta_{s_i}} \begin{bmatrix} \sin\phi_i(t) - \sin\theta_{s_i} \\ -\cos\phi_i(t) + \cos\theta_{s_i} \end{bmatrix} \tag{4}$$

This formula converts the angles into circular arcs and overlaps them.

### 3.2. *Signature duration and estimation of* $t_{0_i}$

Estimating the duration of a signature is a requirement to define its synthetic velocity profile. As the handwritten signature can be seen as a sequence of strokes, we need to define the timing of the individual strokes. Then, the total duration of the signature will be the sum of these single stroke durations. To do so, we assume that each stroke is associated with a set of stimuli ordered by a hypothetical Central Pattern Generator (CPG) [Garipelli (2006)], which produces rhythmic patterned outputs without sensory feedback. In this perspective, it has been suggested that the mammalian locomotor CPG comprises a "timer" which generates step cycles of varying durations and a pattern formation layer which selects and grades the activation of motor pools. However, a certain delay has been observed [Ferrer *et al.* (2020)] between the time of stroke occurrence, $t_{0_i}$, which emulates the timing order issued by the CPG, and the time of end-effector response, $t_{min_{i-1}}$. This delay is related to the human response time [Kandel *et al.* (2000)] which is defined heuristically as: $t_{min_{i-1}} - t_{0_i} \approx 0.5\,sec$. Let $t_{min_0} = 0$ and $t_{min_N} = T$ be the starting and ending timing of the signature and T its total duration. Then, the estimation of the lognormal timing occurrence would result in: $t_{0_i} = t_{min_{i-1}} + 0.5$.

Furthermore, we empirically determined the average time between two consecutive minima, i.e., $t_{min_{i-1}}$ and $t_{min_i}$ by analyzing genuine dynamic signatures from BiosecurID-SONOF [Galbally *et al.* (2015)]. This time is usually between 0.09 and 0.12 sec. Therefore, if the stroke generation was simulated by the CPG step cycle, the duration of each stroke would be established as 0.1 sec. From these hypotheses, we define $\tau_s + r$ as the duration of a single stroke, where $\tau_s = 0.1\,sec$ and $r$ is a random value that follows a normal distribution $\mathcal{N}(0, 0.005)$. This basic operational hypothesis from heuristic observations are assumed to work out a functional model in our procedure.

### 3.3. *Velocity profile*

To exploit this method, we then need to fit a lognormal function between two perceptually important points estimated in the trajectory. However, such lognormals need to be defined in $\tau_s\,sec$ and at a distance $D_i$. Moreover, the lognormal function should be overlapped with the adjacent ones since Sigma-Lognormal model considers the synergy of the neighbour lognormal functions.

As such, the distance of the continuous signature, $d(t)$, travelled at time $t$ could be calculated through the lognormal cumulative function [Diaz *et al.* (2017); Ferrer *et al.* (2017b)]:

$$d(t) = \int_{-\infty}^{+\infty} v_i(t)\, dt = \frac{D_i}{2}\left(1 + \text{erf}\left(\frac{(\ln(t - t_{0_i}) - \mu_i)^2}{\sqrt{2}\sigma_i}\right)\right) \tag{5}$$

Solving for $t$ Eq. (5), we can get the time in terms of distance $t(d)$ as follows:

$$t(d) = t_0 + \exp\left(\sqrt{2}\sigma_i \text{erf}^{-1}\left(\frac{2\, d(t)}{D_i} - 1\right) + \mu_i\right)^{0.5} \tag{6}$$

The following step thus aims to deduce the six lognormal parameters from a dense polyline trajectory, stroke by stroke. Formally, we assumed that the dynamic distance $l_{s_i}$ traveled by a stroke in the signature has a duration $t_{s_i}$. As such, Eq. (5) could be rewritten as follows:

$$l_{s_i} = \frac{D_i}{2}\left(1 + \text{erf}\left(\frac{(\ln(t_{s_i}) - \mu_i)^2}{\sqrt{2}\sigma_i}\right)\right) \tag{7}$$

One operational way [Ferrer *et al.* (2017b)] to solve this non-linear equation is to set $\left(\frac{(\ln(t_{s_i}) - \mu_i)^2}{\sqrt{2}\sigma_i}\right) = 3$. As erf$(3) = 1$, from Eq. (7) it holds that:

$$l_{s_i} = D_i \tag{8}$$

$$\mu_i = \ln(t_{s_i}) - 3\sqrt{2}\sigma_i \tag{9}$$

To solve the last equation with two variables, we obtain a new equation from the calculation of the skewness and kurtosis of the individual lognormals. To make these estimates, we worked out the skewness and kurtosis of the lognormals of the signatures from the BiosecurID-SONOF database [Galbally *et al.* (2015)], which is detailed in Sec. 5.1, extracted using the ScriptStudio software [O'Reilly and Plamondon (2009)]. An averaged skewness of 0.1301 and an averaged kurtosis of 3.082 were observed. This solution leads to lognormal functions, with their maximum values in the middle between strokes. Consequently, since the lognormal mode is analytically defined by $e^{\mu_i - \sigma_i^2}$ and it is approximately occurring at $t_{s_i}/2$ with a slightly left skew, numerically, it holds that [Ferrer *et al.* (2017b)]:

$$\delta_k\, t_{s_i} = e^{\mu_i - \sigma_i^2} \tag{10}$$

where $\delta_k$ denotes a uniform distribution heuristically defined in the $[\delta_k^{min}, \delta_k^{max}] = [0.4, 0.5]$ range. Combining Eqs. (9) and (10), we obtain:

$$\sigma_i^2 + 3\sqrt{2}\sigma_i - \ln(\delta_k) = 0 \tag{11}$$

With this approach, the parameter $D_i$ is assigned the value $l_{s_i}$ (see Eq. (8)), $\sigma_i$ is the positive solution of a simple second-order equation, as can be seen in Eq. (11), and $\mu_i$ is calculated by substituting $\sigma_i$ in Eq. (9).

Once the velocity profiles of individual strokes are obtained, the velocity profile of the signature, $v_i(t)$, is calculated as defined in Eq. (1). As a final check of the synthetic velocity profile of a flowing signature thus obtained, the integral of such a profile must be equal to the total length of the ballistic trajectory.

### 3.4. *Resampling*

Finally, the time at every pixel in the trajectory is calculated as the integral of the velocity from zero up to each multiple of $1/f_m$, where $f_m$ denotes the sampling frequency. Figure 3 visually illustrates an example of a real non-fluent handwritten signature and the synthetic result obtained with this method. Comparing both velocity profiles, we can see that the synthetic profile contains fewer lognormals than the real one. As a direct consequence of increasing the fluency of the signatures, the synthetic velocity is shorter in time and greater in amplitude. Additionally, the similarities between trajectories is observed with a 20.27 dB of Signal-to-noise ratio. A detailed description of this method can be read in [Ferrer *et al.* (2018b)].

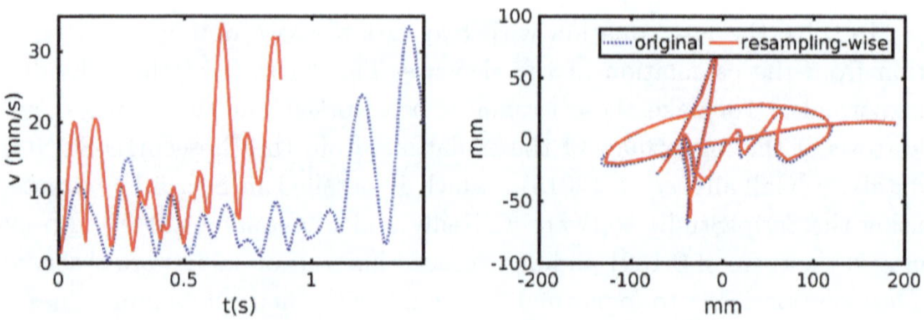


Fig. 3. Original (dotted blue) and reconstructed (solid red) velocity and trajectory obtained with the resampling-wise method.

## 4. Recovering-wise Method: Modifying the Fluency by Signature Recovery

In this method, both the trajectory and the velocity profile of a signature are modified. In opposition to the resampling-wise method, the novelty of this second approach is that it allows an analytical estimation of the lognormal parameters from a static signature, i.e., $\mathbf{P}_i = [D_i, \sigma_i, \mu_i, t_{0_i}, \theta_{s_i}, \theta_{e_i}]$. We can then automatically recover the on-line signature from Eq. (4).

### 4.1. *Positioning the virtual target points*

*Where is the x-y position of a virtual target point?* This question is a matter of debate and studies [Parziale *et al.* (2019)], but according to the Kinematic Theory, it is not necessarily located on the trajectory when the writing process is fluent [Ferrer *et al.* (2017b)]. According to the authors in [Plamondon and Privitera (1996)], the virtual target points are interpreted as a sequence of sorted points. These points are defined in the handwriting engram model of a writer. This engram contains the goal points required to design a signature. Even though they are stored in the long-term motor memory of the writer, their location on the 2D trajectory depends on the neuromotor state of the writer [Plamondon and Privitera (1996)].

Our method carries out an estimation of a possible location of the virtual target points, which should be somewhere near the estimated perceptually important points.

Essentially, each virtual target point, $tp_i$, is associated with a perceptually important point, $pp_i$. Thus, the position of $tp_i$ can be directly estimated from the geometrical relationship of $pp_{i-1}$, $pp_i$ and $pp_{i+1}$. Let $tp_0 = pp_0$ and $tp_N = pp_N$ be the initial and final boundary conditions, with $N$ being the total number of virtual targets or perceptually important points required to define a trajectory. Then, $tp_i$, $i \in 1, \ldots, N-1$ is obtained from the geometric triangle defined by $pp_{i-1}$, $pp_i$ and $pp_{i+1}$. We define the location of the virtual target point $tp_i$ over the bisector angle of the $pp_i$ vertex. Moreover, the virtual target point is defined out of the triangle, at a certain distance $ltp_i$ from the line between $pp_{i-1}$ and $pp_{i+1}$, which is calculated as follows:

$$ltp_i = htp_i \left(1 + \frac{\cos\left(\frac{\varphi_i}{2}\right)}{2}\right) \tag{12}$$

where $htp_i$ is the height of the triangle defined by the above-mentioned mediatrix and $\varphi_i$ is the angle at the vertex $pp_i$. For a better understanding of this calculation, Figure 4(a) illustrates this procedure. It is worth taking into consideration that the Eq. (12) was heuristically obtained according to trigonometrical observations. The hypothesis assumed in this estimation is that the virtual target points is located over the mediatrix, and the more salient a perceptually important point is, the farther is its associated virtual target point. For this reason, and according to the cosine function, the smaller $\varphi_i$, the larger the distance $ltp_i$. Similarly, a less sharpen trajectory would imply a larger $\varphi_i$ and a smaller $ltp_i$.

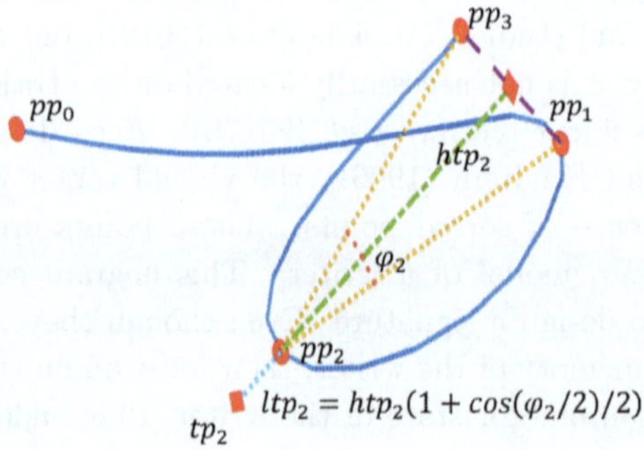


(a) Estimation of the location of the virtual target point $tp_2$ from perceptually important points $pp_1$, $pp_2$ and $pp_3$.

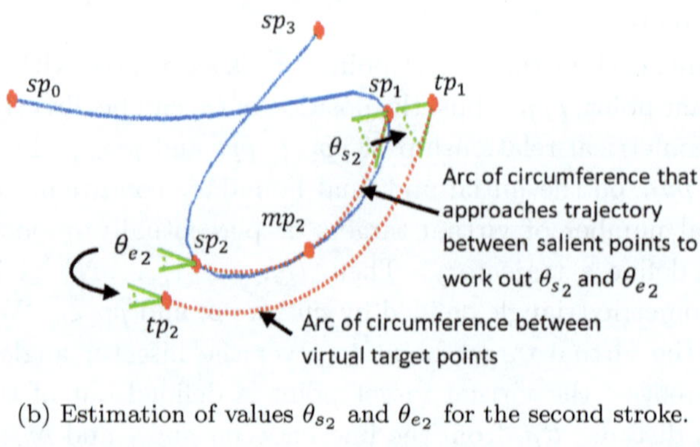


(b) Estimation of values $\theta_{s_2}$ and $\theta_{e_2}$ for the second stroke.

Fig. 4. Overview of the Sigma-Lognormal parameter estimation.

### 4.2. *Calculating the starting and ending angles of a stroke*

Let $\theta_{s_i}$ and $\theta_{e_i}$, $\forall i \in 1, \ldots, N$ be the starting and ending angles. It is important to note that the Sigma-Lognormal model approaches the trajectory produced by a single stroke by a piece of circular arcs. Moreover, the value of the starting and ending angles modifies the trajectory of each arc. In our proposal, these angles are calculated as follows:

- A middle point, $mp_i$ between the two perceptually important points, $pp_{i-1}$ and $pp_i$ is calculated for the $i^{th}$ stroke. The distance on the trajectory between the points $(mp_i, pp_{i-1})$ is the same as $(mp_i, pp_i)$.
- These three points are then used in defining a circumference.
- The starting angle, $\theta_{s_i}$, is computed as the angle between the tangent of the circumference at the point $pp_{i-1}$ and a horizontal line. Similar, the ending point, $\theta_{e_i}$, denotes the angle between the tangent lines to the circumference and the horizontal line.

This procedure is illustrated in Fig. 4(b).

### 4.3. *Calculating $D_i, \sigma_i, \mu_i, t_{o_i}$*

The timing occurrence of each lognormal is calculated following the same procedure as in Sec. 3.2. In the case of $(\sigma_i, \mu_i)$, the previous method estimated these values by adding a lognormal in the middle of two perceptually important points $(pp_{i-1}, pp_i)$. In our present method, we follow the same procedure as in Sec. 3.3 to compute these two pairs of parameters. Instead of using the perceptually important points as references, this second method uses the same equations over the virtual target points.

Furthermore, in our method, the amplitude of the lognormals, $D_i$, which refers to the amplitude of the single impulse movement, can be defined by the position of the virtual target points $(tp_{i-1}, tp_i)$ and the ending and starting angles $(\theta_{s_i}, \theta_{e_i})$ as follows:

$$D_i = r_i \cdot (\theta_{e_i} - \theta_{s_i}),\ i \in 1, \ldots, N \tag{13}$$

where $r_i$ is the radius of the circumference that joins the virtual target points $(tp_{i-1}, tp_i)$. To calculate this radius, we first calculate the center of the circumference as the intersection of the line that transverses $tp_{i-1}$ with slope $-1/\tan(\theta_{s_i})$ and the line that transverses $tp_i$ with slope $-1/\tan(\theta_{e_i})$. Then, $r_i$ is calculated as the distance from the center of the circumference to either $tp_{i-1}$ or $tp_i$.

### 4.4. *Recovering the on-line trajectory and velocity*

The above procedure provides an estimation of the number of strokes, $N$, as well as the six parameters of the Sigma-Lognormal model for each stroke, $D_i, \sigma_i, \mu_i, t_{0_i}, \theta_{s_i}, \theta_{e_i}$.

Finally, to recover the on-line trajectory and, therefore, the velocity profile, we can use Eqs. (3) and (4). This reconstructed trajectory is more rounded than the obtained with the former method, even in the areas where the non-fluent trajectories cause tremor, indecision or more doubts. Therefore, it is expected that the reconstructed signature will be considered more fluid than the seed one.

An example of the recovered velocity and trajectory used with this method can be seen in Fig. 5. Only in the case of non-fluent trajectories is it worth mentioning that the execution time is considerably reduced since the reconstructed specimen represents a fluid and natural style. In the case of trajectory, we can observe the dissimilarities obtained between the original and recovered signatures with a 12.04 dB of signal-to-noise-ratio. More details about this method can be found in our previous works [Ferrer *et al.* (2018a, 2020)].

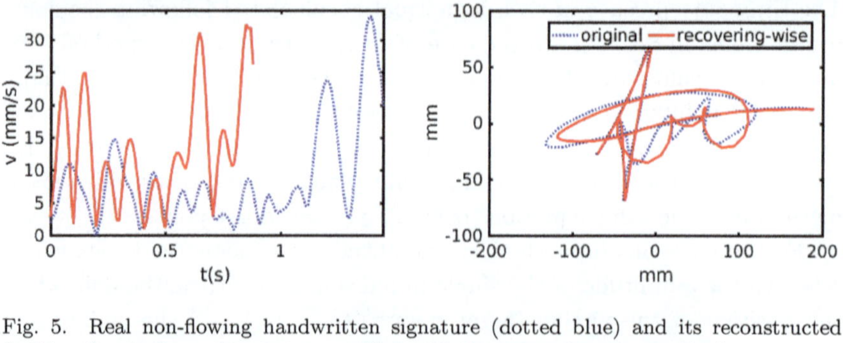


Fig. 5. Real non-flowing handwritten signature (dotted blue) and its reconstructed (solid red) velocity and trajectory profiles with the recovering-wise method.

For a visual comparison of the two proposed methods, Fig. 6 represents on-line handwritten signatures from forgeries, which can be considered non-fluent specimens. It also overlaps both the velocity profiles and trajectories obtained with the resampling-wise and recovering-wise methods. While both methods provide a better flowing velocity, the trajectory with the latter method flows better than that obtained with the former.

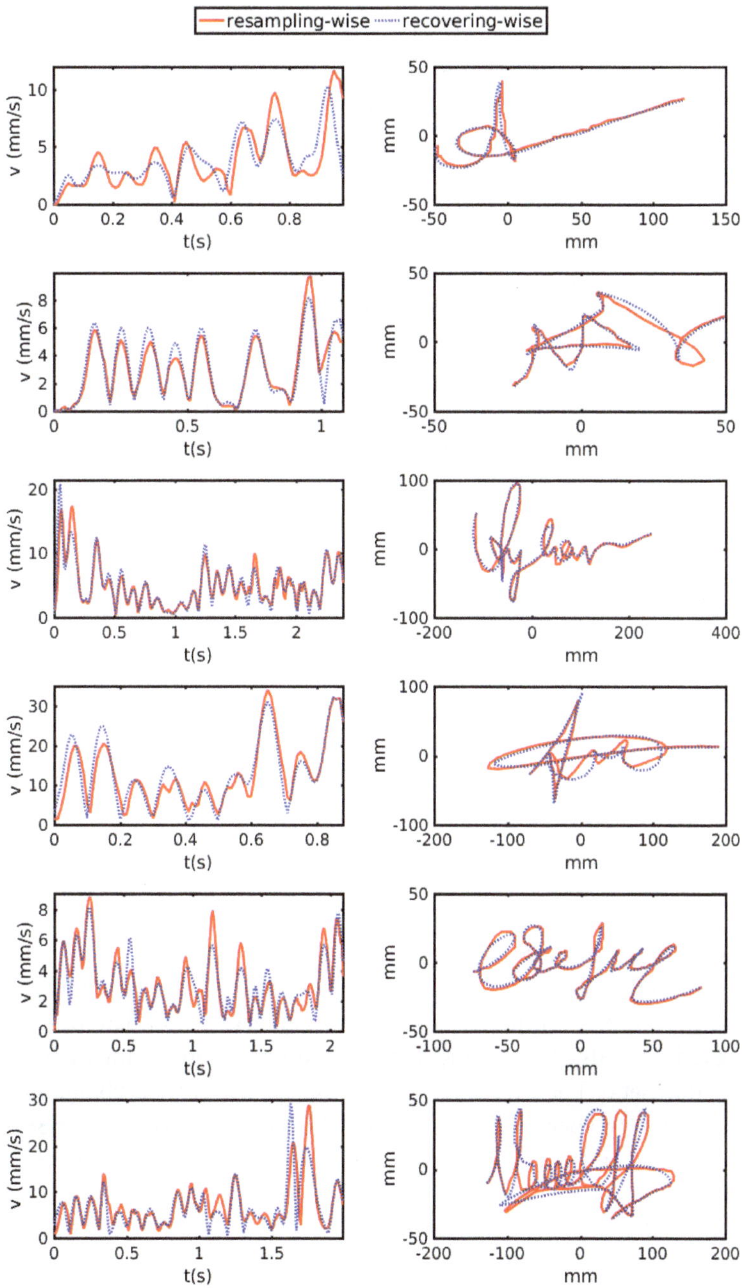


Fig. 6. Visual results on four signatures: solid red line refers to the resampling-wise method whereas dotted blue line the recovering-wise method.

## 5. Experiments

The tests aimed to demonstrate that the proposed methods are capable of improving the fluency of skilled forgeries since they often appear as less fluent gestures. To this end, we conducted several biometric-based experiments with dynamic signature verification.

Signature verification may be considered as a dichotomous problem in pattern recognition: *"does this questioned signature belong to the claimed identity?"* [Diaz *et al.* (2019)] The handwritten signatures produced by an original person are named genuine, and those by an impostor, forgeries. Generally, forged signatures are less fluid than genuine ones. This is especially true for signatures generated by forgers who participate voluntarily to a data collection and who have not practiced the specimen to falsify. In this context, to confirm that the proposed methods improve signature fluency and make forged signatures more genuine [Black (1962)], we must show that the performance of a system in signature verification decreases.

### 5.1. *Databases and experimental protocol*

In this work, a comprehensive evaluation was carried out with the following handwritten signature databases:

- Dynamic real signatures from the BiosecurID-SONOF dataset [Galbally *et al.* (2015)]. It comprises 132 users, 16 genuine signatures and 12 forgeries per user.
- The MCYT-100 corpus [Ortega-Garcia *et al.* (2003)], which is composed of the first 100 users of MCYT-300. It has 25 genuine and 25 forged specimens per signer.
- The SigComp2009 dataset [Blankers *et al.* (2009)]. We use the available subset from the complete dataset used in competitions. It contains 100 users, 932 genuine signatures and 620 forgeries.
- The SUSIG-Visual and Blind [Kholmatov and Yanikoglu (2009)] datasets. The first dataset contains 94 users, with 20 genuine and 10 forged specimens per signer. The second one consists of 88 users, with 8 or 10 genuine repetitions and 10 forged signatures per user.

Additionally, two state-of-the-art automatic signature verifiers (ASV) were used. A Manhattan-based ASV, which consists in describing on-line signatures as histograms of function-based features. Then, a Manhattan distance is used to calculate the membership of a questioned signature. We use our own implementation [Ferrer *et al.* (2017b)] of the ASV detailed in [Sae-Bae and Memon (2014)]. A DTW-based ASV was also used [Ferrer *et al.* (2017b)]. This second system uses function-based signatures to develop a Euclidean distance matrix between a questioned signature and a reference one. Such a matrix is elastically optimized by dynamic programming with DTW. The score of a questioned signature is calculated as the nearest DTW distance obtained from the comparison of all reference signatures with the questioned specimen.

The reference signatures consist of the first five enrolled genuine signatures of each user in the databases. For the False Rejection Rate (FRR), we used the remaining genuine signatures of the same signer, whereas all available skilled forgeries of each signer were used to calculate the False Acceptation Rate (FAR) curve. These experiments were performed for the original signatures (baseline) and for the resampled and the recovered signatures according to the proposed methods, under the same conditions.

### 5.2. *Comparing false acceptance rate and false rejection rate curves*

The experiments in this subsection aim at assessing whether the proposed methods render signatures produced by on-line skilled forgeries more fluid. To this end, we compare the FRR of genuine signatures (baseline) with the FAR curves of *i*) original forgeries available in the database, hereafter seed forgeries (baseline); *ii*) forged signatures with the resampling-wise method, and *iii*) forged signatures with the recovering-wise method.

It is expected that if the forged signatures can be produced more fluently, they should be more genuine and difficult to detect. As such, the FAR curve would be closer to the FRR than the FAR curve obtained with the seed forgeries (baseline).

Figure 7 shows the FAR and FRR curves with the baseline, i.e., genuine and seed forgery signatures and the skilled forgery signatures obtained with resampling-wise and recovering wise methods. In all cases, with both ASVs, the FAR curves are moved to the left. This means that both the resampled and recovered forgeries are closer to the genuine signatures than the seed forgeries. The more beneficial effect of this methodology is seen in the case of the SigComp2009 dataset, where we can observe a greater leftward displacement in the FAR curves, especially in the case of the Manhattan (MAN)-based ASV.

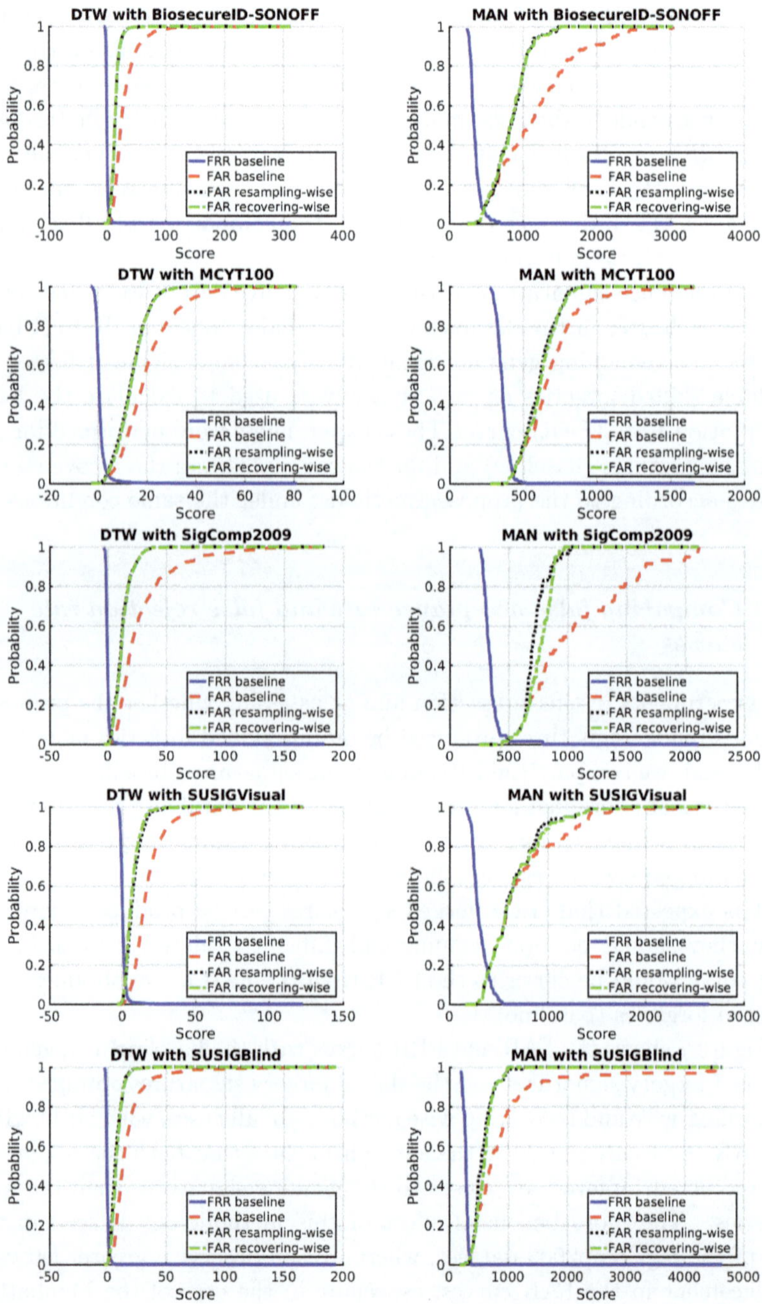


Fig. 7. FAR and FRR curves with the original signatures (baseline) and FAR curves with both resampling-wise and recovering-wise methods.

Even though these methods improve the fluent appearance of the signatures, it was hard to improve the EER of the systems. It suggests that more efforts are required to get proper genuine realism from the forgeries. In the meantime, these methods can be seen as two initial proposals towards the forgeries transformation into genuine signatures regarding the obtained results in the FRR curves.

Overall, it is seen then that both methods improve the flowing velocity of the forgeries. However, the trajectory with the recovery-wise method is smoother and more genuine than that obtained with the resampling-wise. Moreover, this comprehensive evaluation suggests that both automatic signature verifiers are not able to detect the synthetic forgeries as accurately as the seed ones. Furthermore, this can be seen as a biometric-case attack that should be studied to be detected.

## 6. Conclusions

Two methods have been proposed to improve the fluency of a given on-line skilled forgery signature by generating more fluent synthetic on-line specimen. The central control of the fluidity in the signature is performed by estimating its perceptually important points. These points can be worked out from the 8-connected, smoothed Bresenham's trajectory, which is obtained by linking the real on-line signature sampling points. This comprehensive estimation is crucial in this work since estimating fewer points leads to a very fluid, but a distorted specimen. Conversely, adding too many points creates non-fluent artificial-looking trajectories.

For this reason, we used a two-step perceptually important point estimator (TS-PIPE) [Ferrer *et al.* (2017a)]. Once the perceptually important points are estimated, two methods are proposed based on the lognormal properties of the rapid movements: resampling-wise and recovering-wise. The former method improves the velocity profile by resampling the trajectories, while the latter estimates the six parameters of all lognormals hidden in the signal from the estimated virtual target points. Next, the signature is recovered using the Sigma-Lognormal equations. The capacity to make the signature flow more natural is confirmed with the experimental results. Specifically, several signature verification experiments were carried out. In all cases, the false acceptance rate, from non-fluent signatures that claimed to be genuine, were closer to the false rejection rate curves. This indicates that there are a number of non-fluent signatures, which are more similar to genuine specimens.

Regarding future research, apart from its direct application to the design and testing of ASV systems, as indicated in the experiments, the methods can improve the difficulty of a particular database by increasing the quality of its skilled forgeries. It can result in a family of more robust signature verifiers capable of detecting more challenging forgeries [Ferrer *et al.* (2018a,b)].

Moreover, although this study has focused on signatures, other examples of fluidity control are typically observed in the production of cursive handwriting, drawings, doodles or further trajectories that involve rapid handwriting movements [O'Reilly and Plamondon (2009)]. In this context, the present study sheds the basis for other investigations on gesture fluency. For example, fluent handwriting can be seen as representing basic characteristics of the human neuromotor system. However, this characteristic could be seen as a difficult challenge for people who have not developed a mature neuromotor system or have any neurological disease [Fischer *et al.* (2014)]. This observation and characterization of fluid and non-fluid handwriting regarding the lognormality principle might have many implications in e-health, e-learning apart from the e-security applications associated with signatures [Plamondon *et al.* (2018)].

In e-health, for example, the Sigma-Lognormal model has widely assisted practitioners in discriminating between healthy and unhealthy handwriting: analyses of Sigma-Lognormal parameters have led to promising results in the estimation of early-stage Alzheimer's disease [Pirlo *et al.* (2015)]. In future e-health applications involving handwriting, these methods could serve as a cornerstone in rehabilitating neurological disorders. These methods could improve the natural handwriting of patients who suffer from any neuromotor impairment. This could open up opportunities to design neurological screening tests that motivate the recovery process.

In e-learning applications, children's graphomotor performance can be analyzed by using Sigma-Lognormal parameters. This approach has demonstrated statistically significant differences in kindergarten children's handwriting abilities aged 3, 4, and 5 [Duval *et al.* (2015)]. In the future, in e-learning applications, these methods could encourage children to improve their calligraphy [Simonnet *et al.* (2019)], and show their ideal handwriting with their personalised styles. If such sequels occur, this article will have fulfilled its mission.

## References

Black, D. A. (1962). Forged signatures more skillfully written than the true signatures, *J. Crim. L. Criminology & Police Sci.* **53**, p. 109.

Blankers, V. L., van den Heuvel, C. E., Franke, K. Y. and Vuurpijl, L. G. (2009). Icdar 2009 signature verification competition, in *10th International Conference on Document Analysis and Recognition, ICDAR'09*, pp. 1403–1407.

Brault, J.-J. and Plamondon, R. (1993). Segmenting handwritten signatures at their perceptually important points, *IEEE Transactions on Pattern Analysis & Machine Intelligence*, 9, pp. 953–957.

De Stefano, C., Garruto, M. and Marcelli, A. (2004). A saliency-based multiscale method for on-line cursive handwriting shape description, in *9th International Workshop on Frontiers in Handwriting Recognition*, pp. 124–129.

Diaz, M., Ferrer, M. A., Impedovo, D., Malik, M. I., Pirlo, G. and Plamondon, R. (2019). A perspective analysis of handwritten signature technology, *ACM Computing Surveys (CSUR)* **51**, 6, p. 117.

Diaz, M., Ferrer, M. A., Parziale, A. and Marcelli, A. (2017). Recovering western on-line signatures from image-based specimens, in *14th IAPR International Conference on Document Analysis and Recognition*, pp. 1204–1209.

Diaz-Cabrera, M., Ferrer, M. A. and Morales, A. (2015). Modeling the lexical morphology of western handwritten signatures, *PloS one* **10**, 4, p. e0123254.

Duval, T., Rémi, C., Plamondon, R., Vaillant, J. and O'Reilly, C. (2015). Combining Sigma-Lognormal modeling and classical features for analyzing graphomotor performances in kindergarten children, *Human movement science* **43**, pp. 183–200.

Ferrer, M. A., Diaz, M., Carmona, C. and Plamondon, R. (2020). iDeLog: Iterative dual spatial and kinematic extraction of Sigma-Lognormal parameters, *IEEE Trans. on Pattern Analysis and Machine Intelligence (PAMI).* **42**(1): 114–125.

Ferrer, M. A., Diaz, M., Carmona, C. and Plamondon, R. (2018a). A biometric attack case based on signature synthesis, in *52nd IEEE International Carnahan Conference on Security Technology (ICCST)*, pp. 1–6.

Ferrer, M. A., Diaz, M., Carmona-Duarte, C. and Plamondon, R. (2018b). Improving on-line signature skillfulness, in *Workshop on Lognormality Principle and its Applications in Int. Conf. on Pattern Recognition and Artificial Intelligence (ICPRAI 2018)*, pp. 768–773.

Ferrer, M. A., Diaz, M. and Carmona-Duarte, C. (2017a). Two-steps perceptual important points estimator in 8-connected curves from handwritten signature, in *Seventh International Conference on Image Processing Theory, Tools and Applications (IPTA)*, pp. 1–5.

Ferrer, M. A., Diaz, M., Carmona-Duarte, C. and Morales, A. (2017b). A behavioral handwriting model for static and dynamic signature synthesis. *IEEE Trans. Pattern Anal. Mach. Intell.* **39**, 6, pp. 1041–1053.

Fischer, A. and Plamondon, R. (2017). Signature verification based on the kinematic theory of rapid human movements, *IEEE Transactions on Human-Machine Systems* **47**, 2, pp. 169–180.

Fischer, A., Plamondon, R., O'Reilly, C. and Savaria, Y. (2014). Neuromuscular representation and synthetic generation of handwritten whiteboard notes, in *14th International Conference on Frontiers in Handwriting Recognition (ICFHR)*, pp. 222–227.

Galbally, J., Diaz-Cabrera, M., Ferrer, M. A., Gomez-Barrero, M., Morales, A. and Fierrez, J. (2015). On-line signature recognition through the combination of real dynamic data and synthetically generated static data, *Pattern Recognition* **48**, 9, pp. 2921–2934.

Garipelli, G. (2006). *A neuromotor model of handwriting generation highlighting the role of basal ganglia*, Ph.D. thesis, Department of Electrical Engineering, Indian institute of technology.

Gomez-Barrero, M., Galbally, J., Fierrez, J., Ortega-Garcia, J. and Plamondon, R. (2015). Enhanced on-line signature verification based on skilled forgery detection using Sigma-Lognormal features, in *International Conference on Biometrics (ICB)*, pp. 501–506.

Kandel, E. R., Schwartz, J. H., Jessell, T. M., of Biochemistry, D., Jessell, M. B. T., Siegelbaum, S. and Hudspeth, A. (2000). *Principles of neural science*, Vol. 4, McGraw-hill New York.

Kholmatov, A. and Yanikoglu, B. (2009). SUSIG: an on-line signature database, associated protocols and benchmark results, *Pattern Anal. and Applications* **12**, 3, pp. 227–236.

O'Reilly, C. and Plamondon, R. (2009). Development of a Sigma–Lognormal representation for on-line signatures, *Pattern Recognition* **42**, 12, pp. 3324–3337.

Ortega-Garcia, J., Fierrez-Aguilar, J. *et al.* (2003). MCYT baseline corpus: A bimodal biometric database, *IEE Proceedings - Vision, Image and Signal Processing* **150**, 6, pp. 395–401.

Parziale, A., Senatore, R. and Marcelli, A. (2019). Looking at the principle of motor equivalence through the "lognormal-glasses", in *19th International Graphonomics Conference (IGS 2019 - Your brain on art)*, pp. 1–4.

Pirlo, G., Diaz, M., Ferrer, M. A., Impedovo, D., Occhionero, F. and Zurlo, U. (2015). Early diagnosis of neurodegenerative diseases by handwritten signature analysis, in *International Conference on Image Analysis and Processing*, pp. 290–297.

Plamondon, R. and Djioua, M. (2006). A multi-level representation paradigm for handwriting stroke generation, *Human movement science* **25**, 4-5, pp. 586–607.

Plamondon, R., Pirlo, G., Anquetil, E., Rémi, C., Teulings, H.-L. and Nakagawa, M. (2018). Personal digital bodyguards for e-security, e-learning and e-health: A prospective survey, *Pattern Recognition* **81**, pp. 633–659.

Plamondon, R. and Privitera, C. M. (1996). A neural model for generating and learning a rapid movement sequence, *Biological cybernetics* **74**, 2, pp. 117–130.

Sae-Bae, N. and Memon, N. (2014). Online signature verification on mobile devices, *IEEE Transactions on Information Forensics and Security* **9**, 6, pp. 933–947.

Simonnet, D., Girard, N., Anquetil, E., Renault, M. and Thomas, S. (2019). Evaluation of children cursive handwritten words for e-education, *Pattern Recognition Letters* **121**, pp. 133–139.

https://doi.org/10.1142/9789811226830_0003

# Chapter 3

# Modeling the Complexity of Signature and Touch-Screen Biometrics using the Lognormality Principle

Ruben Vera-Rodriguez, Ruben Tolosana, Javier Hernandez-Ortega, Alejandro Acien, Aythami Morales, Julian Fierrez and Javier Ortega-Garcia

*BiDA Lab, School of Engineering, Universidad Autonoma de Madrid*
*Madrid, Spain*

*(ruben.vera, ruben.tolosana, javier.hernandezo, alejandro.acien, aythami.morales, julian.fierrez, javier.ortega)@uam.es*

This paper focuses on modeling the complexity of biomechanical tasks through the usage of the Sigma-LogNormal model of the Kinematic Theory of rapid human movements. The Sigma-LogNormal model has been used for several applications, in particular related to modeling and generating synthetic handwritten signatures in order to improve the performance of automatic verification systems. In this paper we report experimental work for the usage of the Sigma-LogNormal model to predict the complexity of biomechanical tasks on two case studies: 1) on-line signature recognition in order to generate user-based complexity groups and develop specific verification systems for each of them, and 2) detection of age groups (children from adults) using touch screen patterns. The results achieved show the benefits of using the Sigma-LogNormal model for modeling the complexity of biomechanical tasks in the two case studies considered.

## 1. Introduction

On-line signature verification and other handwritten tasks (drawings, touch patterns, etc.) are experiencing a high development recently due to the technological evolution of digitizing devices, including smartphones and tablets. Such handwritten data can be applied to many applications in different sectors such as security, e-government, healthcare, education, user profiling, advertising or banking [Plamondon *et al.* (2014); Guest (2006); Fierrez *et al.* (2018b); Tolosana *et al.* (2019a, 2020a); Acien *et al.* (2020d)].

This paper focuses on modeling the complexity of handwritten information, which can be a very important factor in different applications related to handwriting. We propose to model the complexity of handwritten tasks through the usage of the Sigma-LogNormal model of the Kinematic Theory of rapid human movements [O'Reilly and Plamondon (2009)]. The Sigma-LogNormal model has been used in the past for several applications. One of the most successful ones has been the synthetic generation of handwriting, in particular signatures (two examples in [Galbally *et al.* (2012)] and [Diaz *et al.* (2017)]). This model has recently been used in [Gomez-Barrero *et al.* (2015)] and [Fischer and Plamondon (2017)] not to generate synthetic signature samples, but to improve the performance of traditional signature verification systems. In [Gomez-Barrero *et al.* (2015)] the authors proposed a skilled forgery detector using some features extracted from the Sigma-LogNormal model whereas in [Fischer and Plamondon (2017)], a new set of features based on the Sigma-LogNormal model was proposed achieving very good performance.

In this paper we report experimental work for the usage of the Sigma-LogNormal model to predict the complexity of biomechanical tasks on two case studies: 1) The first one describes its application to on-line signatures in order to generate user-based complexity groups (as there are users with very complex signatures and others with very simple ones). Then, a specific signature verification system is developed for each complexity group achieving very significant improvements of verification performance [Tolosana *et al.* (2017b, 2020c)]. 2) On the other hand, the second one describes its application to detect age groups (children from adults) in touch dynamic tasks performed on smartphones or tablets [Hernandez-Ortega *et al.* (2017)], as the difference between adults and children is mainly caused by the different maturity of their anatomy and neuromotor system. These are less mature in children, so they have worse manual dexterity causing rougher movements [O'Reilly and Plamondon (2009); Piaget and Inhelder (1969)].

The remainder of the paper is organized as follows. Section 2 describes the Sigma-LogNormal model, used in this work to model the complexity of handwritten tasks. Section 3 describes the first case study focused on modeling the complexity of on-line signatures and its experimental results. Section 4 describes the second case study focused on modeling the complexity of touch dynamic information in order to detect age groups and its experimental results. Finally, Sec. 5 draws the final conclusions and points out some lines for future work.

## 2. The Sigma-LogNormal Model

Many models have been proposed to analyze human movement patterns in general and handwriting in particular. These models allow the analysis of features

Table 1. Sigma-LogNormal parameters description.

| Parameter | Description |
|---|---|
| $D_i$ | Input command: covered distance when executed in isolation. |
| $t_{0i}$ | Time occurrence of the input command. |
| $\mu_i$ | Logtemporal delay. |
| $\sigma_i$ | Logresponse time of the neuromotor system. |
| $\theta_{si}$ | Initial angular position of the stroke. |
| $\theta_{ei}$ | Final angular position of the stroke. |

related to motor control processes and the neuromuscular response, providing complementary features to the traditional *X* and *Y* coordinates related to handwriting tasks. One of the most well known writing generation models is the Sigma-LogNormal model [O'Reilly and Plamondon (2009); Djioua and Plamondon (2009)].

The Sigma-LogNormal model decomposes the complex signals that describe the speed of muscular movements into simpler ones that can be explained by a few parameters. These parameters contain information about the activity itself and about the neuromotor skills of the person [Duval *et al.* (2015)]. In particular, the Sigma-LogNormal model states that the velocity profile of human hand movements can be decomposed into strokes. Moreover, the velocity of each of these strokes, $i$, can be described with a speed signal $v_i(t)$ that has a lognormal shape:

$$|v_i(t)| = \frac{D_i}{\sqrt{2\pi}\sigma_i(t - t_{0i})} \exp\left(-\frac{(\ln(t - t_{0i}) - \mu_i)^2}{2\sigma_i^2}\right) \tag{1}$$

where each of the parameters are described in Table 1. The complete velocity profile is modelled as a sum of the different individual stroke velocity profiles as:

$$v_r(t) = \sum_{i=1}^{N} v_i(t) \tag{2}$$

where *N* is the number of lognormals of the entire movement. A complex action, like a handwritten signature or touch task, is a summation of these lognormals, each one characterized by different values for the six parameters in Table 1. Figure 1 shows an example of the lognormal velocity profiles extracted for each stroke of one signature. In this work the software ScriptStudio provided by Scribens Lab[a] has been used to reconstruct the signatures and touch screen patterns.

## 3. Case Study 1: On-Line Signature Complexity

Signature verification systems have been shown to be highly sensitive to signature complexity [Fierrez, Ortega-Garcia and Gonzalez-Rodriguez (2005)].

[a]https://www.polymtl.ca/expertises/en/scribens-laboratory

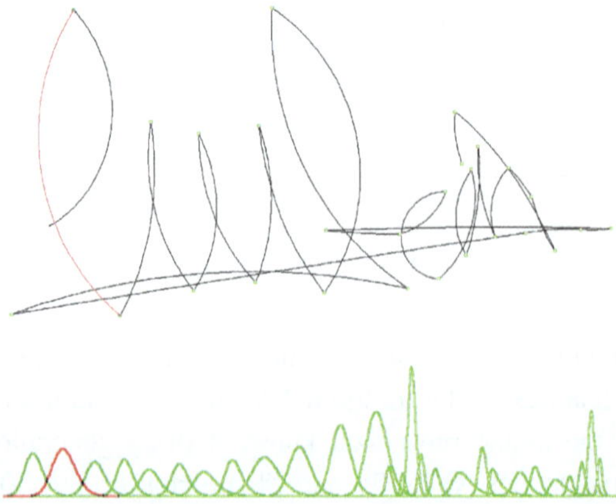

Fig. 1. Trace and velocity profile of one reconstructed on-line signature using the Sigma-LogNormal model. A single stroke of the signature and its corresponding lognormal profile are highlighted in red colour. Individual strokes are segmented within the Script-studio software algorithm [O'Reilly and Plamondon (2009)].

In [Alonso-Fernandez, Fairhurst, Fierrez and Ortega-Garcia (2007)], Alonso-Fernandez *et al.* evaluated the effect of the complexity and legibility of the signatures for off-line signature verification (i.e. signatures with no available dynamic information) pointing out the differences in performance for several matchers. Signature complexity has also been associated to the concept of entropy, defining entropy as the inherent information content of biometric samples [Lim and Yuen (2016); Zhou (2009)]. In [Houmani, Garcia-Salicetti and Dorizzi (2008)] a "personal entropy" measure based on Hidden Markov Models (HMM) was proposed in order to analyse the complexity and variability of on-line signatures regarding three different levels of entropy. In addition, the same authors have recently proposed in [Houmani and Garcia-Salicetti (2016)] a new metric known as "relative entropy" for classifying users into animal groups where skilled forgeries are also considered. Despite all the studies performed regarding on-line signature as a biometric trait, none of them have exploited, as far as we are aware, the concept of complexity in order to develop more robust and accurate on-line signature verification systems.

### 3.1. *Proposed system*

The architecture of our proposed system is shown in Fig. 2. Based on the parameters of the Sigma-LogNormal model, we propose to use the number of lognormals ($N$) that models each signature as a measure of the complexity level of the

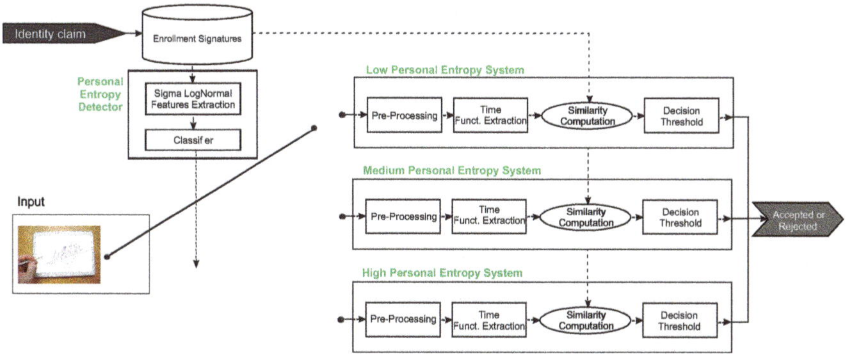


Fig. 2. Architecture of our proposed methodology focused on the development of an on-line signature verification system adapted to the signature complexity level.

signature. It is worth noting that lognormals with amplitude value lower than a threshold were discarded. Once this parameter is extracted for all available genuine signatures of the enrolment phase, the user is classified into a complexity level using the majority voting algorithm (low, medium and high complexity levels). Only genuine signatures are considered in our proposed approach for measuring the complexity level. The advantage of this approach is that the signature complexity detector can be performed off-line thereby avoiding time consuming delays and making it feasible to apply in real time scenarios. It is worth noting that before extracting the number of lognormals, a normalization based on the mean and standard deviation was applied to the X, Y and pressure coordinates of the signatures as a preprocessing stage.

Then, after having classified a given user into a complexity group, a specific on-line signature verification module based on time functions (a.k.a. local system) [Martinez-Diaz *et al.* (2009a)] has been adapted to each signature complexity level. For each signature acquired, signals related to $X$ and $Y$ pen coordinates are used to extract a set of 23 time functions, similar to [Martinez-Diaz *et al.* (2014)] (see Table 2). The most discriminative and robust time functions of each complexity level are selected using the Sequential Forward Feature Selection algorithm (SFFS) enhancing the signature verification system in terms of EER.

The local system considered in this work for computing the similarity between the time functions from the input and training signatures is based on DTW algorithm [Martinez-Diaz *et al.* (2009b)]. Scores are obtained as:

$$score = e^{-D/K} \tag{3}$$

where $D$ and $K$ represent respectively the minimal accumulated distance and the number of points aligned between two signatures using DTW algorithm.

Table 2. Set of time functions considered in this work.

| # | Feature |
|---|---|
| 1 | x-coordinate: $x_n$ |
| 2 | y-coordinate: $y_n$ |
| 3 | Pen-pressure: $z_n$ |
| 4 | Path-tangent angle: $\theta_n$ |
| 5 | Path velocity magnitude: $v_n$ |
| 6 | Log curvature radius: $\rho_n$ |
| 7 | Total acceleration magnitude: $a_n$ |
| 8-14 | First-order derivate of features 1-7: $\dot{x_n}, \dot{y_n}, \dot{z_n}, \dot{\theta_n}, \dot{v_n}, \dot{\rho_n}, \dot{a_n}$ |
| 15-16 | Second-order derivate of features 1-2: $\ddot{x_n}, \ddot{y_n}$ |
| 17 | Ratio of the minimum over the maximum speed over a 5-samples window: $v_n^r$ |
| 18-19 | Angle of consecutive samples and first order difference: $\alpha_n, \dot{\alpha_n}$ |
| 20 | Sine: $s_n$ |
| 21 | Cosine: $c_n$ |
| 22 | Stroke length to width ratio over a 5-samples window: $r_n^5$ |
| 23 | Stroke length to width ratio over a 7-samples window: $r_n^7$ |

### 3.2. *Database and experimental protocol*

In this case, BiosecurID database [Fierrez, Galbally, Ortega-Garcia, *et al.* (2010)] is considered. Signatures were acquired from a total of 400 users using a Wacom Intuos 3 pen tablet with a resolution of 5080 dpi and 1024 pressure levels. The database comprises 16 genuine signatures and 12 skilled forgeries per user, captured in 4 separate acquisition sessions. Each session was captured leaving a two month interval between them, in a controlled and supervised office-like scenario. Signatures were acquired using a pen stylus. The available information within each signature is: *X* and *Y* pen coordinates and pressure. In addition, pen-up trajectories are available.

The experimental protocol has been designed to allow the study of different signature complexity levels in the system performance. Two main experiments are carried out: 1) evaluation of the signature complexity detector proposed in this work in order to classify users into different complexity levels, and 2) evaluation of the proposed approach based on a separate on-line signature verification system adapted to each signature complexity level.

For the first experiment, our proposed signature complexity detector is analyzed using all available users from BiosecurID. For the second experiment, the BiosecurID database is split into development dataset (40% of the users) and evaluation dataset (the remaining 60% of the users). The development dataset is considered in order to select the most discriminative and robust time functions for each signature complexity level using the SFFS algorithm whereas the evaluation dataset is considered for the evaluation of the proposed system. Both skilled and random forgeries are considered using the 4 signatures from the enrolment session as reference signatures and the remaining 12 genuine signatures and 12 skilled forgeries signatures as the test. The final score is obtained after performing the average score of the four one-to-one comparisons.

## 3.3. *Results*

### 3.3.1. *Analysis of the signature complexity detector*

The first experiment was designed to evaluate the proposed approach for signature complexity detection. For this, the signature complexity detector was performed in two different steps. First, each user of the BiosecurID database was manually labelled in a signature complexity level (low, medium, high). This process was carried out by manually labelling the image of just one genuine signature per user. This was performed by two annotators and two times each in order to keep consistency on the results. Three different complexity levels were considered based on previous works [Houmani and Garcia-Salicetti (2016)]. Users with signatures longer in writing time and with an appearance more similar to handwriting were labelled as high-complexity users whereas those users with signatures shorter in time and with generally simple flourish with no legible information were labelled as low-complexity users. This first stage served as a ground truth. Following this stage, the Sigma-LogNormal parameter $N$ was extracted for each available genuine signature of the BiosecurID database (i.e. a total of $400 \times 16 = 6400$ genuine signatures). Then, we represented for each complexity level their corresponding distribution of lognormals according to the ground truth performed during the first stage. Figure 3 shows the distributions of the number of lognormals obtained for each complexity level using all genuine signatures of the BiosecurID database. The three proposed complexity-dependent decision thresholds are highlighted by black dashed lines and were selected in order to minimize the number of misclassifications between different signature complexity levels. Signatures with lognormal values equal or less than 17 are classified as low-complexity signatures whereas those signatures with more than 27 lognormals are classified into the high-complexity group. Otherwise, signatures are categorized into medium-complexity level. Figure 4 shows some of the signatures classified into each complexity level.

We now analyse each resulting complexity level following the same procedure proposed in [Houmani and Garcia-Salicetti (2016)]: analysing the system performance for different complexity groups considering only $X$ and $Y$ pen coordinates. It is important to remark that each user is classified into a complexity

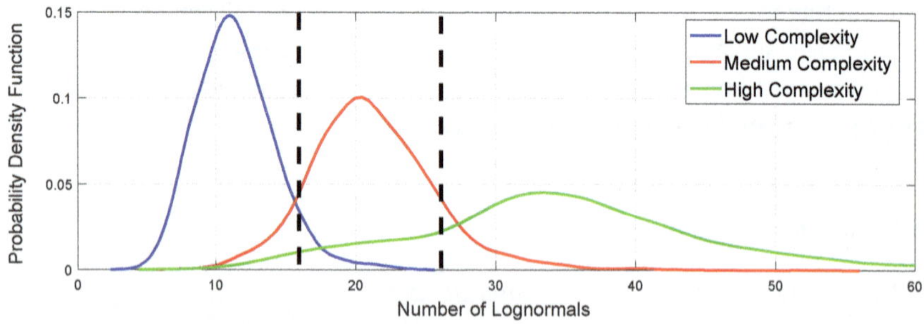


Fig. 3. Probability density function of the number of lognormals for each complexity level using all genuine signatures of the BiosecurID database. The three proposed complexity-dependent decision thresholds are highlighted by black dashed lines.

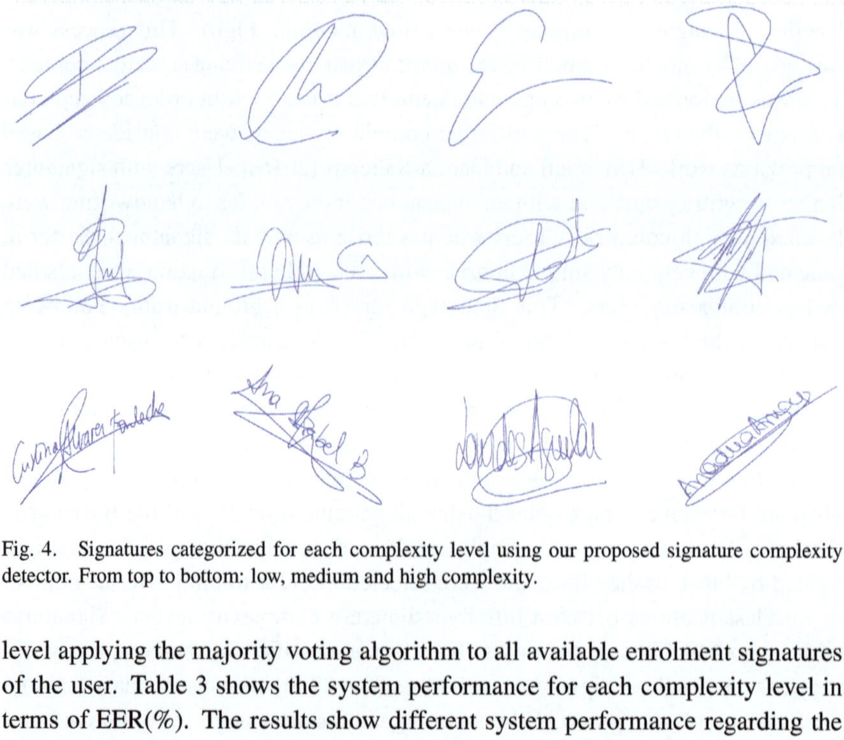

Fig. 4. Signatures categorized for each complexity level using our proposed signature complexity detector. From top to bottom: low, medium and high complexity.

level applying the majority voting algorithm to all available enrolment signatures of the user. Table 3 shows the system performance for each complexity level in terms of EER(%). The results show different system performance regarding the signature complexity level. Users with a high complexity level have an absolute improvement of 4.3% compared to users categorized into a low complexity level for skilled forgeries. Therefore, the idea of considering a different optimal on-line signature verification system for each signature complexity level is analysed in next sections in order to select the most discriminative and robust time functions for each complexity group and reduce the system performance.

### 3.3.2. *Time-functions selection for the complexity-based signature verification system*

First we analyse which are the most discriminative and robust time functions for each signature complexity level using the SFFS algorithm over the development dataset. The following three cases are studied:

(1) Time functions selected for all three signature complexity levels.
(2) Time functions selected only for medium and high signature complexity levels.
(3) Time functions selected only for low and medium signature complexity levels.

For the first case, the time functions $\dot{z_n}$, $\dot{a_n}$ and $v_n^r$ (see Table 2) have been selected in all systems as robust time functions regardless of the signature complexity level. These time functions are the variation of pressure, variation of acceleration and ratio of the minimum over the maximum speed and provide general and valuable information to all signature verification systems about the knowledge and speed of the users performing their signatures. For the second case, the time functions $\dot{v_n}$, $\ddot{y_n}$ and $\dot{\alpha_n}$ have been selected for both medium and high signature complexity levels. These time functions provide information related to the variation of the velocity, vertical acceleration and variation of angle, time functions more related to the geometry of characters and therefore, with the handwriting. Finally, the time function $c_n$ is the only one selected for the third case and provides information related to the angles as signatures with low and medium complexity level are usually categorized for having simple flourishes with no legible information. It is important to highlight that the time function $\ddot{y_n}$ is not selected for users with low signature complexity level. In other studies such as [Martinez-Diaz *et al.* (2014)], this time function was selected in most optimal systems. However, the vertical acceleration seems not to be very discriminative for users with low signature complexity level as their signatures are usually simpler and not related to handwriting.

### 3.3.3. *Experimental results of the complexity-based signature verification system*

The second part of the experimental work was focused on developing a specific verification system for each group of signature complexity. For this, the SFFS algorithm was applied to the development dataset in order to find the most discriminative time functions for each complexity group. Then, the evaluation of the proposed system was compared to a baseline system based on DTW and the same

Table 3. Experiment 1: System performance results (EER in %) of the BiosecurID database of each personal complexity level.

| | Low C. | Medium C. | High C |
|---|---|---|---|
| Skilled forgeries | 22.2 | 21.7 | 17.9 |
| Random forgeries | 3.6 | 2.4 | 2.6 |

Table 4. Experiment 2: System performance results (EER in %) on the evaluation dataset for each signature complexity level.

| | Low C. | | Medium C. | | High C. | |
|---|---|---|---|---|---|---|
| | Baseline | Proposed | Baseline | Proposed | Baseline | Proposed |
| Skilled forgeries | 13.8 | 10.1 | 7.5 | 5.2 | 6.2 | 4.6 |
| Random forgeries | 1.5 | 1.3 | 0.7 | 0.5 | 0.9 | 0.9 |

system (same time functions) for all complexity groups, similar to the baseline system presented in [Gomez-Barrero *et al.* (2015)].

Table 4 shows the evaluation results achieved considering our proposed approach based on personal entropy on-line signature verification systems. Analysing the results obtained, our Proposed Systems achieve an average absolute improvement of 2.5% EER compared to the Baseline System for the case of skilled forgeries. It is important to note that for the most challenging users (users with high personal entropy level), our proposed approach achieves an absolute improvement of 3.7% EER compared to the Baseline System. Analysing the results obtained for the random forgery cases, our Proposed Systems also achieves improvements for all personal entropy levels. For this case, the improvement has been lower than for skilled forgery cases due to its low values and the way that the SFFS algorithm was applied during the training of the systems (focused on skilled forgery cases). Results obtained after applying our proposed approach based on personal entropy on-line signature verification systems outperform the results of the state-of-the-art for the BiosecurID database. In [Gomez-Barrero *et al.* (2015)], the authors achieved an absolute improvement of 1.0% EER for skilled forgery cases whereas our proposed approach achieves an average absolute improvement of 2.5% EER compared to the same Baseline System.

For completeness, Fig. 5 shows the performance of the Baseline and Proposed Systems considering all personal entropy levels together in terms of the false rejection rate (FRR) at different values of false acceptance rate (FAR). Our Proposed Systems achieve a final value of 5.8% FRR for a FAR = 5.0% and 3.9% FRR for a FAR = 10.0%. These results show the importance of considering different signature verification systems for each personal entropy level in order to enhance the verification systems with more robust time functions.

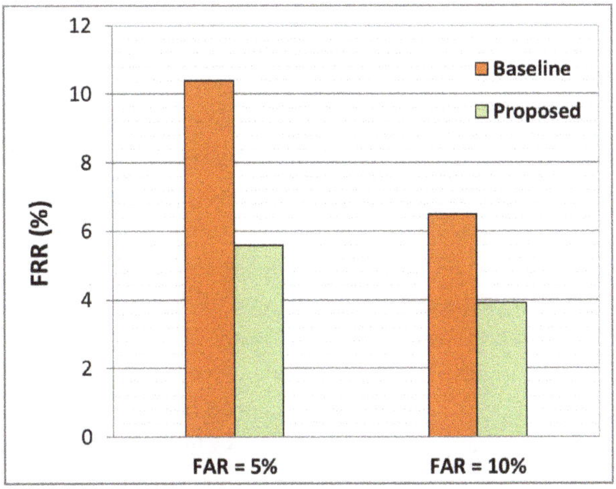


Fig. 5. **Experiment 2:** Analysis of the False Rejection Rate (FRR) at different values of False Acceptance Rate (FAR) for both Proposed and Baseline Systems on the whole evaluation dataset.

## 4. Case Study 2: Predicting Age Groups from Touch Patterns

Age groups prediction based on handwritten touch patterns acquired from touch-screen devices such as smartphones or tables is a recent and important challenge. Touchscreen devices provide mobile access to an unlimited number of digital contents and services (e.g. more than a half of YouTube visits come from mobile devices and this percentage is increasing[b]). Digital services are used by people from everywhere, all ages, all ethnicities and all socioeconomic status. In this context, the classification of users according to geographic and demographic attributes is crucial for service personalization (e.g. recommender systems, parental control, security[c]). Some of these attributes can be obtained from metadata associated to the device (e.g. IP address, language selection, GPS location) or can be inferred from the user behavior (e.g. browsing history, social network contents, and keystroke dynamics) [Morales *et al.* (2016)]. We want to highlight the spread of the use of this kind of devices by young children. The study in [Kabali *et al.* (2015)] reveals that 97% of US children under the age of four use mobile devices, regardless of family income.

In this case study we analyze a way to classify users of touch panels according to two age groups (children and adults). The age is a key attribute in user

---

[b]Youtube statistics. https://about.twitter.com/company Retrieved September 13, 2016.
[c]https://www.bba.org.uk/news/press-releases/mobile-phone-apps-become-the-uks-number-one-way-to-bank Retrieved September 13, 2019.

profiling with direct application on several automatic systems (e.g. parental control, recommender systems, advertising). Three examples of use cases are: i) locking content and/or applications: locking some services in tablets and smartphones when children are using them, i.e. buying new applications or sensitive content; ii) user's age study by service providers: this way service providers could develop new content that fits better to their actual audience; iii) real-time interface adapting: as children have worse control of their fine movements than adults, changing default interfaces to special tailored ones could be beneficial.

The most popular method to reveal the age of the user is based on an online questionnaire in which the user directly answers questions about his age. However, this solution assumes: i) honesty on the response of the users, and ii) users can read. Both assumptions cannot be guaranteed because of many practical reasons. Besides the fact that people lie, nowadays children start to use digital platforms and services before learning to read.

In the existing literature, there are many experiments exploring the use of technology by children, seeking how to improve the design of adapted interfaces and applications [Cassidy and McKnight (2010)]. However, modeling and characterizing mathematically how children interact with touch devices and how their conduct differs from the adult's one is a field that has not been studied deeply enough. A work related to this topic is [Abdul Aziz *et al.* (2013)] where they analyzed different types of touching tasks like tap, rotate or drag and drop, and they found that children have different success rates when trying to perform different tasks. Simple tasks — for example tapping — can be done by all children without any problem, but the more complex ones are very difficult to be completed by very young children.

In [Anthony *et al.* (2012)], they measured the touch patterns of children and compared it to patterns from adults. They discovered that children have a larger miss rate than adults when trying to hit small targets. In [Vatavu *et al.* (2015a)] tap tasks are used to extract time and precision-based features. They designed two different approaches, using only one tap for classification and using 7 consecutive tasks. They get high accuracy rates: 86.5% in the one tap approximation, and 99% of accuracy using 7 consecutive taps to combine their scores. Even though they get good results using tap tasks, we decide to use drag and drop tasks because the differences between the neuromotor development of users can be manifested in a better way. The direct comparison between approaches is not fair because we are using different tasks/information to classify users. However, in our work we demonstrate that using a very common and fast action (e.g. unlock screen based on drag and drop) we can achieve higher classification rates that those achieved in [Vatavu *et al.* (2015a)] for the one task approach (the second approach has not been implemented yet). In our opinion, both approaches are complementary, have very different nature, and can be combined to achieve higher performances.

The difference between adults and children is mainly caused by the different maturity of their anatomy and neuromotor system. These are less mature in children, so they have worse manual dexterity causing rougher movements [Piaget and Inhelder (1969); O'Reilly and Plamondon (2009)].

In order to characterize the interaction of children and adults with touchscreen devices, we propose to use a model of the human neuromotor system. The Sigma-LogNormal theory of rapid human movements represents complex movements with an analytic model that describes some physical and cognitive features of human beings [Fischer and Plamondon (2015); Plamondon *et al.* (2013)]. Studies like [Duval *et al.* (2015)] have proved that the Sigma-LogNormal model can be used to characterize children handwriting. They conclude that there are two main groups of children separable by looking at their learning stage. Children's neuromotor skills become more similar to the adults' skills when they grow up, namely, when they finish their preoperational stage. At age 10 children know how to activate each little muscle properly to produce determinate fine movements [Vatavu *et al.* (2015b)]. As they are based on the same neuromotor skills, the principles applied to handwriting models can be also used to model touchscreen patterns.

In this case study we propose to exploit the lognormality principle and use the Sigma-Lognormal model to detect age groups as simple application of the model to drag and drop touch tasks showed large differences between adults and children velocity profiles. In particular, this case study is focused in age classification of users into two groups: children under 6 years old and adults. We use information of simple touch tasks collected from 119 people (89 children and 30 adults) using two different types of devices: a smartphone and a tablet. Single-sensor and cross-sensor scenarios have been evaluated. The results show accuracies over 90% in several scenarios with top correct classification rate of 96% for the data obtained from tablets.

### 4.1. *Proposed system*

In this case, a more complex system was developed compared to Case Study 1 in order to predict age groups from drag and drop touch tasks, as the main focus here was to optimize the final classification result.

The parameters of the Sigma-LogNormal model (as described in Sec. 2) were used to calculate 18 different features per lognormal (see Table 5) as described in [Fischer and Plamondon (2015)]. These features can be classified into two groups: space-based and time-based. Space-based features are those that give information

Table 5. Sigma-LogNormal features extracted.

| Space-based features | Time-based features |
|---|---|
| $f_1 = D_i$ | $f_6 = \mu_i$ |
| $f_2 = \sin(\theta_{si})$ | $f_7 = \sigma_i$ |
| $f_3 = \cos(\theta_{si})$ | $f_8 = \Delta t_0 = t_{0i} - t_{0i-1}$ |
| $f_4 = \sin(\theta_{ei})$ | $f_9 = v_2 = \lvert v_i(t_{2i})\rvert$ |
| $f_5 = \cos(\theta_{ei})$ | $f_{10} = v_3 = \lvert v_i(t_{3i})\rvert$ |
| | $f_{11} = v_4 = \lvert v_i(t_{4i})\rvert$ |
| | $f_{12} = \delta t_{05} = t_{5i} - t_{0i}$ |
| | $f_{13} = \delta t_{15} = t_{5i} - t_{1i}$ |
| | $f_{14} = \delta t_{13} = t_{3i} - t_{1i}$ |
| | $f_{15} = \delta t_{35} = t_{5i} - t_{3i}$ |
| | $f_{16} = \delta t_{24} = t_{4i} - t_{2i}$ |
| | $f_{17} = \Delta t_1 = t_{1i} - t_{1i-1}$ |
| | $f_{18} = \Delta t_3 = t_{3i} - t_{3i-1}$ |

about the spatial distribution of the strokes, such as $D_i$, $\mu_i$, $\sigma_i$, and other features based in $\theta_{si}$ and $\theta_{ei}$ (see Table 1). Time based features are composed by the values of speed at some relevant points of the strokes like their maximum or inflexion points; and the time-offsets between those points. The task time and the number of lognormals in each task have been added as additional features.

It is worth noting that the lognormals with amplitude value lower than a threshold were discarded. Then, the 18 features from [Fischer and Plamondon (2015)] are computed for each stroke, and each parameter is averaged across strokes. The 18 averaged parameters are augmented with the task time and the number of strokes to generate the final feature vector of size 20.

Regarding the classification of the age of the user, quite often it is possible to differentiate between children and adults by simply looking at the velocity profile of a touch screen task. In Fig. 6, an example of these types of profiles is presented, consisting in performing a drag and drop task in both cases. A visual comparison between children and adults velocity profiles shows that children's signals are usually composed by a higher number of strokes than the adults' ones, and therefore have a higher degree of complexity.

Figures 7(a) and 7(b) show the histograms of two features (Covered distance $f_1$, and Logtemporal delay $f_6$) for children and adults. These two features are highly discriminative as their histograms are clearly separated, showing differences between both classes and therefore suggesting the potential for the classification task.

As a classifier we use a SVM (Support Vector Machine) with a RBF (Radial Basis Function) kernel because of its good general performance in binary classification tasks and the few number of parameters to configure.

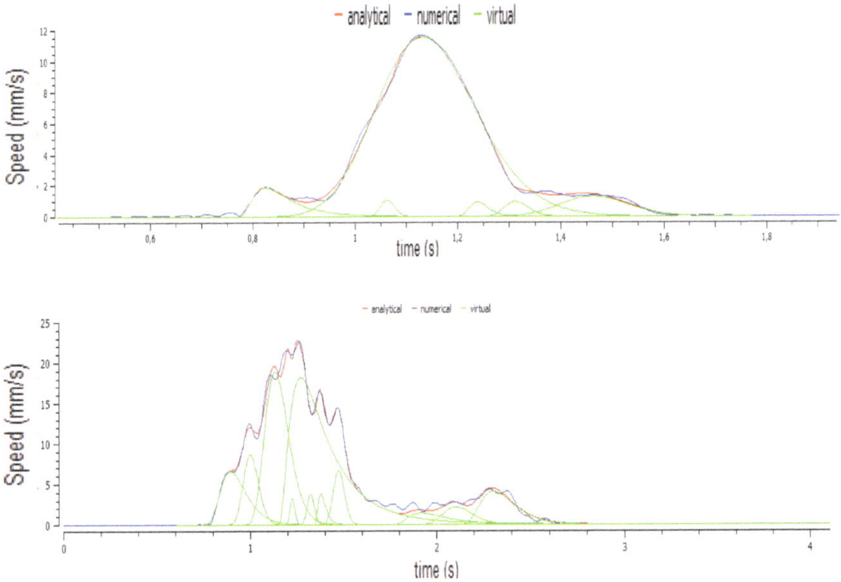


Fig. 6. Comparison between Sigma LogNormal speed profiles for (a) an adult and (b) a child following the same task.

### 4.2. *Database and experimental protocol*

The database used is publicly available and was presented in [Vatavu *et al.* (2015b)]. It is comprised with data from touchscreen activity of both children and adults performing predesigned tasks in an ad-hoc app. In the present work, we have used the data from singletouch and multitouch drag and drop activities. Drag and drop activities consist of picking one object on the device screen and moving it to a target area. Multidevice information is available as the users have completed the tasks both in a smartphone and in a tablet. Both single-sensor and cross-sensor tasks are analyzed.

The dataset is composed by 89 children between 3 and 6 years old and 30 young adults under 25 years old. The mean age of the children is 4.6 years. The total number of drag and drop tasks is 2912 for children and 1157 for adults (see [Vatavu *et al.* (2015b)] for more details).

As the experimental protocol, the database was divided randomly into training (60%) and testing (40%). The random selection was repeated 50 times and the final performance is presented in terms of averaged correct classification accuracy.

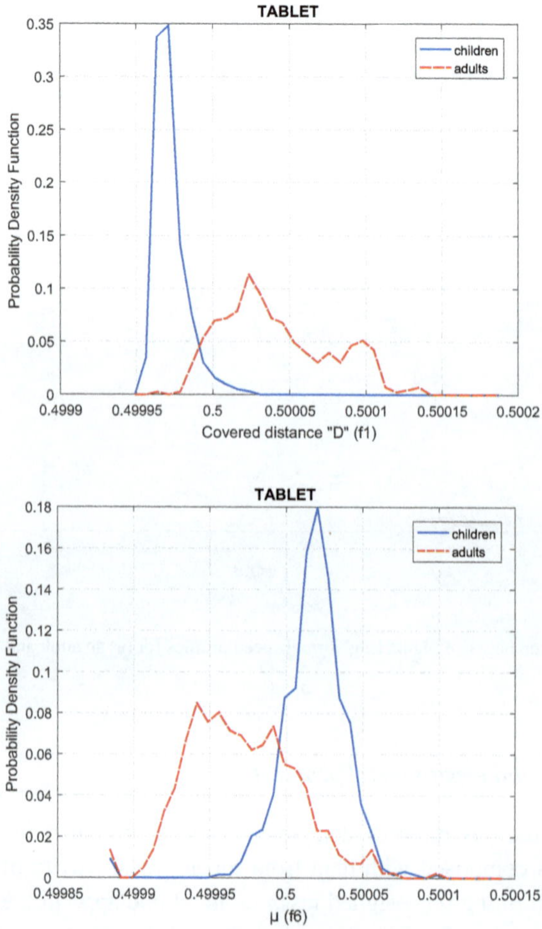


Fig. 7. Probability Density Functions for two features ($f_1$ and $f_6$). These are highly discriminative features as histograms are separated.

### 4.3. *Results*

Table 6 shows the accuracies obtained according to the different scenarios. They are presented in terms of correct classification accuracy (percentage of samples from both classes correctly classified).

The mean value of accuracy having into account all the evaluated scenarios is 92.8%. The classification rates are over 96% in a single-sensor setting and over 95% in a cross-sensor scenario. The best results are obtained with tablets as sensors, while using smartphone's data slightly degrades the results.

Table 6. Accuracy results for the 20 lognormal features. The accuracy is measured as the rate of correct classifications considering both classes.

| | | Testing samples | | | |
|---|---|---|---|---|---|
| | | Phone Singletouch | Tablet Singletouch | Phone Multitouch | Tablet Multitouch |
| Traning samples | Phone Singletouch | 93.6% | 95.0% | 88.0% | 92.1% |
| | Tablet Singletouch | 93.7% | 96.3% | 88.9% | 94.0% |
| | Phone Multitouch | 94.1% | 95.9% | 88.0% | 92.8% |
| | Tablet Multitouch | 93.0% | 96.3% | 87.9% | 94.6% |

Compared with [Vatavu *et al.* (2015a)] where they get an accuracy rate of 86.5% using one tap task for classification and with a single-sensor approximation (using smartphone's data), our system performs better, getting a 93.6% of accuracy using only data from smartphones, and over 96% using data from tablets. Another conclusion that can be extracted from Table 6 is that the data obtained from multitouch tasks get worse results than the singletouch cases. The best multitouch scenario is obtained using tablet's data for both training and testing, with a 94.6% of accuracy, compared with its singletouch counterpart that gets a 96.3%. This may be caused by the less developed control of the left hand by right-handed people and vice versa. The main reason for using the Sigma-LogNormal model is that adults have a better control of fine movements than children, what is translated to different values for the model parameters [Vatavu *et al.* (2015b)].

The cross-sensor scenarios get results not too far from the single-sensor scenarios. The results obtained using smartphone singletouch data for training, and tablet singletouch data for testing (95.9% of accuracy) are quite similar to those obtained using only tablet singletouch data (96.3% of accuracy). This fact makes this type of systems very suitable for real applications due to its high independence of the device used.

Due to the higher number of children in the database compared to adults, selecting a percentage of the total users make the two scenarios unbalanced. Experiments balancing the number of both classes in training and testing have been made. The results show small variations around 1% of accuracy with respect to the presented results.

Figure 8 shows histograms of the scores calculated in the classification process. It can be seen that the scores from children and adults are visibly separated into two different zones, making possible to obtain high accuracy rates (over 96%). There are also other zones where the scores distributions overlap. These regions are the source of incorrect classifications. Combining scores from several tasks of the same user could make possible to reduce the overlap areas, increasing even more the accuracy rate.

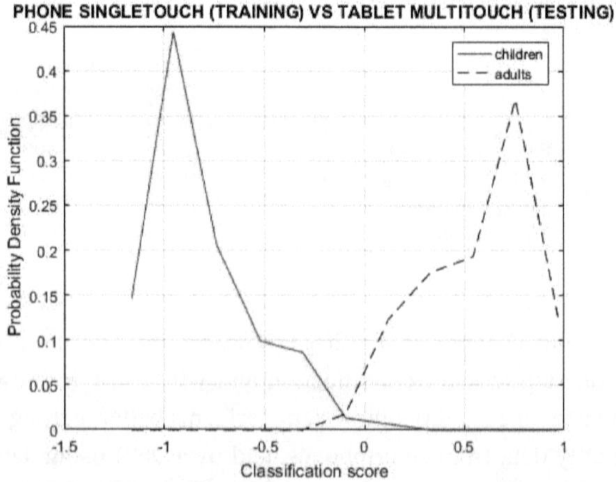


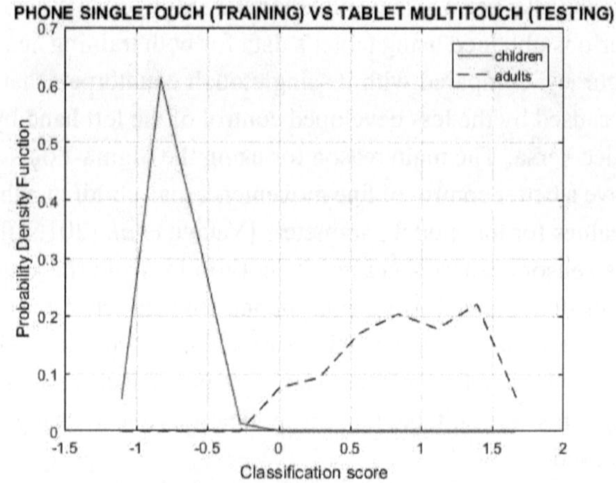


Fig. 8. Histograms of scores using the Sigma-LogNormal model features. Top figure represents the scores for single-sensor scenario, using tablet singletouch data for both training and testing. Bottom figure shows the histogram for a cross-sensor scenario, using phone singletouch data for training and tablet multitouch data for testing the classifier.

## 5. Conclusions

This work has reported experimental results on modeling the complexity of biomechanical tasks through the usage of the Sigma-LogNormal model of the Kinematic Theory of rapid human movements. Two different case studies have been analyzed.

The first case study has focused on applying the Sigma-LogNormal model to develop an on-line signature complexity detector. Just by using the number of strokes of the signatures was enough to obtain very good results differentiating between three different signature complexity groups (low, medium and high). As a second stage, a specific signature verification system was developed for each signature complexity group by carrying out a time functions selection process. Very significant improvements of recognition performance have been shown when comparing the proposed system with a baseline, being both based on DTW and time functions as features. For future work, the approach considered in this work will be further analysed using the e-BioSign public database [Tolosana *et al.* (2017a)] in order to consider new scenarios such as the case of using the finger as the writing tool. It would be possible to define fuzzy areas between two consecutive levels of complexity and this might improve the final results. Novel systems based on the usage of Recurrent Neural Networks (RNNs) [Tolosana *et al.* (2018a, 2019b, 2020b); Vera-Rodriguez *et al.* (2019)] and the fusion of different systems [Fierrez *et al.* (2018a); Tolosana *et al.* (2019c)] will be considered. Also, different types of presentation attacks to signature recognition systems [Tolosana *et al.* (2018b)] will be considered analysing how signatures with different complexity levels are affected.

On the other hand, the second case study has focused on age group prediction (children from adults) from handwritten touch patterns acquired from touchscreen devices such as smartphones or tables. Applying the Sigma-LogNormal model to some examples of drag and drop tasks from children and adults showed that children had a more complex velocity profiles with a larger number of lognormals, which is consistent with the lognormality principle. The proposed approach is based on 20 features extracted from the model, and results achieved were very promising with classification rates over 96% in a single-sensor setting and over 95% in a cross-sensor scenario. Further work includes performing feature selection to find out the most discriminative features and also the analysis of touchscreen data to continuously monitor the user behaviour [Acien *et al.* (2019, 2020b,d)] or to detect whether there is a person or a bot interacting [Acien *et al.* (2020a,c)]

**Acknowledgements**

This work has been supported by projects: IDEA-FAST (H2020-IMI2-2018-853981), PRIMA (H2020-ITN-2019-860315), TRESPASS-ETN (H2020-ITN-2019-860813), BIBECA (RTI2018-101248-B-I00 MINECO/FEDER), Bio-Guard (Ayudas Fundación BBVA a Equipos de Investigación Científica 2017) and by

UAM-CecaBank Project. Alejandro Acien is supported by a FPU Fellowship from Spanish MECD, and Javier Hernandez by a FPI Fellowship from UAM.

## References

Abdul Aziz, N. A., Batmaz, F., Stone, R. and Paul Wai Hing Chung (2013). Selection of touch gestures for children's applications, in *Proc. Science and Information Conference (SAI), 2013*, pp. 721–726.

Acien, A., Morales, A., Fierrez, J. and Vera-Rodriguez, R. (2020a). Becaptcha-mouse: Synthetic mouse trajectories and improved bot detection, *arXiv preprint arXiv:2005.00890*.

Acien, A., Morales, A., Fierrez, J., Vera-Rodriguez, R. and Bartolome, I. (2020b). BeCAPTCHA: Detecting Human Behavior in Smartphone Interaction using Multiple Inbuilt Sensors, in *AAAI Workshop on Artificial for Cyber Security (AICS)*.

Acien, A., Morales, A., Fierrez, J., Vera-Rodriguez, R. and Delgado-Mohatar, O. (2020c). BeCAPTCHA: Bot Detection in Smartphone Interaction using Touchscreen Biometrics and Mobile Sensors, *arXiv preprint arXiv:2005.13655*.

Acien, A., Morales, A., Vera-Rodriguez, R. and Fierrez, J. (2020d). *Mobile Active Authentication based on Multiple Biometric and Behavioral Patterns*, pp. 161–177, Springer.

Acien, A., Morales, A., Fierrez, J., Vera-Rodriguez, R. and Hernandez-Ortega, J. (2019). Active detection of age groups based on touch interaction, *IET Biometrics* **8**, 1, pp. 101–108.

Alonso-Fernandez, F., Fairhurst, M.C., Fierrez, J. and Ortega-Garcia, J. (2007). Impact of Signature Legibility and Signature Type in Off-Line Signature Verification, *In Proc. IEEE Biometrics Symposium*, pp. 1–6.

Anthony, L., Brown, Q., Nias, J., Tate, B. and Mohan, S. (2012). Interaction and recognition challenges in interpreting children's touch and gesture input on mobile devices, in *Proc. ACM Intl. Conference on Interactive tabletops and surfaces*, pp. 225–234.

Cassidy, B. and McKnight, L. (2010). Children's interaction with mobile touch-screen devices: Experiences and guidelines for design, *Int. J. Mob. Hum. Comput. Interact.* **2**, 2, pp. 1–18.

Diaz, M., Fischer, A., Ferrer, M. A. and Plamondon, R. (2017). Dynamic signature verification system based on one real signature, *IEEE Trans. on Cybernetics*, 99, pp. 1–12.

Djioua, M. and Plamondon, R. (2009). A new algorithm and system for the characterization of handwriting strokes with delta-lognormal parameters, *IEEE Transactions on Pattern Analysis and Machine Intelligence* **31**, 11, pp. 2060–2072.

Duval, T., Rémi, C., Plamondon, R., Vaillant, J. and O'Reilly, C. (2015). Combining sigma-lognormal modeling and classical features for analyzing graphomotor performances in kindergarten children, *Human Movement Science* **43**, pp. 183–200.

Fierrez, J., Morales, A., Vera-Rodriguez, R. and Camacho, D. (2018a). Multiple classifiers in biometrics. Part 2: Trends and challenges, *Information Fusion* **44**, pp. 103–112.

Fierrez, J., Pozo, A., Martinez-Diaz, M., Galbally, J. and Morales, A. (2018b). Benchmarking touchscreen biometrics for mobile authentication, *IEEE Trans. on Information Forensics and Security* **13**, 11, pp. 2720–2733.

Fierrez, J., Galbally, J., Ortega-Garcia, J. *et al.* (2010). BiosecurID: A Multimodal Biometric Database, *Pattern Analysis and Applications* **13**, 2, pp. 235–246.

Fierrez, J., Ortega-Garcia, J. and Gonzalez-Rodriguez, J. (2005). Target Dependent Score Normalization Techniques and Their Application to Signature Verification, *IEEE Transactions on Systems, Man, and Cybernetics. Part C* **35**, 3, pp. 418–425.

Fischer, A. and Plamondon, R. (2017). Signature verification based on the kinematic theory of rapid human movements, *IEEE Trans. on Human-Machine Systems* **47**, 2, pp. 169–180.

Fischer, A. and Plamondon, R. (2015). A dissimilarity measure for on-line signature verification based on the sigma-lognormal model, in *Proc. 17th Biennial Conference of the International Graphonomics Society*.

Galbally, J., Plamondon, R., Fierrez, J. and Ortega-Garcia, J. (2012). Synthetic on-line signature generation. Part I: Methodology and algorithms, *Pattern Recognition* **45**, pp. 2610–2621.

Gomez-Barrero, M., Galbally, J., Fierrez, J., Ortega-Garcia, J. and Plamondon, R. (2015). Enhanced on-line signature verification based on skilled forgery detection using Sigma-LogNormal Features, in *Proc. 2015 International Conference on Biometrics (ICB)*, pp. 501–506.

Guest, R. (2006). Age Dependency in Handwritten Dynamic Signature Verification Systems, *Pattern Recognition Letters* **27**, 10, pp. 1098–1104.

Hernandez-Ortega, J., Morales, A., Fierrez, J. and Acien, A. (2017). Detecting age groups using touch interaction based on neuromotor characteristics, *IET Electronics Letters*, pp. 1–2.

Houmani, N. and Garcia-Salicetti, S. (2016). On Hunting Animals of the Biometric Menagerie for Online Signature, *PLOS ONE* **11**, 4, pp. 1–26.

Houmani, N., Garcia-Salicetti, S. and Dorizzi. B. (2008). A Novel Personal Entropy Measure Confronted to Online Signature Verification Systems' Performance, *In Proc. Intl. Conf. on Biometrics: Theory, Applications and System, BTAS*, pp. 1–6.

Kabali, H. K., Irigoyen, M. M., Nunez-Davis, R., Budacki, J. G., Mohanty, S. H., Leister, K. P. and Bonner, R. L. (2015). Exposure and use of mobile media devices by young children, *Pediatrics* **136**, 6, pp. 1044–1050.

Lim, M. and Yuen, P. (2016). Entropy Measurement for Biometric Verification Systems, *IEEE Transactions on Cybernetics* **46**, 5, pp. 1065–1077.

Martinez-Diaz, M., Fierrez, J., Krish, R. P. and Galbally, J. (2014). Mobile Signature Verification: Feature Robustness and Performance Comparison, *IET Biometrics* **3**, 4, pp. 267–277.

Martinez-Diaz, M., Fierrez, J. and Hangai, S. (2009a). *Signature Features*, pp. 1185–1192, Springer US, Boston, MA, ISBN 978-0-387-73003-5.

Martinez-Diaz, M., Fierrez, J. and Hangai, S. (2009b). *Signature Matching*, pp. 1192–1196, Springer US, Boston, MA, ISBN 978-0-387-73003-5.

Morales, A., Fierrez, J., Tolosana, R., Ortega-Garcia, J., Galbally, J., Gomez-Barrero, M., Anjos, A. and Marcel, S. (2016). Keystroke Biometrics Ongoing Competition, *IEEE Access* **4**, pp. 7736–7746.

O'Reilly, C. and Plamondon, R. (2009). Development of a Sigma-Lognormal Representation for On-Line Signatures, *Pattern Recognition* **42**, 12, pp. 3324–3337.

Piaget, J. and Inhelder, B. (1969). *The psychology of the child*, Vol. 5001, Basic books.

Plamondon, R., Pirlo, G. and Impedovo, D. (2014). *Online Signature Verification*, pp. 917–947, Springer London, London.

Plamondon, R., O'Reilly, C., Rémi, C. and Duval, T. (2013). The lognormal handwriter: learning, performing, and declining, *Frontiers in psychology* **4**, p. 945.

Tolosana, R., Vera-Rodriguez, R., Fierrez, J. and Ortega-Garcia, J. (2020a). BioTouchPass2: Touchscreen Password Biometrics Using Time-Aligned Recurrent Neural Networks, *IEEE Transactions on Information Forensics and Security*.

Tolosana, R., Vera-Rodriguez, R., Fierrez, J. and Ortega-Garcia, J. (2020b). DeepSign: Deep On-Line Signature Verification, *arXiv preprint arXiv:2002.10119*.

Tolosana, R., Vera-Rodriguez, R., Guest, R., Fierrez, J. and Ortega-Garcia, J. (2020c). Exploiting Complexity in Pen-and Touch-based Signature Biometrics, *International Journal on Document Analysis and Recognition*.

Tolosana, R., Vera-Rodriguez, R. and Fierrez, J. (2019a). BioTouchPass: Handwritten Passwords for Touchscreen Biometrics, *IEEE Transactions on Mobile Computing*.

Tolosana, R., Vera-Rodriguez, R., Fierrez, J., Morales, A. and Ortega-Garcia, J. (2019b). Do You Need More Data? The DeepSignDB On-Line Handwritten Signature Biometric Database, in *Proc. 15th International Conference on Document Analysis and Recognition, ICDAR*.

Tolosana, R., Vera-Rodriguez, R., Fierrez, J. and Ortega-Garcia, J. (2019c). Reducing the Template Aging Effect in On-Line Signature Biometrics, *IET Biometrics*.

Tolosana, R., Vera-Rodriguez, R., Fierrez, J. and Ortega-Garcia, J. (2018a). Exploring Recurrent Neural Networks for On-Line Handwritten Signature Biometrics, *IEEE Access*, pp. 1–11.

Tolosana, R., Vera-Rodriguez, R., Fierrez, J. and Ortega-Garcia, J. (2018b). *Presentation Attacks in Signature Biometrics: Types and Introduction to Attack Detection*, Springer.

Tolosana, R., Vera-Rodriguez, R., Fierrez, J., Morales, A. and Ortega-Garcia, J. (2017a). Benchmarking Desktop and Mobile Handwriting across COTS Devices: the e-BioSign Biometric Database, *PLOS ONE* **5**, 12.

Tolosana, R., Vera-Rodriguez, R., Guest, R., Fierrez, J. and Ortega-Garcia, J. (2017b). Complexity-based Biometric Signature Verification, in *Proc. 14th IAPR Int. Conference on Document Analysis and Recognition, ICDAR*.

Vatavu, R.-D., Anthony, L. and Brown, Q. (2015a). Child or adult? inferring smartphone users' age group from touch measurements alone, in *Human-Computer Interaction*, pp. 1–9.

Vatavu, R.-D., Cramariuc, G. and Schipor, D. M. (2015b). Touch interaction for children aged 3 to 6 years: Experimental findings and relationship to motor skills, *International Journal of Human-Computer Studies* **74**, pp. 54–76.

Vera-Rodriguez, R., Tolosana, R., Caruana, M., Manzano, G., Gonzalez-Garcia, C., Fierrez, J. and Ortega-Garcia, J. (2019). Deepsigncx: Signature complexity detection using recurrent neural networks, in *Proc. 15th International Conference on Document Analysis and Recognition, ICDAR*.

Zhou, Z.H. (2009). *Entropy, Biometric*, pp. 273–274, Springer US, Boston, MA, ISBN 978-0-387-73003-5.

https://doi.org/10.1142/9789811226830_0004

# Chapter 4

# Could Sigma-Lognormal Modeling Help Teachers to Characterize the Kinematic Efficiency of Pupils' Cursive Procedures of Handwriting?

Céline Rémi[1,*], Jimmy Nagau[1,†], Jean Vaillant[1,‡] and Réjean Plamondon[2,§]
[1]*LAMIA, Université des Antilles*
*Pointe-à-Pitre, Guadeloupe*
[2]*Laboratoire Scribens, Département de génie électrique,*
*Polytechnique Montréal, Canada*
[*]*celine.remi@univ-antilles.fr,* [†]*Jimmy.nagau@univ-antilles.fr,*
[‡]*jean.vaillant@univ-antilles.fr,* [§]*rejean.plamondon@polymtl.ca*

Teachers need some reliable, objective and simple approaches to follow the various aspects of children's graphomotor progress involved in handwriting learning all along their scholarship. This paper deals with the relevance of the Sigma-Lognormal modeling as a tool to identify reliable objective parameters in such a perspective. It focuses on the search of variables for the representation of the kinematic and procedural efficiency of children's procedures for cursive isolated words handwriting. This research was performed with French pupils in elementary school. The main results show the capacity of the Sigma-Lognormal modeling to reproduce cursive writing movements produced by children since the early stage of their scholarship at primary school when they write isolated words. Furthermore, we underline the potential interest of a new offline variable that could be estimated by teachers: the *deltaPenlift*, and of the relevance of the signal to noise ratio (SNR), provided by the online Sigma-Lognormal modeling of children's isolated word handwriting. Both factors help to characterize pupils' level of graphomotor expertise and their capacity to respect the cursive handwriting procedure taught at elementary school.

## 1. Introduction

Studies have shown that a certain number of children labeled as poor hand writers see their access to the written language and literacy strongly restrained by graphomotor difficulties [Chartrel *et al.*, 2004; Zesiger, 2003]. Indeed, handwriting remains even today the main way of capture and expression of knowledge for pupils. So, at school, to be a proficient hand writer, pupils have to know and to be able to automatically mobilize some efficient perceptual-motor handwriting procedures [Graham *et al.*, 1998; Rosenblum, 2018]. In school, as soon as possible, they must deal with this major challenge that will make them able to focus their attention on the linguistic dimension of their handwritten productions and to avoid the risk of compromising the success of all their apprenticeship [Dinehart, 2014].

According to most school systems, in conformity with research literature like [Duval *et al.*, 2014; Julius *et al.*, 2016; Medwell and Wray, 2007; Rosenblum, 2006], an efficient handwriting is defined as a perceptual-motor process which allows balancing aesthetic features like formal quality and legibility of handwritten productions with characteristics like rapidity, fluidity and regularity of the handwriting movements. Rapidity is defined as the capacity to write words, sentences and texts with sufficient speed to meet academic expectations. According to [Duval *et al.*, 2015], fluidity refers to the capacity of writing without sudden stops such that the writing movements can be clearly decomposed into kinematic premises. This allows observing the strategy that has been used by the children to accomplish this task. Regularity is defined as the capacity to produce written patterns that minimize the total distance of the generated trajectory by avoiding tremors that result in jerky writing. Regularity also takes into account the writer capacity to have similar gestures when repeating the same task over time.

In this context, since kindergarten, activities dealing with handwriting learning that are proposed and driven by teachers, must have as essential and complementary goals to strengthen each pupil's kinematic and spatio-graphic efficiency and to respect the procedural performance. For the kinematics, this means consolidating his ability to execute rapid, fluid and regular handwriting movements when he has to handwrite a particular simple or complex sequence of graphical shapes or signs. For the spatio-graphics, this requires reinforcing his capacity to comply with all the shapes or signs that he has to write to produce a given word or a sentence with the conventional formal requirements of the type of handwriting taught in school.

For the procedural performance, this means to respect the degree of conformity to the continuity, the orientation and direction of the pen-tip movements when the pupil is handwriting according to the conventional scholar procedural rules taught in school.

All over the academic cursus, teachers at least have to evaluate each of these three efficiency dimensions as well as each schoolchild's capacity to reach them, simultaneously promoting the readability and the speed of the handwritten productions during scholar activities like dictation, writing sentences or texts, note taking.

Thanks to his own perceptual abilities, on the basis of an offline visual analysis of the final result of a handwriting activity, a teacher can evaluate the spatio-graphic efficiency of all his pupils looking at the formal conformity of the letters that have been handwritten. In the same offline way, the teacher, by counting the number of letters of words handwritten during a given time, can estimate each pupil's velocity. However, this offline method provides a partial appreciation of the pupil's kinematic capacity through a rough estimation of rapidity. In the case of low rapidity, it is difficult to determine if it is a fluidity default, a lack of regularity or if this is due to the limitations of the teacher offline visual analysis of the handwritten production. Moreover, the challenge of a reliable and complete evaluation of the kinematic efficiency and of the procedural performance for each pupil in the classroom seems very difficult for teachers, based on their own perceptual abilities only. For such a purpose, it would be better if they could observe and analyze the handwriting procedures, their fluency and regularity, online, that is during each pupil's activity. Digital devices like digitizing tablets, thanks to their capacity of online data acquisition during the handwriting process all over its execution introduce useful new possibilities to analyze the kinematic dimension of handwriting. Such ability has allowed the development of numerous digital tools dedicated to handwriting learning and interactive offline or automated online evaluation like those described in deDiego-Cottinelli and Barros [2010], Plamondon *et al.* [2018], Rosenblum *et al.* [2003] and Simonnet *et al.* [2017]. One of the most sophisticated of these digital tools, Intuiscript, achieves an automated analysis of the procedure of cursive word handwriting such that it provides an online feedback on the spatio-graphic and procedural conformity to the pupil and teacher [Simonnet *et al.*, 2019]. However, none of these tools achieves an explicit evaluation of all the aspects of the procedural efficiency of handwriting. So, they cannot help teachers to completely appreciate each of the three dimensions of the strategy efficiency that pupils must master at a particular moment of their academic courses to deal with the requirements of their handwriting productions. In fact, there are no studies concerning which online variables should be relevant, from the beginning stage of the handwriting learning experience at kindergarten up to the end of elementary school, to estimate the kinematic efficiency of a pupil's handwriting procedure.

We have chosen to deal with this first question using the cursive procedure because it is taught and practiced during primary schooling in France and in

Canadian provinces like the Province of Québec or the Prince Edwards Island [Morin *et al.*, 2017; Prince Edward Island, 2010]. Previously, we have shown the interest of the Sigma-Lognormal modeling to express and analyze young children's kinematic efficiency on non-linguistic but continuous handwritten productions like scribbles [Rémi *et al.*, 2015, 2017], straight lines or pseudo-letters [Duval *et al.*, 2015, 2013]. We aim in the present paper to show its potential interest for characterizing the efficiency of the cursive procedures of isolated word handwriting. For such purpose, we will verify thanks to an experimental study the following assumptions:

(1) The Sigma-Lognormal model can reconstruct the velocity profiles of cursive handwriting movements for pupils of elementary school.
(2) According to the Lognormality Principle, some online global variables that are estimated through the Sigma-Lognormal modeling of the velocity profile of the pupils' cursive handwriting movements have a behavior coherent, according to the scholar grades, with the trends that can be observed for usual offline classical variables.
(3) The signal to noise ratio SNR is a relevant online variable that can help to synthetically characterize the kinematic efficiency of the pupil's handwriting procedures with respect to the academic set of cursive rules taught at school and that can help to validate the relevance of deltaPenlift, an original offline variable defined in this paper.

In the second section of this chapter, we review the lognormal paradigm in the specific context of this study. Among other things, we report some of our previous experimental results which have shown that over time, thanks to graphomotor practice and training, pupil handwriting tends to respect more and more the Lognormality Principle.

The third section deals with the cursive procedure of handwriting. Among other things, we expose its fundamental rule and the set of academic cumulative cursive rules that decline it through the teaching methods used at school. On the basis of these rules, we finally deduce a numerical easy-to-compute differential variable, from the classical offline number of pen lifts realized during handwriting. This new variable is then investigated for the estimation of the procedural performances achieved by the pupils with respect to the set of academic cursive rules taught at school.

The rest of the chapter presents a new experimental study, extending the one reported in Rémi *et al.* [2018]. After the presentation of the experimental protocol and the methodology that is used for data collection and the variable extraction, each of our three assumptions is studied.

## 2. Overview of the Lognormal Paradigm

The lognormal paradigm is based on the Kinematic Theory that describes a neuromuscular network involved in handwriting production as a linear system which controls the velocity of the pen.

This paradigm assumes that the Sigma-Lognormal modeling [Plamondon and Djioua, 2006] constitutes a relevant way to approximate, with at least a satisfying quality, continuous, fast and more or less complex handwriting movements like those involved in adults who execute signatures [O'Reilly and Plamondon, 2009] or triangles [Plamondon *et al.*, 2013a] and those used by 4–6 years old children to draw straight lines, bridges [Plamondon *et al.*, 2013b], pseudo-letters [Duval *et al.*, 2015] or scribbles [Rémi *et al.*, 2015]. Indeed, according to the lognormal handwriting generation paradigm, such movements take their origins in the execution, by the neuromuscular system, of action plans made up of a sequence of NbLog virtual targets. These targets are linked pairwise with circle arcs forming virtual trajectories, as shown in Fig. 1 for a cursive lowercase letters "b", and the speed profile for each virtual trajectory follows a scaled time-shifted lognormal function:

$$v_i(t) = \frac{D_i}{\sigma_i (t - t_{0i})\sqrt{2\pi}} \exp\left( -\frac{1}{2\sigma_i^2}[\ln(t - t_{0i}) - \mu_i]^2 \right) \tag{1}$$

The velocity $v_i(t)$ is defined by a set of six sigma-lognormal parameters $\{t_{0i}, D_i, \theta_{si}, \theta_{ei}, \mu_i, \sigma_i\}$ which are assumed to ultimately reveal the central and peripheral motor control process involved in the neuromuscular response [Plamondon and Djioua, 2006].

As illustrated in Fig. 1(b), the velocity profile for the complex movement is pretty well approximated with the extracted set of neuromuscular commands used to generate it.

In other words, the tangential velocity signal is approximated by the vector summation of the set of NbLog lognormals $\vec{v}_i(t)$ defined by the equation:

$$\vec{v}_r(t) = \sum_{i=1}^{nbLog} \vec{v}_i(t) \sum_{i=1}^{nbLog} \frac{\overrightarrow{D_i}}{\sigma_i (t - t_{0i})\sqrt{2\pi}} \exp\left( -\frac{1}{2\sigma_i^2}[\ln(t - t_{0i}) - \mu_i]^2 \right);$$

$$NbLog \geq 2 \tag{2}$$

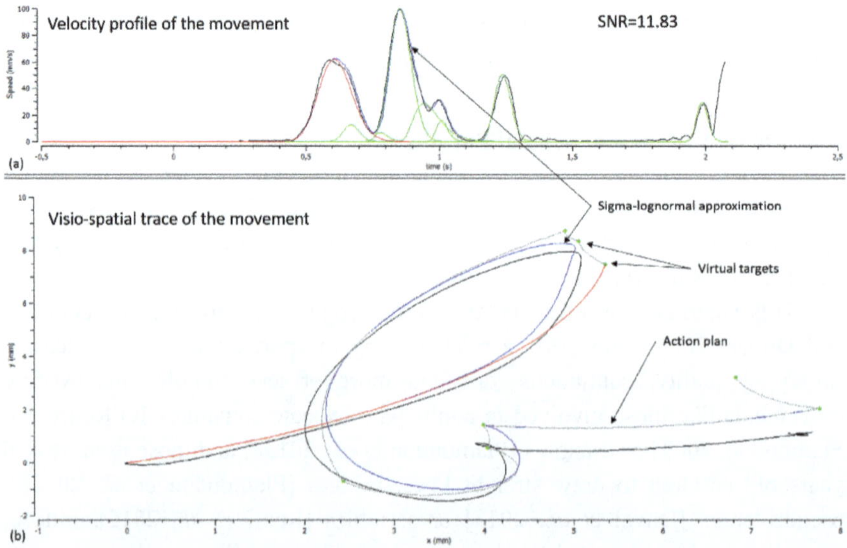


Fig. 1. Examples of the trace (a) and the tangential velocity profile (b) of the Sigma-Lognormal modeling with 8 virtual targets of a handwritten cursive lowercase letter "b" produced by a healthy young adult in good control his neuromuscular system.

This approximation $\vec{v}_r(t)$ is estimated with a controllable value of the signal-to-noise ratio (SNR), between the original [$v_o(t)$] and the reconstructed [$v_r(t)$] velocity profiles, as defined by:

$$SNR = 10\log\left(\frac{\int (v_r(t))^2\,dt}{\int (v_o(t) - v_r(t))^2\,dt}\right) \tag{3}$$

For a more detailed description, from the mathematical and conceptual points of view of the Kinematic Theory and the lognormal handwritten paradigm, the reader is referred to Plamondon *et al.* [2013a]. What is of interest here are the principal conclusions of our previous experimentations dealing with the lognormality paradigm.

First, according to the Lognormality Principle, we have shown that for constrained intentional and voluntary continuous drawing or scribbling, the reconstruction of the pen tip trajectory and its velocity profile all over the execution time both for adults [O'Reilly *et al.*, 2009; O'Reilly and Plamondon, 2011, 2012] as well as for young children [Rémi *et al.*, 2015] who have a high level of control of their handwriting movements during the task, is achieved with a higher value of the SNR than for subjects that have a lesser control.

In addition, for performing tasks involving kinematic and procedural efficiency constraints in terms of speed, orientation and direction of strokes and management of the pen lifts, it has also been experimentally shown that, at similar performance levels in terms of movement control, there is a relationship between the programming and coordination capacity of the commands of the movements and the number of lognormals used for the reconstruction of these movements [Plamondon *et al.*, 2013a; O'Reilly *et al.*, 2013]. This was particularly evident in the case of older adults performing a triangular plotting task in two different directions of rotation without pen lifts. It was suggested that the deterioration of motor control with aging was associated with the development of compensatory strategies such as emitting more motor commands to generate an adequate movement for a given task. A reverse phenomenon has been observed in the case of pen tasks that consist in rapid continuous plotting of oblique straight lines and bridges executed by kindergarten pupils aged from 3 to 6 years who were, a priori, less experimented than expert healthy adults in full possession of their graphomotor capacity. Indeed, more lognormals were needed for the reconstruction with similar motor control levels as expressed by the SNR, of movements produced by young children. The younger, the lower the scholar grade and the lesser their graphomotor training resulted in lower SNR, as predicted by the Lognormality Principle. This phenomenon was interpreted in [Duval *et al.*, 2015] as reflecting the progressive improvement performances in terms of fluidity and movement control. By analogy to what was assumed in [Plamondon *et al.*, 2013b] for adults, from the point of view of neuromotor programming capabilities, this phenomenon could also be interpreted as reflecting the development of more effective motor control strategies. In this case, for similar motor control levels, the efficiency was implicitly online expressed in terms of the difference of the numbers of virtual targets extracted by the Sigma-Lognormal modelling [Rémi *et al.*, 2017] or in terms of the reduction of the overlapping rates of the lognormal strokes [O'Reilly *et al.*, 2013].

## 3. Cursive Rules Taught and Propositions to Quantify the Procedural Performances

The adjective cursive is derived from the Medieval Latin *cursivus* (running), from the Latin *cursus* (a running) from the past participle of *currere* (to run). The cursive procedure of handwriting learning taught and practiced at school is based on the use of the cursive Modern Roman Alphabet which is illustrated in Fig. 2 [Primus, 2004].

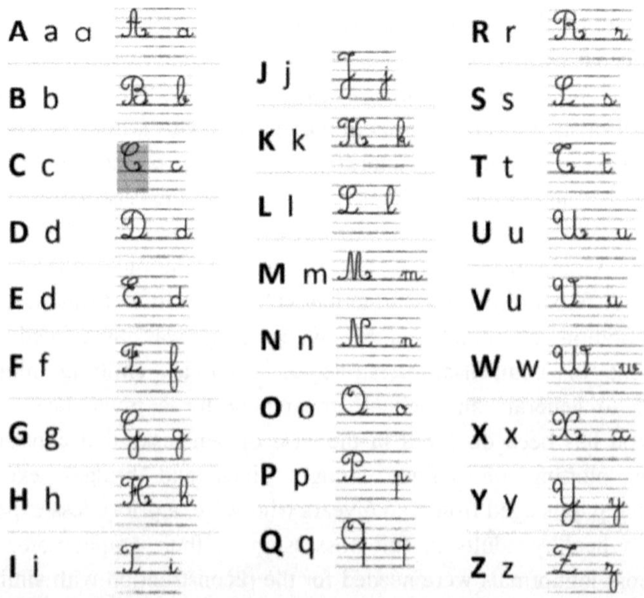

Fig. 2. Uppercase, script, uppercase and lowercase cursive handwritten representations of each letter of Roman alphabet [Primus, 2004].

Whatever the word, the expert cursive handwriting procedure consists to write and join a maximum of cursive letters, like those depicted in Fig. 2, contained in this word by lifting the pen as little as possible.

So, an expert writer respecting this fundamental cursive principle will tend to minimize the number of pen lifts during the handwriting process.

Consequently, we assume that if the cursive movement patterns are mastered for all the required letters and for their junctions, an expert will also tend to optimize the kinematic efficiency, that is to say the rapidity, fluidity and regularity, of his handwriting gesture.

### 3.1. *Regarding the methods for teaching the cursive procedures of handwriting*

An analysis of the recommendations made for teaching the cursive procedure of handwriting like in Deltour [2011], Ministère de l'Éducation nationale et de la Jeunesse, Eduscol [2018], shows that the academic methods are essentially based on the application of four cumulative cursive rules which guide the orientations, directions of the trajectory of the pen tip on the sheet of paper and manage the pen tip lifts. These cumulative cursive rules are: 1) with the exception of lowercase x, uppercase A, B, F, K, H, P, Q, R, X, Z and letters with diacritics, any other cursive letter that compose a word has to be entirely written in a single and continuous movement tracing a sequence of up and down strokes without generating any pen lift. 2) when tracing a round letter (a, o, g, d, c, q) that is inserted within a cursive word, a preliminary pen lift is required to begin the movement at the conventional starting point. 3) any letter can be joined with any consecutive non-round cursive letter without lifting the pen. 4) all diacritical marks (accents, dots, dieresis, bars). of the diacritical letters have to be placed, one after the other from the far left to the right, after the cursive writing of the complete set of letters of the word that have to be joined according to the joining rules; each of these diacritical marks generates a pen lift.

### 3.2. *Proposition of an offline variable to quantify the pupil's procedural performances in cursive handwriting*

Let us remark that, whatever the word, the strict application of the four cumulative cursive rules previously described provides a single integer that we shall call from now on the academic minimal number of pen lifts. As examples, Table 1 provides the academic minimal number of pen lifts that require the pseudo-word *tintin*, the first names: *Tintin*, *Jean*, *Alexandre*, the French words *ananas,* which means pineapple, *Ville* ,which means city, and the Creole word *woulo* which means bravo.

Table 1. Examples of academic minimal number of pen lifts for a selection of cursive words.

| word | academic minimal number of pen lifts |
|---|---|
| *tintin* | 5 |
| *Tintin* | 5 |
| Jean | 2 |
| Alexandre | 6 |
| ananas | 3 |
| Ville | 3 |
| woulo | 3 |

The pen lifts done along the cursive handwriting of these words according to the four cumulative cursive rules are presented by Fig. 3. If one considers the case of the pseudo-word *tintin*, according to an academic expert cursive procedure, in accordance with the four cumulative cursive rules, it is necessary to proceed by writing all the bodies of the 6 successive letters t, i , n , t , i, n without lifting the pen. Then, at the end of the pen-lift ending after the second letter n, the four diacritical marks of the t, i, t and i letters are successively placed from the left to the right before the final pen lift (Fig. 3(a)). So, the strict application of the academic cumulative cursive rules will lead to the execution of a minimum of 5 pen lifts, as shown in Table 1. The same number is requested for the first name *Tintin* because, the beginning uppercase T obliges to the execution of a pen lift before the continuous writing of the sequence *intin* and then the last three diacritical marks (Fig. 3(b)).

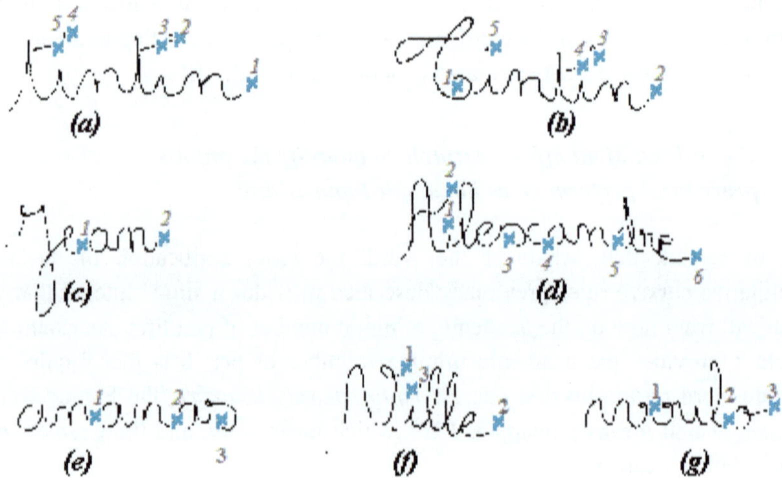

Fig. 3. Pen lifts done according to the four cumulative cursive rules all along the cursive handwriting of words. They are illustrated by x in blue and numbered according to their temporal order on each word.

As illustrated in Table 2, teachers start to introduce the rules and procedures of cursive writing to pupils from 5 to 6 years old in French school systems, that is to say since the last year of kindergarten: the Great section (GS). In the province of Québec it is during the second grade of primary school that cursive writing is introduced [Bara *et al.*, 2011; Morin *et al.*, 2012] when pupils are 6 years old. They usually begin with the cursive lowercase letters, either of the student's first name or of usual simple words. Table 2 provides a comparison of the teaching periods of cursive handwriting and the types of exercises that are practiced in France and the province of Québec.

Cursive word plotting exercises are opportunities for pupils to learn and train the four academic cumulative procedural rules that are presented and demonstrated to them by their teachers. All over a pupil's scholar learning and practice period, the number of pen lifts he executes when he writes words can be visually counted by the teacher or online estimated through pen-based digital devices or tablets.

Table 2. Partial comparative vision of the educational systems in France and in the province of Québec. * indicates in which grade cursive handwriting is introduced to pupils in France (GS) and in the Province of Québec (2nd primary).

| **Educational system in France** | | | **Educational system in Québec** | |
|---|---|---|---|---|
| **Schools** | **Grades** | **Starting age** | **Schools** | **Grades** |
| Middle school | 3th | 14 | Secondary | 3rd secondary |
| | 4th | 13 | | 2nd secondary |
| | 5th | 12 | | 1rst secondary |
| | 6th | 11 | *Primary* | *6th primary* |
| *Elementary school* | *CM2* | 10 | | *5th primary* |
| | *CM1* | 9 | | *4th primary* |
| | *CE2* | 8 | | *3rd primary* |
| | *CE1* | 7 | | *2nd primary** |
| | *CP* | 6 | | *1st* |
| Kindergarten | Great section (GS)* | 5 | Pre-school | Kindergarten |
| | Middle Section (MS) | 4 | | |
| | Small Section (SS) | 3 | | |

On this basis, we propose a new differential offline variable to help to evaluate, with respect to cursive rules, the procedural performance achieved by pupils when they write a particular word. This offline variable, which we shall call from now on *deltaPenlift*, is defined as a single integer which measures, comprehensively but synthetically, the difference between the number of pen lifts realized during the handwriting of a given word and the academic minimal number of pen lifts estimated for this word thanks to the four cumulative cursive rules.

We assume that, for a given cursive word, the value taken by *deltaPenlift* can help to give a global picture of the procedural differences between the academic cursive procedure of handwriting taught and the procedure of handwriting really used by the writer.

### 3.3. *Task*

The Mellier's task [Mellier *et al.*, 1999] chosen for this study was conceived to clinically test the automation of the procedure of handwriting for 6 to 12-year-old children. In Mellier's task, the child is asked to write his first name and a control pseudoword *tintin* without models, from memory on a sheet of paper or a digitizer. These two handwriting activities have to be executed according to two different modalities: with visual control [VC] and with no visual control [NVC]. According to the VC modality, the child can see what he is doing. So, he can visually control all the effects of his handwriting movements while he produces the word that is required by the experimenter. For the NVC modality, the child's eyes are blindfolded while he is writing the requested word. So, he cannot visually control the effects of his handwriting movements. This test assumes that automation is reflected by invariance, similar values have to be observed for the same online variable whatever the modality of control of the task of the same word.

We have chosen this Mellier's task because we suppose that by observing if the principle of invariance is verified, it shall be possible to evaluate the relevance of the kinematic and procedural variables investigated. Indeed, a pertinent variable should remain invariant whatever modality [VC vs. NVC] for pupils who already master the procedure of cursive handwriting taught at school. So, we expect that this task should help to discriminate the good handwriters from the less experienced or the poor ones, according to the measure of their procedural performances and the kinematic efficiency of their handwriting movements.

## 4. Experimental Study

### 4.1. *Goals and steps of the experimental study*

The purpose of this experimental study is to verify whether some online variables estimated through the Sigma-Lognormal modeling and the variable *deltaPenlift* contributes significantly to the characterization of the kinematic efficiency of a pupil's handwriting with respect to the procedure which is still most common in the teaching and practice of handwriting in France and Quebec today [Morin *et al.*, 2017]. The period considered extends from the beginning of handwriting learning experience after the end of kindergarten to the end of elementary school.

First, we verify if the Sigma-Lognormal modeling can reconstruct at least satisfactorily the velocity profiles of cursive handwriting for a majority of the words produced by pupils in the Mellier's task [Mellier *et al.*, 1999] regardless of the modalities of control of their movements.

If the first step shows such capacity of reconstruction of the Sigma-Lognormal modeling, in a second step, we analyze and discuss the behaviors of the Sigma-Lognormal global online variables in comparison with the classical online variables usually extracted from the Mellier's task, according to factors like the type of-word produced and the modality of control.

We assume that the pupils' procedures of handwriting a priori involve the production of cursive movements that can be characterized by the offline variable *deltaPenlift,* with respect to the four academic cumulative procedural rules previously described. Thus, if the second step shows relationships between some global Sigma-Lognormal online variables and some of the classical ones, including the number of pen lifts executed during the handwriting task, in the third and last step we will try to find a link between the signal-to-noise ratio, SNR, that evaluates the quality of the Sigma-Lognormal reconstruction of a velocity curve and the *deltaPenlift*.

The following is a description of the participants and the methodological aspects of the experimental study.

## 4.2. *Participants*

One hundred and eleven (111) pupils of 5 grades of French primary school passed the Mellier's test in a transversal experimentation [Rémi *et al.*, 1997]. Eighteen of these pupils were from CP grade, 17 from CE1, 24 from CE2, 24 from CM1 too and 28 were from CM2 grade. The number of the pupils and their average age is indicated in Table 3 for each grade.

Table 3. Number of pupils and their average age for each grade.

| Grade | Number of pupils | Average age of the pupils |
|---|---|---|
| CP | 18 | 6 years and 10 months |
| CE1 | 17 | 7 years and 8 months |
| CE2 | 24 | 8 years and 6 months |
| CM1 | 24 | 9 years and 9 months |
| CM2 | 28 | 10 years and 10 months |

Each participant had to handwrite his first name under two conditions: first with visual control [P_VC] and next without visual control of his production [P_NVC]. All the pupils also had to handwrite the cursive pseudo-word *tintin* twice: at first with visual control [T_VC] and next while they could not visually control their production [T_NVC].

## 5. Method for the Acquisition of the Words and the Estimation of the Classical Online Variables

The Mellier's task was performed by each student on a clean sheet of white A4 paper on the active surface of a Calcomp DrawingBoardIII digitizer. The acquisition units was placed in front of the child while the screen of the control and extraction unit was hidden from him using an opaque partition. Children's productions were recorded online at 100 Hz with a spatial resolution of 0.001 inches owing to the Anatrace software [Rémi *et al.*, 2002].

Finally, 222 first names and 220 pseudowords tintin were collected: 111 first names for the VC modality [P_VC] and 111 for the NVC modality [P_NVC]. In the case of tintin, 110 handwritten productions were collected for the VC modality [T_VC] and 110 for the NVC modality [T_NVC].

Whatever the control modality of the handwriting movements, with vision [VC] or without vision [NVC], all words were processed with the Dekattras software [Duval *et al.*, 2015] for the extraction of the following classical global online variables:

- mean speed of handwriting [*WritingSpeed*],
- length of the trajectory [*Length*],
- mean duration of the pen-lifts [*meanPenliftDuration*],
- number of pen lifts [*nbPenlifts*],
- mean pressure applied on the pen [*Pressure*],
- duration of the pen-down [*WritingDuration*],
- total time with the pen lift [*TotalTimeWithPenlift*],
- total duration of handwriting [*TotalDuration*],
- ratio between pen-lift duration and total duration of handwriting [*PenliftDuration/WritingDuration*]

In the Mellier's task, these global online variables are usually analyzed to evaluate the level of automation of the handwriting processes [Rémi *et al.*, 2000; Frélicot *et al.*, 2002].

## 6. Study of the Relevance of the Sigma-Lognormal Processing of the Children's Words

We used the ScriptStudio software [O'Reilly and Plamondon, 2009] to compute the reconstructed velocity vector for each isolated cursive first name and each pseudoword *tintin* handwritten by the participants.

To determine if the Sigma-Lognormal reconstructions of the children's movements were satisfying, we have performed a descriptive analysis on the quality of the reconstructions of the trajectories of the pen. As in Duval *et al.* [2015], we used a categorization of the values of the SNR into four levels as follows: SNR $\geq 25$ = excellent quality, $20 \leq$ SNR $< 25$ = good quality, $15 <$ SNR $< 20$= correct quality, SNR $\leq 15$ = unsatisfactory quality.

As illustrated in Fig. 4, according to the SNR, 2.3%, 40.5% and 50.7% of the recorded samples of cursive handwriting movements were reconstructed respectively with excellent, good, and correct quality for the four tasks: P_VC, P_NVC, T_VC and T_NVC produced by all participants. Only 6.6% of the reconstructions were evaluated as unsatisfactory.

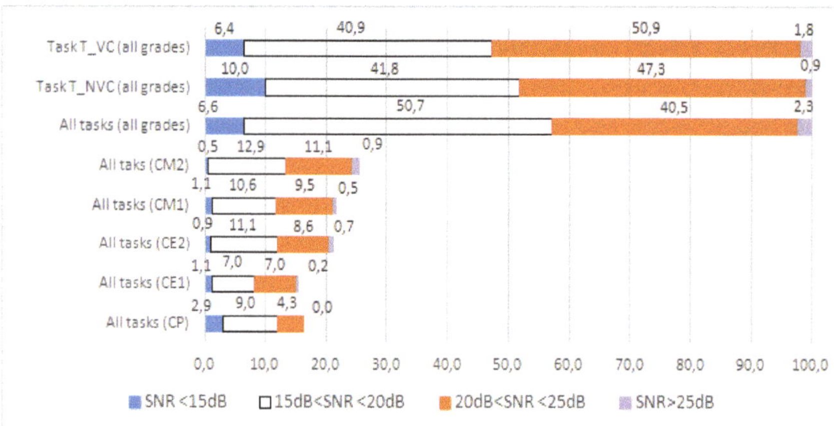


Fig. 4. Percentages of the reconstructed trajectories within the four SNR intervals for each grade for all tasks. The global percentages for all grades within the four SNR intervals are also provided for task T_VC and task T_NVC.

This global distribution clearly confirms that the Sigma-Lognormal model can accurately reproduce most of the children's trajectories when they write familiar and simple cursive isolated words. In this context, it makes sense to use this model to characterize the kinematic efficiency of the children's writing gesture.

It must also be noted that 40.5% of the required trajectory was reproduced with a SNR higher than 20 dB. This means that these trajectories were made by children who have almost good neuromotor control and good procedural performances. The rates provided by this step of our study, in conjunction with those already obtained on kindergarten children's production [Rémi *et al.*, 2015] reveals that the Sigma-Lognormal modeling can be used to satisfyingly reproduce children's more or less complex cursive handwriting movements from scribbles up to cursive isolated words. Furthermore, these results suggest that in scholar usual context of handwriting learning, the SNR estimated by the Xzero robust algorithm of the ScriptStudio Software could be used as a first filter for the graphomotor control evaluation of the pupils to make an early detection of children with a potential graphomotor deficit.

Figure 5 shows the Sigma-Lognormal reconstruction of the tangential velocity profile and the cursive spatial pen-tip trajectory produced by a pupil when he writes his first name. Figure 6 shows similar data extracted when a pupil writes the pseudoword *tintin*.

With the ScriptStudio Software, the 3 global Sigma-Lognormal online variables: nbLog, SNR and SNR/nbLog can then be estimated for each of the 4 tasks: P_VC, P_NVC, T_VC and T_NVC.

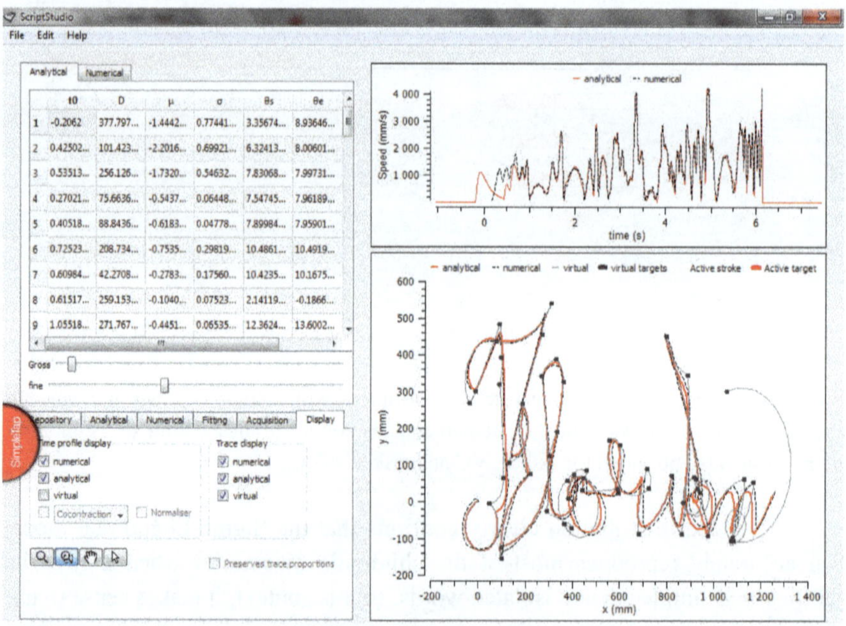


Fig. 5. Sigma-Lognormal reconstruction of the tangential velocity profile and the cursive spatial pen-tip trajectory observed for a pupil when he writes his first name Florian during task P_VC.

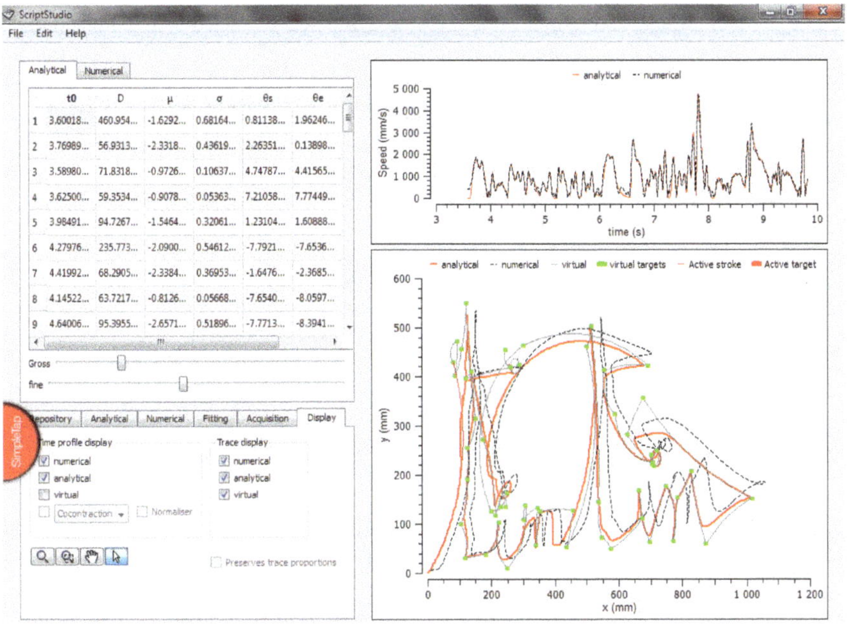


Fig. 6. Sigma-Lognormal reconstruction of the tangential velocity profile and the cursive spatial pen-tip trajectory observed for a pupil when he writes the pseudo-word *tintin* during task T_NVC.

## 7. Selection of Online Variables that Can Better Reflect the Effects of Automation Through Learning

For each of the 4 tasks: P_VC, P_NVC, T_VC and T_NVC, twelve online variables, summarized in Table 4, were estimated at each of the words collected. These online variables are related to four graphomotor criteria: procedural performance, rapidity, fluidity and regularity with which performances in handwriting are usually analyzed and described in Duval *et al.* [2015], Rosenblum *et al.* [2006], Vinter and Chartrel [2010], Zesiger *et al.* [2000].

For each of these 12 online variables, the ratio of their values observed for the two conditions VC and NVC were also included, as in Rémi *et al.* [2000]. We have introduced these ratios on the basis that they were supposed to be informative about the consistency of the graphomotor performance criteria with respect to perturbations like the lack of visual feedback. So, finally, this is 72 variables coded as Xi that have to be a priori analyzed thanks to various native functions from the R programming environment [Release 3.4.0].

Table 4. Definitions and types of the online variables extracted and corresponding graphomotor criteria.

| Definition of the variable | Type of the variable | Graphomotor criteria |
|---|---|---|
| Pressure | Postural | Fluidity |
| Length | Procedural | Regularity |
| nbPen-lifts | Spatial | Procedural performance |
| meanPen-lift Duration | Kinematic | Procedural performance |
| WritingVelocity | Temporal | Rapidity |
| WritingDuration | Temporal | Rapidity |
| TotalDuration | Temporal | Rapidity |
| Pen-liftDuration | Temporal | Fluidity, Regularity |
| SNR | Sigma-Lognormal | Rapidity |
| nbLog | Sigma-Lognormal | Fluidity |
| SNR/nbLog | Sigma-Lognormal | Regularity |

By means of the R package FactoMineR, we performed a principal component analysis [PCA] with respect to the scholar grades on this set of 72 variables. Sixteen variables, presented in Fig. 7(a), were found highly correlated with the first two principal components that explain 31% of the variance. A second PCA was performed with these sixteen variables. These results are illustrated in Figs. 7(b) and 7(c).

Figure 7(b) shows that the first principal component is highly positively correlated with WritingDuration and Pen-liftDuration for P_NVC and T_NVC [X27, X28, X63 and X64] and with TotalDuration and meanPenliftDuration [X26, X33] for P_NVC and highly negatively correlated with writingVelocity for P_VC, P_NVC, T_VC and T_NVC [X19, X31, X55, X67]. Therefore, the horizontal axis in Fig. 7(b) can be considered as a scale of scale of execution

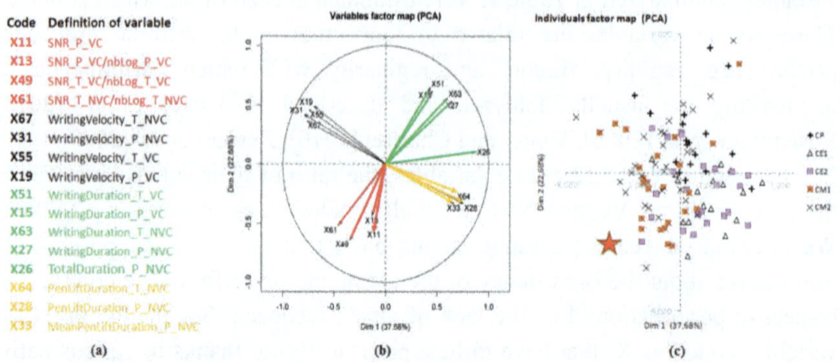


Fig. 7. (a) list of the sixteen most important variables; (b) correlation circle for these 16 variables; (c) individual profiles according to the grade.

duration and as an inverse scale of regularity and fluidity which explains 38% of the variability. The second principal component is highly positively correlated with WritingDuration for P_VC and T_VC [X15 and X51] and highly correlated negatively with SNR/nbLog for T_VC and T_NVC [X49 and X61]. This axis is opposing slowness and irregularity for T_VC and T_NVC.

Children from CE1 and CE2 write slowly and have longer pause duration as compared to those of CM1 and CM2 whatever the type of isolated word and the nature of feedback, visio-spatial or kinesthetic, involved in the task. Moreover, children from CE1, CE2 and CM1 take less time to write their first name or *tintin* than the younger and less experienced pupils of CP and the elder pupils of CM2. Furthermore, a majority of the children from the intermediary grades of CE1, CE2 and CM1 have a greater regularity in their movements: they write with more precision than a majority of children from CP and CM2 when they are writing *tintin* whatever the modality of control of the execution: with [VC] or without vision [NVC]. To try to explain such results we can refer to three aspects that contribute to different strategies of handwriting. The first aspect is related to the progressive changes of the type of control used by pupils from proactive and visual to retroactive and kinesthetic and its consequences on the nature of the movements they execute [Chartrel and Vinter, 2004]. The second aspect is related to the change of the status of handwriting that pupils build on the basis of the teaching goals that they perceive all over their scholar cursus. At the beginning in CP, handwriting is presented and so perceived by pupils as a calligraphic skill with an essential academic challenge of legibility without great temporal constraints, while its communicative dimension becomes more evident and crucial in higher grades where productivity requirements are greater [Morin *et al.*, 2017]. So, a compromise, depicted as personalization [Zesiger, 2003], have to be made by pupils before the end of primary school to try to achieve simultaneously the productivity and the legibility that are the norm required at middle school. The third aspect is related to the fact that in the same time, the procedural constraints related to the reduction of the number of pen-lifts are not usually evaluated by teachers. This is partially a consequence of the fact that teachers have no online or offline way for such purpose. However, such procedural constraints should perhaps be a condition to increase the kinematic efficiency of the gesture of handwriting.

Next, we performed Kruskal-Wallis tests [Kruskal and Wallis, 1952] to investigate if there were significant behavioral differences among the 5 learning grades with respect to some of the 16 variables. The results are presented in Table 5. They show that 13 of the 16 variables that were extracted contribute significantly to discriminate between at least 2 of the 5 grades.

Let us remark that the variable X19, defined as the rate SNR/nbLog, computed on the handwriting of the pseudo-word tintin without vision [NVC], which is a priori the less usual task and the more difficult for pupils, has a significant behavior according to the scholar grade. This fact and the results obtained with the principal component analysis suggest that by computing the SNR/nbLog ratio, which is related to regularity, on non-usual isolated words or pseudowords handwritten with and without vision control, it might be possible to monitor pupil's lognormality all along primary school since 5 years old.

Table 5. Variables Xi with respect to each task and graphomotor criteria for which the Kruskal-Wallis test was performed. Significant tests [p-value < 0.003 with Bonferroni correction] are in bold.

| Modality | Task | Definition of the Variable | Name of the Variable | Kruskal-Wallis |
|---|---|---|---|---|
| NVC | *tintin* | SNR/nbLog | X19 | **< 0,0001** |
| | | WritingVelocity | X26 | **0,000** |
| | | WritingDuration | X51 | **< 0,0001** |
| | | Pen-liftDuration | X63 | **0,001** |
| | first name | WritingVelocity | X27 | **0,001** |
| | | WritingDuration | X55 | **0,000** |
| | | TotalDuration | X61 | **0,000** |
| | | Pen-liftDuration | X64 | **< 0,0001** |
| | | MeanPen-liftDuration | X67 | **< 0,0001** |
| VC | *tintin* | SNR/nbLog | X15 | 0,029 |
| | | WritingVelocity | X28 | **< 0,0001** |
| | | WritingDuration | X33 | **< 0,0001** |
| | first name | SNR | X11 | 0,003 |
| | | SNR/nbLog | X13 | 0,295 |
| | | WritingVelocity | X31 | **< 0,0001** |
| | | WritingDuration | X49 | **< 0,0001** |

## 8. Can Lognormality Help to Characterize the Kinematic Efficiency of the Procedures Used to Write Cursive Words

We have defined the notion of academic minimal number of pen lifts and the variable *deltaPenlift* that can be estimated easily for each word. Our purpose here is to study the relevance of such compact representation. Can it help in all likelihood to consistently measure the procedural conformity and the kinematic efficiency of the procedure of cursive word handwriting with respect to its recommended version?

To answer such a question, for the 6 to 11-year-old participants, we have studied experimentally if there is a relationship between the degree of lognormality of pupils' handwriting expressed with the SNR and the procedural performance estimated by the variable *deltaPenlift* relatively to the cursive rules taught in school.

To answer such a question, for the 6 to 11-year-old participants, we have studied experimentally if there is a relationship between the degree of lognormality of pupils' handwriting expressed with the SNR and the procedural performance estimated by the variable *deltaPenlift* relatively to the cursive rules taught in school.

We have chosen to use the SNR because it was one of the sixteen variables selected by the PCA. Moreover, as previously depicted in the section that deals with the review of the lognormal paradigm, the SNR is assumed to measure the similarity of the levels of motor control and implicitly the efficiency. More precisely, to determine whether the Lognormality Principle can help to analyze the procedures used by pupils when they write cursive words, we investigated the following question:

When the number of pen lifts executed by the pupil as he is writing a cursive word is closer to its academic minimal number of pen lifts, is the higher is the SNR value?

To deal with this question, we have proceeded in two steps. First, we considered only children that seem to be efficient hand writers according to the Lognormality Principle through the Mellier's task. Next, we took all pupils.

## 8.1 *Data selection*

Let us recall that the choice of the Mellier's task as a basis for initiating a study on children's cursive handwriting strategies is justified by the fact that the writing of their first name and the procedure for carrying out the cursive trace are a priori taught to each student from the preschool period. This fact is assumed to be a factor of automation of the cursive procedure used by pupils to write their first name.

In order to neutralize or control certain factors related to the structural characteristics of first names that could bias our study, we have chosen to consider only first names with at least 4 characters and 2 to 4 syllables. This selection procedure seemed appropriate given that for 98% of the French first names from 1900 up to the present day are composed of at least 4 characters and 2 syllables.

Moreover, given the results described in the previous sections, as a method to select only the best hand writers according to the Lognormality Principle through the Mellier's task, we used the cursive handwriting of *tintin* as the reference. We only considered pupils for whom we have at least one Sigma-Lognormal reconstruction of the samples *tintin* they wrote with at least a correct quality, so with SNR > 15 dB as in Plamondon *et al.* [2013].

From this two-step selection, 104 students were first selected because 7 of them had a first name with less than 4 characters. Of these 104 students, only those who had been able to write the pseudo-word *tintin* completely in cursive letters with and without vision and for whom at least one SNR value was greater than or equal to 15 dB was retained. This ensures that in at least one of the imposed conditions the production was of sufficient quality for its Sigma-Lognormal reconstruction. From this second step, we finally kept 67 students who were distributed according to their grades, ages expressed in years and genders as indicated by Table 6.

Table 6. Characteristics of the selected sample of pupils.

| **Grade/** | **Age** | | | | | | | | | | | **Total** |
|---|---|---|---|---|---|---|---|---|---|---|---|---|
| **Gender** | **6,5** | **7** | **7,5** | **8** | **8,5** | **9** | **9,5** | **10** | **10,5** | **11** | **12** | |
| **CP** | **4** | **5** | **1** | | | | | | | | | **10** |
| **Girls** | 3 | 2 | | | | | | | | | | 5 |
| **Boys** | 1 | 3 | 1 | | | | | | | | | 5 |
| **CE1** | **1** | **1** | **4** | **1** | | **1** | | | | | | **8** |
| **Girls** | - | 1 | 3 | | | 1 | | | | | | 5 |
| **Boys** | 1 | | 1 | 1 | | | | | | | | 3 |
| **CE2** | | | **1** | **3** | **6** | **7** | | | | | | **17** |
| **Girls** | | | 1 | 1 | 2 | 5 | | | | | | 9 |
| **Boys** | | | | 2 | 4 | 2 | | | | | | 8 |
| **CM1** | | | | | | **2** | **3** | **8** | | **1** | | **14** |
| **Girls** | | | | | | | 1 | 7 | | 1 | | 9 |
| **Boys** | | | | | | 2 | 2 | 1 | | | | 5 |
| **CM2** | | | | | | | | **4** | **5** | **8** | **1** | **18** |
| **Girls** | | | | | | | | 1 | 2 | 4 | 1 | 8 |
| **Boys** | | | | | | | | 3 | 3 | 4 | | 10 |
| **All Grades** | **5** | **6** | **6** | **4** | **6** | **10** | **3** | **12** | **5** | **9** | **1** | **67** |
| **Total Girls** | 3 | 3 | 4 | 1 | 2 | 6 | 1 | 8 | 2 | 5 | 1 | 36 |
| **Total Boys** | 2 | 3 | 2 | 3 | 4 | 4 | 2 | 4 | 3 | 4 | 0 | 31 |

The structural compositions of the first names for these 67 selected pupils are summarized in Table 7. Most of the first names are composed of 2 or 3 syllables and have between 5 and 7 letters.

Table 7. A cross-tabulation of the number of letters (Rows) and the number of syllables (Columns) for the 67 selected pupils' first names.

| | | NbSyllables | | | |
|---|---|---|---|---|---|
| | | **2** | **3** | **4** | **Total** |
| NbLetters | **4** | 2 | - | - | **2** |
| | **5** | 6 | 1 | - | **7** |
| | **6** | 16 | 13 | - | **29** |
| | **7** | 8 | 11 | - | **19** |
| | **8** | 1 | 4 | - | **5** |
| | **9** | - | 3 | 2 | **5** |
| | **Total** | **33** | **32** | **2** | **67** |

## 8.2. *Cross-analysis of the behaviors of deltaPenlift and SNR*

The next three tables summarize what we have observed while studying the relationship between the *deltaPenlift* and the SNR. Table 8 cross-references the mean values of *deltaPenlift* for the 67 selected children for the cursive writing of their first names for each modality of control, VC and NVC, and each SNR interval. There is no reconstruction of the first name with SNR < 15 dB for these 67 selected children.

Table 8. Mean deltaPenlift for the cursive writing of the first name of the 67 children selected according to the SNR value.

| SNR interval | VC | NVC |
|---|---|---|
| **15 dB < SNR < 20 dB** | 1,45 | 1,24 |
| **20 dB <SNR < 25 dB** | 1,48 | 1,27 |
| **> 25 dB** | 1,55 | 1,30 |
| **Overall total** | **1,47** | **1,25** |

The Table 9 cross-references the mean values of *deltaPenlift* for all the 104 pupils for the cursive writing of *tintin* for each modality of control, VC and NVC, and each interval of the SNR.

Table 9. Mean deltaPenlift for the word tintin with respect to the SNR value (Row) and the modality VC/NVC of the task (Column) for the 104 all-comers pupils.

| SNR Interval | VC | NVC |
|---|---|---|
| **SNR < 15 db** | 4,14 | 1,90 |
| **15 dB < SNR < 20 dB** | 3,19 | 1,95 |
| **20 dB < SNR < 25 dB** | 2,65 | 1,29 |
| **> 25 dB** | 3,50 | 3,00 |
| **Overall total** | **2,99** | **1,63** |

The Table 10 cross-references the mean values of *deltaPenlift* for all the 104 pupils for the cursive writing of their first names for each modality of control, VC and NVC, and each interval of the SNR.

Table 10. Mean deltaPenlift for the cursive writing of the first name with respect to the SNR value (Row) and the modality VC/NVC of the task (Column) for the 104 all-comers pupils.

| SNR Interval | VC | NVC |
|---|---|---|
| **SNR < 15 db** | 3,40 | 4,40 |
| **15 dB < SNR < 20 dB** | 3,46 | 2,44 |
| **20 dB < SNR < 25 dB** | 3,03 | 2,22 |
| **> 25 dB** | 2,67 | 2,33 |
| **Overall total** | **3,32** | **2,45** |

The Tables 8 to 10 show the same trend for the cursive writing of *tintin* and the first name, with or without vision: the higher the SNR, the smaller is *deltaPenlift*, The less coherent mean values observed for the range > 25 dB, are due to the very low number of writers [1 to 2] with such a SNR value.

In addition, we observed that the individual values of the SNR for the 67 students are all above 15 dB when writing the first name with and without vision. This validates our choice of basing our selection on the observation of the SNR for the writing of *tintin*. Moreover, the mean value of *deltaPenlift* in the case of writing without vision is systematically smaller than that with vision. This is probably due to the fact that in the case of the NVC mode, most pupils prefer a strategy consisting of maintaining contact with the sheet in order to guarantee the best respect of the other constraints [e.g., binding letters, etc.] imposed by the cursive writing rules taught. Finally, it appears that the *deltaPenlift*s observed for the 67 selected writers are smaller than for the all-coming students when writing the first name. This suggests that students with a good SNR are in majority those who better respect the rules of pen lifting during cursive writing that are taught in school. The offline variable *deltaPenlift* seems to follow the same trend as the online variable SNR. This parameter could contribute to the evaluation of the efficiency of students' cursive writing in addition to the traditional formal parameters, as well as helping the pre-screening of students with learning disabilities.

## 9. Conclusion

Cursive handwriting plays an essential role in the child's learning in school. In the French school system, the use of this process is mandatory for any activity requiring a written record. It is a standard precondition. In this context, modalities for representing its implementation have emerged in the French school system, such as the number of pen-lifts that are required when writing words. The efficiency of cursive handwriting when learning the graphomotor gestures for letters and words can eventually predict the child's attention and engagement in school. Indeed, many studies [Zesiger *et al.*, 2000; Bara *et al.*, 2011; Morin *et al.*, 2017] show that the activity of handwriting, if poorly mastered, can require an intense cognitive effort on the part of the students. This focuses their cognitive resources on the procedural execution leaving other activities such as comprehension, orthography or listening in the background.

Because of the importance of cursive handwriting skill, we chose to determine whether the use of the Sigma-Lognormal model, widely tested in the literature for other types of handwriting activities, in the context of designing Personal Digital Bodyguards [Plamondon *et al.*, 2018], would provide information on the efficiency of such procedure that is involved in pupils when they write words in elementary school. For this purpose, we first confirmed that the Sigma-Lognormal model was able to reconstruct with a level of quality that satisfied the speed profiles used by students aged 6 to 12 when writing more or less complex and familiar words. It is important to underline that this is the first time that a study on the power of Sigma-Lognormal modeling is realized in complex handwriting procedures of isolated words produced by such groups of children. So, on the basis of this positive result, we then conducted an exploratory study on the behavior of classical and Sigma-Lognormal variables for isolated words in two different modalities of visual motor control. The variables we considered had already been previously identified as reflecting, for other handwritten drawing situations, the degree of procedural and kinematic efficiency of writing as well as the general tendency of writers to move towards lognormality during the learning period of their writing skill. In particular, this analysis revealed a trend towards more frequent observation of high SNR levels and the SNR/nbLog rate for pupils who had a longer learning period than for younger pupils. This confirmed the potential interest of using Sigma-Lognormal analysis to help characterize the effectiveness of cursive handwriting in school-aged children.

In this study we also wanted to observe the behavior of an original criterion, the *deltaPenlift* for the characterization of the procedural efficiency of cursive handwriting. As estimated by the difference between the number of pen-lifts executed during the cursive handwriting of a given word and its corresponding academic minimal numbers according to the set of cursive rules taught at school, the *deltaPenlift* is easy and quick to estimate automatically. It can also be evaluated by a teacher himself. We also wanted to determine if there was a relationship between the degree of lognormality of pupils' handwriting expressed with the SNR and the procedural performances estimated by the variable *deltaPenlift* relatively to the set of cursive rules taught at school. For this purpose, we analyzed the writing of the pseudo-word and first name *tintin*/*Tintin* as well as the handwriting of their first name through the Mellier's Task. The results obtained show that this new criterion, *deltaPenlift*, tends to decrease as the SNR calculated on these words increases. From these initial results, we can conclude that this parameter makes sense in the characterization of the efficiency and procedural quality of cursive writing gestures in children.

This new study adds isolated cursive words to the panel of the various usual handwriting movements like those involved in scribbling tasks [Rémi *et al.*, 2015, 2017], movement patterns used to sketch rapid straight lines, isolated cursive pseudo-letters and letters [Duval *et al.*, 2013, 2015] that we have already analyzed thanks to Sigma-Lognormal modeling. It seems to confirm that this modeling methodology is relevant, with regard to the corpus of movement patterns that is taught and used in school, to qualify children kinematic handwriting efficiency all over their scholarship from kindergarten up to the end of primary school.

Thanks to this study we have tried to identify variables that could be used to characterize, to evaluate and to monitor the kinematic efficiency of pupils' handwriting at school. The results provided by the present experimental study tend to show that the online sigma-lognormal global variables SNR and SNR/nbLog and the offline variable deltaPenlift can play such role when cursive handwriting is the reference taught and practiced at school. Now, thanks to these variables we will be able to question the differences of pupils' performances that can exist for example according to gender or according to the pedagogical approaches used by teachers. To go further on this topic, we have started a vast campaign where this time nearly 1300 children of 3 to 12 years old are involved, from the beginning kindergarten to middle school [Rémi *et al.*, 2018]. Its results will be the subject of a next paper.

## References


Bara, F., Morin, M.-F., Montésinos-Gelet, I. and Lavoie N. (2011). Graphomotricity conceptions and practices among primary school teachers in France and in Quebec, *Revue française de pédagogie*, 176, pp. 41–56 (in French).

Chartrel, E. and Vinter, A. (2004). L'écriture, une activité longue et complexe à acquérir, *J. A.N.A.E.*, 78, pp. 174–180.

deDiego-Cottinelli A. and Barros, B. (2010). TRAZO: a tool to acquire handwriting skills using tablet-PC devices. *Proc. of the 9th International Conference on Interaction Design and Children (IDC '10)*, pp. 278–281.

Deltour, R. (2011). *Du graphisme à l'écriture*, ed. Circonscription de Lens, France (in French).

Dinehart, L. H. (2014). Handwriting in early childhood education: Current research and future implications. *Journal of Early Childhood Literacy,* 15(1), pp. 97–118.

Duval, T., Rémi, C., Plamondon, R., O'Reilly, C. (2013). On the use of the Sigma-Lognormal model to study children handwriting, *Proc. of the 16th Biennial Conference of the Graphonomics Society (IGS 2013)*, pp. 26–29.

Duval, T., Rémi, C., Prevost, L., Dorville, A., Plamondon, R. (2014). Etude de la faisabilité de l'évaluation de l'efficacité des mouvements de tracé du jeune apprenti-scripteur. *Proc. 26e conférence francophone sur l'Interaction Homme-Machine (IHM'14)*, pp. 29–37 (in French).

Duval, T., Rémi, C., Plamondon, R., Vaillant, J., O'Reilly, C. (2015). Combining Sigma-Lognormal modeling and classical variables for analyzing graphomotor performances in kindergarten children, *J. Human Movement Science*, 43, pp. 183–200.

Frélicot, C., Rémi, C., Courtellemont, P, (2002). School level recognition from children's drawings and writing, *Proc. of the 16th Int. Conf. on Pattern Recognition (ICPR* 2002), pp. 493–496.

Graham, S., Berninger, V., Weintraub N. and Schafer, W. (1998). Development of handwriting speed and legibility in grades 1-9, *J. Educ. Res.*, 92, pp. 42–52.

Julius, M. S., Meir, R., Shechter-Nissim, Z., Adi-Japha, E. (2016). Children's ability to learn a motor skill is related to handwriting and reading proficiency, *J. Learning and Individual Differences*, 51, pp. 265–272.

Kruskal, W.H. and Wallis, W.A. (1952). Use of ranks in one-criterion variance analysis, *J. Am. Stat. Assoc.* 47, 583–621 and errata, ibid. 48, 907–911.

Medwell, J. and Wray, D. (2007). Handwriting: what do we know and what do we need to know? *J. Literacy,* 41(1), pp. 10–16.

Mellier, D., Fernandez-Berani, L. and Fessard, C. (1999). Devenir à 6 ans d'enfants grands prématurés. *J. Enfance*, 1, pp. 67–78 (in French).

Ministère de l'Éducation nationale et de la Jeunesse, Eduscol (2018). *Informer et accompagner les professionnels de l'éducation CYCLE 2: FRANÇAIS: le geste d'écriture et la copie, La mise en œuvre de l'enseignement de l'écriture cursive*. ed. Eduscol: Ministère de l'Éducation nationale et de la Jeunesse, France. Retrieved from https://cache.media.eduscol.education.fr/file/Gestes_et_ecritures/61/0/3-RA16_C2_FRA_3_mise_en_oeuvre_ens_ecriture_cursive_635610.pdf (in French).

Morin, M.-F., Lavoie, N. and Montésinos-Gelet, I. (2012). The Effects of Manuscript, Cursive or Manuscript/Cursive Styles on Writing Development in Grade 2. *J Language and Literacy*, 14(1), pp. 110–124.

Morin, M.-F.,Bara, F. and Alamargot, D. (2017). Apprentissage de la graphomotricité à l'école: Quelles acquisitions? Quelles pratiques? Quels outils? *J. Scientia Paedagogica Experimentalis*, 54 (1-2), pp. 47–84 (in French).

O'Reilly, C. and Plamondon, R. (2009). Development of a Sigma-Lognormal Representation for On-Line Signatures, *Pattern Recognition,* 42(12), pp. 3324–3337.

O'Reilly, C., Plamondon, R., Lebrun, L.-H., Clément, B., and Mathieu, P. A. (2009). Sigma-Lognormal Analysis of a Complex Movements Neuromuscular Test., *Proc. of the 14th Conference of the International Graphonomics Society (IGS 2009),* pp. 48–51.

O'Reilly, C. and Plamondon, R. (2011). Impact of the principal stroke risk factors on human movements. *J. Human Movement Science*, 30(4), pp. 792–806.

O'Reilly, C. and Plamondon, R. (2012). Design of a Neuromuscular Disorders Diagnostic System Using Human Movement Analysis. *Proc. of the 11th International Conference on Information Science, Signal Processing and Their Applications (ISSPA 2012)*, pp. 787–792.

O'Reilly, C., Plamondon, R., Landou, MK. and Stemmer, B. (2013). Using kinematic analysis of movement to predict the time occurrence of an evoked potential associated with a motor command. *European Journal of Neuroscience*, 37(2), pp. 173–180.

Plamondon, R. and Djioua, M. (2006). A Multi-Level Representation Paradigm for Handwriting Stroke Generation, *J Human Movement Science*, 25(4-5), pp. 586–607.

Plamondon, R., Djioua, M. and Mathieu, P.-A. (2013). The Kinematic Theory of Rapid Human Movements: Experimental Observation of the Proportional Effect Hypothesis, *Human Movement Science*, 32(5), pp. 1026–1039.

Plamondon, R., O'Reilly, C., Rémi, C., Duval, T. (2013). The lognormal handwriter: Learning, performing, and declining, *J. Frontiers in Psychology*, 4, 945.

Plamondon, R., Pirlo, G., Anquetil, E., Rémi, C., Teulings, H.-L. and Nakagawa M. (2018) Personal digital bodyguards for e-security, e-learning and e-health: A prospective survey. *Pattern Recognition*, 81, pp. 633–659.

Primus, B. (2004). A featural analysis of the Modern Roman Alphabet, J. *Written Language and Literacy* 7(2), pp. 235–274.

Prince Edward Island, Department of Education and Early Childhood Development (DEECD) (2010). *Guidelines for handwriting instruction: printing and cursive, kindergarten to grade 6.* ed. DEECD: Prince Edward Island, Canada. Retrieved from http://www.gov.pe.ca/photos/original/eecd_printcurk6.pdf

Rémi, C., Amara, M., Courtellemont, M. and de Brucq, D. (1997). An experimental program for the automatic study of handwriting learning at elementary school, *Proc. of the Eighth Conference of the International Graphonomics Society,* pp. 75–76.

Rémi, C., Frélicot, C., Courtellemont, P. and de Brucq, D. (2000). Analyse de productions textuelles et graphiques pour l'aide à la détection du retard scolaire, *Proc. of Colloque International Francophone sur l'Ecrit et le Document (CIFED'2000)*, pp. 141–150 (in French).

Rémi, C., Frélicot, C., and Courtellemont, P. (2002). Automatic analysis of the structuring of children's drawing and writing, *Pattern Recognition*, 35(5), pp. 1059–1069.

Rémi, C., Vaillant, J., Plamondon, R., Prévost L. and Duval, T. (2015) Exploring the Kinematic Dimensions of Kindergarten Children's Scribbles. *Proc. of the 17th Biennial Conference of the International Graphonomics Society (IGS 2015)*, pp. 79–82.

Rémi, C., Nagau, J., Vaillant, J. and Plamondon, R. (2017). Preliminary Study of t0, a Sigma-Lognormal Parameter Extracted from Young Children's Controlled Scribbles, *Proc. of the 18th International Graphonomics Society Conference (IGS 2017)*, pp. 109–113.

Rémi, C., Vaillant, J., Nagau, J., Plamondon, R. (2018). Can the Sigma-Lognormal modeling help to monitor Child's graphomotor skills? *Proc. of the first International Conference on Pattern Recognition and Artificial Intelligence (ICPRAI2018)*, pp. 780–783.

Rémi, C., Nagau, J., Vaillant, J., Dorville, A. and Plamondon, R. (2018). Multimodal Acquisition and Analysis of Children Handwriting for the Study of the Efficiency of Their Handwriting Movements: The @MaGma Challenge, *Proc of the 16th International Conference on Frontiers in Handwriting Recognition (ICFHR 2018)*, pp. 459–464.

Rosenblum, S., Parush, S. and Weiss, P. L. (2003). Computerized temporal handwriting characteristics of proficient and poor handwriters, *Am. J. Occup. Ther.*, 57, pp. 129–138.

Rosenblum, S., Goldstand, S. and Parush, S. (2006). Relationships among biomechanical ergonomic factors, handwriting product quality, handwriting efficiency, and computerized handwriting process measures in children with and without handwriting difficulties. *Am. J. Occup. Ther.*, 60(1), pp. 28–39.

Rosenblum, S. (2018). Inter-relationships between objective handwriting variables and executive control among children with developmental dysgraphia. *PloS one*, 13(4), e0196098. doi:10.1371/journal.pone.0196098.

Simonnet, D., Anquetil, E. and Bouillon, M. (2017). Multi-Criteria Handwriting Quality Analysis with Online Fuzzy Models. *Pattern Recognition*, 69, pp. 310–324.

Simonnet, D., Girard, N., Anquetil, E., Renault, M. and Thomas, S. (2019). Evaluation of children cursive handwritten words for e-education, *Pattern Recognition Letters*, 121, pp. 133–139.

Vinter, A. and Chartrel, E. (2010) Effects of Different Types of Learning on Handwriting Movements in Young Children, *Learning and Instruction*, 20, pp. 476–486.

Zesiger, P., Deonna, T. and Mayor, C. (2000). L'acquisition de l'écriture. *Enfance*, (3), pp. 295–304 (in French).

Zesiger, P. (2003). Acquisition et troubles de l'écriture, *Enfance*, (55), pp. 56–64 (in French).

https://doi.org/10.1142/9789811226830_0005

# Chapter 5

# Classifying the Kinematics of Fast Pen Strokes in Children with ADHD using Different Machine Learning Models

Nadir Faci[1,*], Hung Tuan Nguyen[2,†], Patricia Laniel[3,‡], Bruno Gauthier[3,§], Miriam H. Beauchamp[3,4,ǂ], Masaki Nakagawa[2,**] and Réjean Plamondon[1,††]
[1]*Laboratoire Scribens, Département de génie électrique, Polytechnique Montréal*
[2]*Tokyo University of Agriculture and Technology*
[3]*Département de Psychologie, Université de Montréal*
[4]*Ste-Justine Hospital Research Center*
[*]*nadir.faci@polymtl.ca*, [†]*ntuanhung@gmail.com*,
[‡]*patricia.laniel@umontreal.ca*, [§]*bruno.gauthier@umontreal.ca*,
[ǂ]*miriam.beauchamp@umontreal.ca*, [**]*nakagawa@cc.tuat.ac.jp*,
[††]*rejean.plamondon@polymtl.ca*

This exploratory study examines whether the sigma-lognormal model derived from the Kinematic Theory of rapid human movements discriminates between the handwriting strokes produced by children with and without Attention Deficit Hyperactivity Disorder (ADHD). Twelve children with ADHD and 12 controls aged 8–11 years were asked to produce handwriting strokes on a digitizing tablet. The strokes were analyzed using the sigma-lognormal model. Strokes made by children with ADHD reflected poorer motor control, action planning and execution than strokes made by controls. Different Machine learning models were trained to classify the subjects according to the discriminatory parameters used as features. Although the sample size and data are modest and will require replication in a larger forthcoming study, promising preliminary results are obtained, suggesting that the sigma-lognormal model may be a useful tool in the assessment of ADHD.

## 1. Introduction

Attention deficit/hyperactivity disorder (ADHD) is a neurodevelopmental disorder marked by patterns of inattention, hyperactivity, and/or impulsivity that is inconsistent with developmental expectations and that interferes with daily functioning in a variety of settings (e.g., school, family) [American Psychiatric Association, 2013]. Many studies suggest that the motor skills of children with ADHD are poorer than those of their peers (e.g., [Kaiser *et al.*, 2015; Rosa Neto *et al.*, 2015]). A delay in maturation of the prefrontal brain regions involved in executive functioning, attention, and motor planning has been observed in children with ADHD [Shaw *et al.*, 2007] and may underlie motor deficits. These deficits could also result from altered fronto-striato-cerebellar connectivity patterns, including dopamine pathways, which are responsible for the planning and modulation of movements [Farone, 2015; Langmaid *et al.*, 2016]. Moreover, abnormal connectivity between the primary motor cortex and several other regions involved in motor and sensorimotor functioning (e. g., lower frontal gyrus, insular cortex, amygdala, putamen, pallidum) has been observed in children with ADHD and/or developmental coordination disorder (DCD) and may lead to both motor and attention deficits [Mcleod *et al.*, 2014].

A recent meta-analysis indicates that children with ADHD have less legible and slower handwriting than their peers without ADHD [Graham *et al.*, 2016]. Poor handwriting may impact on academic achievement, self-esteem, and quality of life [Brossard-Racine *et al.*, 2011; Goulardins *et al.*, 2011]. It is thus essential to assess the motor and graphomotor skills of children with ADHD. Early detection of ADHD and of the problems associated with it may reduce their negative impact on development and help guide intervention strategies.

The study of graphomotor behavior through the kinematic analysis of handwriting movements yields online, precise, and objective quantitative measures of handwriting [Tucha *et al.*, 2008]. The results of the few studies that have used kinematic analysis with children with ADHD show that these children have greater difficulty planning and programming movements effectively than do typically developing children [Eliasson *et al.*, 2004; Rosenblum *et al.*, 2008; Yan and Thomas, 2002], as manifested by slower [Shen *et al.*, 2012; Yan and Thomas, 2002] and less consistent handwriting [Adi-Japha *et al.*, 2007; Eliasson *et al.*, 2004; Langmaid *et al.*, 2014; Pitcher *et al.*, 2002], and less precision and fluidity [Adi-Japha *et al.*, 2007; Eliasson *et al.*, 2004; Schoemaker *et al.*, 2005; Yan and Thomas, 2002]. However, previous studies have been limited to the study of global kinematic variables such as maximum speed, reaction time, total duration of movement or degree of fluidity, which provide only partial information on the state of the neuromotor system.

Plamondon and colleagues [Plamondon *et al.*, 2013] recently developed a novel approach based on the Kinematic Theory of rapid human movements [Plamondon, 1995a, 1995b] to study fine motor skills. They used the lognormal speed profile obtained from pen strokes executed on a digitizer as the model of ideal motor control. This technique can be used to assess graphomotor impairment caused by neurological problems [Van Gemmert *et al.*, 2013]. It can be used in children to assess the general state of the neuromotor system using global parameters (SNR, nbLog, SNR/nbLog). It can also provide new parameters describing the internal functioning of the neuromotor system, that is, a representation of its action plan ($t_0$, D, $\theta_s$, $\theta_e$) and execution ($\mu$, $\sigma$, Mode (M), Median (m), Time delay ($\bar{t}$), Response time (s), Asymmetry (Ac)) (see below for a description of each parameter).

To produce a pen stroke, the representation of the desired movement and its motor program are first activated and selected and then converted into motor commands before the movement is finally produced adequately by the neuromuscular system [Plamondon *et al.*, 1991]. Effective execution of a fast pen stroke therefore requires the generation of motor commands to contract the agonist and antagonistic muscles involved in the movement [Plamondon *et al.*, 2013]. This results in an asymmetric bell-shaped velocity profile, representing a rising phase (acceleration) at the beginning of the movement and a slower decreasing phase (deceleration) at the end of the movement [Eliasson *et al.*, 2004; Plamondon *et al.*, 2013]. According to the Kinematic Theory, lognormality occurs when rapid and spontaneous movements are sufficiently controlled and automated to require little or no visual feedback [Plamondon, 1995a, 1995b]. The generation of more complex movements thus amounts to concatenating and superimposing lognormal functions over time, exploiting the sigma-lognormal model [Plamondon and Djioua, 2006; Plamondon *et al.*, 2013].

In a typical experiment, the position of the pen is first recorded by a digitizer. Software programs calculate the speed, then an optimization algorithm breaks down the movement into its lognormal components and extracts the parameters that best represent the pen's trajectory. Among the parameters obtained, the number of lognormal functions required to reconstruct the signal (nbLog) and the signal-to-noise ratio between the original profile and the reconstructed profile (SNR) are used to calculate the SNR/nbLog ratio. These variables are an index of the lognormality of the movements produced, which can be interpreted as an index of the fluidity of movement [Duval *et al.*, 2015]. Using the lognormal approach, recent studies have shown that the signal-to-noise ratio (SNR) declines among older individuals [Plamondon *et al.*, 2013], that nbLog is significantly higher for individuals with Parkinson's disease than for controls [Van Gemmert *et al.*, 2013], and that young children (3 to 5 years old) tend toward lognormal patterns of pen strokes over time (SNR increases and nbLog decreases as fine motor control improves) [Duval *et al.*, 2015]. Together, these results indicate that lognormal parameters (SNR and nbLog) are useful to describe the state of motor control in a variety of contexts.

In the present study we applied the sigma-lognormal model to examine the motor control of children with ADHD. We sought to determine whether the parameters obtained from sigma-lognormal modeling of fast pen stroke speed profiles would discriminate between strokes produced by children with and without ADHD. We predicted that children with ADHD would show poorer handwriting movement quality than controls and that this would be reflected in the lognormal parameters. Moreover, we exploited various pattern recognition techniques to evaluate their classification potential, that is, their ability to discriminate between ADHD subjects and controls. These different techniques are useful in determining the sensitivity and specificity of the fast stroke task. This study is an extension and complement to a previous study using the same sample/data [Laniel *et al.*, 2018, 2019].

## 2. Global Methodology

### 2.1. *Participants*

Twenty-five children 8 to 11 years old (Mean age = 9.77 years) participated in the study [Laniel *et al.*, 2019]. Children in the ADHD group (n = 12, M = 9.57, SD = 1.18 years, 8 males) received a medical diagnosis of ADHD and all took psychostimulant medication, except on the day of the evaluation. They were recruited through ads posted on bulletin boards at the University of Montreal and on groups for parents of children with ADHD on social networks (e.g., Facebook). Children taking Strattera (atomoxetine) were excluded because of its long-lasting effect. One control participant was withdrawn from the analyses because of a problem with data collection. Thus, 12 children without ADHD recruited via day camps and through ads posted on social networks (e.g., Facebook) formed the control group (n = 12, M = 9.97, SD = 1.33 years, 6 males). Exclusion criteria for the control group were: a diagnosed neurological, neurodevelopmental, psychiatric or genetic condition, intellectual disability, developmental coordination disorder (DCD), psychotropic medication.

### 2.2. *Procedures*

Children were asked to rapidly draw 30 straight single strokes on a digitizer with a stylus, one at a time, from a starting point to a large target area printed on a guide sheet. Strokes had to be more than 13 cm long. The go signal for the child to draw a stroke was a sound cue (at 1 kHz for 500 ms) emitted by an audio simulator. Movements were digitized at 200 Hz and recorded on a digitizer (Wacom Intuos2). The digitizer and simulator were synchronized with an in-house program called Sign@medic (for more details on the test, see [O'Reilly *et al.*, 2014]).

### 2.3. *Extraction of the sigma-lognormal parameters*

Invalid strokes were removed (automatically by the software if the stroke started before the cue or manually by the examiner if the stroke was executed improperly). These behavioral rejects (versus software rejects, see below) represented 17.5% of the total number of strokes (157/897) (21% (33) in the control group and 79% (124) in the ADHD group). The ADHD group produced a significantly higher number of strokes leading to rejection, M = 10.33, SD = 8.08, than the control group, M = 2.75, SD = 5.01, $t(22) = -2.76$, $p = 0.013$. Of note, children with ADHD often began their movement before the start cue.

The data were then filtered. The pressure exerted on the tablet indicated when participants lifted the pen, marking the end of data collection for a particular stroke. Disturbances in the signal due to lifting of the pen were eliminated. The speed profiles were then reconstructed with the sigma-lognormal model. This method allowed the extraction of eight lognormal parameters, namely, the number of lognormals per stroke (nbLog), SNR, $t_0$, D, $\theta_s$, $\theta_e$, $\mu$, $\sigma$, Six other parameters were derived based on these indicators (SNR/nbLog, Mode (M), Median (m), Time delay ($\bar{t}$), Response time (s) and Asymmetry (Ac)) [Plamondon *et al.*, 2003].

After parameter extraction, some strokes were rejected due to software processing errors (software rejects). A total of 12 strokes (control group = 3; ADHD group = 9) were rejected because the SNR was less than 15 dB (meaning that the software was unable to reconstruct the trajectory adequately [Plamondon and Djioua, 2006]) and 24 (control group = 17; ADHD group = 7) were rejected when $t_0$ was negative and $\mu$ was positive, caused by a software error. Finally, 43 (control group = 24; ADHD group = 19) were rejected because D (the predicted movement amplitude) exceeded the maximum distance a child could reach. Strokes could be rejected for more than one reason. In the end, a total of 669 strokes (on 897) were kept for analysis (control group = 341; ADHD group = 328). On average, 3.17 strokes, SD = 2.25, were rejected by software for each child in the control group and 2.75 strokes, SD = 2.01, for each child with ADHD. The software reject difference was not significant, $t(22) = 0.46$, $p = 0.637$.

### 2.4. *Sigma-lognormal parameters*

For each group, the mean value of the following parameters was used in the analysis [Plamondon, 1995a, 1995b; Plamondon *et al.*, 2003, 2013]. The first three parameters are global curve fitting parameters and describe the general state of the neuromotor system:

- SNR: The signal-to-noise ratio between the original speed profile and the reconstructed speed profile, computed in decibels (dB). This is a measure of the quality of the sigma-lognormal reconstruction. Higher SNR indicates better reconstruction.
- nbLog: The number of lognormal functions required to reconstruct the signal. This parameter represents fluidity of movement by the participant. Higher nbLog indicates less fluid movement.
- SNR/nbLog: SNR/nbLog is used as a performance criterion. The capacity to reconstruct a movement's speed profile with lognormals can be interpreted as an indicator of the quality of motor control since the lognormal speed profile corresponds to perfect motor control [Plamondon *et al.*, 2013]. According to the lognormality principle, higher SNR/nbLog indicates a movement closer to ideal behavior.

Other parameters describe the internal functioning of the neuromotor system. Four parameters, considered as central, represent the neuromotor action plan:

- $t_0$: Time required (in seconds) for the brain to produce a motor command. $t_0$ should not be confused with the moment a sound cue is emitted ($t = 0$), nor with the response time (RT) measured by moment of movement onset. $t_0$ is the moment when a population of neurons sends a motor command. In other words, $t_0$ occurs after the start cue, given that the central nervous system must perceive the sound cue before sending the command, but before RT, given that a delay is necessary for the motor command to be transmitted and for enough muscle cells to be recruited to begin the movement effectively [O'Reilly *et al.*, 2013]. Longer $t_0$ are indicative of longer delays before a motor command is emitted and, therefore, longer time to perceive the stimulus and prepare the command.
- D: The amplitude of the movement associated with each motor command, in millimeters. Greater D indicates a greater amplitude of the movement associated with a command.
- $\theta_s$: The starting angular direction of each stroke in radian.
- $\theta_e$: The ending angular direction of each stroke in radian.

Two parameters are related to motor program execution and are referred to as peripheral:

- μ: Logtime delay, the time required to reach the middle of movement distance on a logarithmic scale. It can be associated with the general response speed of the neuromuscular system to a command.
- σ: Logresponse time, the response time on a logarithmic scale of the neuromuscular system's response to a command. Sigma (σ) is used to compute an estimate of movement duration.

The following parameters are also related to motor program execution and are intrinsic characteristics of each lognormal [Plamondon *et al.*, 1991, 2003].

- Mode: The Mode (M) is the time at which the amplitude of the impulse response reaches its maximum value.

$$M = t_0 + e^{(\mu-\sigma^2)} \tag{1}$$

- Median: The Median (m) is the time when the integral under curve reaches its half-value. This indicates when half the action is executed. In the current study, the Median means half the distance. A smaller Median means that half of the movement is reached faster.

$$m = t_0 + e^{\mu} \tag{2}$$

- Time delay: The time Delay ($\bar{t}$) is a global evaluation of the rapidity of a given system to react to a command. A smaller Time delay means that the reaction to the command is faster.

$$\bar{t} = t_0 + e^{(\mu - \frac{\sigma^2}{2})} \tag{3}$$

- Response time: The response time (s) represents the impulse response of a neuromuscular system. s is a measure of the spread of the impulse response over time. A smaller response time means that the movement time is smaller.

$$s = (\bar{t} - t_0)\sqrt{(e^{\sigma^2} - 1)} \tag{4}$$

- Asymmetry: Asymmetry (Ac) is related to the shape of the curve. Higher Asymmetry means that the Mode (M) moves to the left with a smaller velocity peak, but without changing the Median position. It means that the peak will be reached faster, but it will be smaller and movement time will be longer.

$$A_c = 1 - e^{-\sigma^2} \tag{5}$$

## 2.5. *Statistical analysis*

First, we compared the quality of motor control of single strokes between children with and without ADHD, conducting t-tests for independent measures on the parameters obtained from the sigma-lognormal analysis. The analysis was conducted on all strokes by group (and not on the average for each child). Bonferroni corrections for multiple comparisons were applied. Since six parameters (SNR/nbLog, Mode, Median, Time delay ($\bar{t}$), Response (s) time and Asymmetry (Ac)) were calculated from the other eight ($t_0$, D, $\theta_s$, $\theta_e$, $\mu$, $\sigma$, SNR, nbLog), Bonferroni corrections were applied for eight parameters (significance threshold was $p < 0.00625$).

## 3. Results

Two-sided chi-square tests revealed no significant difference between the clinical and control groups on sex, $X^2(2, N = 24) = 0.686$, $p = 0.680$, and handedness, $X^2(2, N = 24) = 3.429$, $p = 0.217$. Independent measures t-tests showed no significant difference in age (ADHD group M = 9.57, SD = 1.18; control group M = 9.97, SD = 1.33), $t(22) = 0.77$, $p > 0.05$).

The results revealed a significant between-group difference on SNR/nbLog, indicating significantly poorer quality of motor control in the ADHD group (see Table 1). Mean nbLog was significantly higher for the ADHD group, which indicates that more lognormals were required to reconstruct the pen stroke signal and that the movements were less fluid. A significant between-group difference was found in terms of $t_0$, which was greater for the ADHD group, indicating a longer delay before motor command instantiation (see Table 1). Figure 1 presents a reconstruction of a stroke executed by a control subject using the Sigma-Lognormal model. It used two lognormals to reconstruct the velocity of the stroke with a SNR of 29.77 dB.

In addition, the D parameter was significantly smaller for the ADHD group, which reflects a smaller movement amplitude in their action plans. Sigma ($\sigma$) and Response time (s) parameters were also significantly greater in the ADHD group, suggesting that strokes made by children with ADHD were slower than those of the children in the control group. Time delay ($\bar{t}$) was significantly greater in the ADHD group, indicating that the neuromotor system of children with ADHD took more time to react to the command than children in the control group. No significant difference (with Bonferroni correction) was observed on the other parameters (see Table 1). Figure 2 presents a reconstruction of a stroke executed by an ADHD subject using the Sigma-Lognormal model. It used six lognormals to reconstruct the velocity of the stroke with a SNR of 26.54dB.

Table 1. Results of t-test for independent measures Between ADHD Group and Control group on sigma-lognormal parameters

| **Parameters** | **ADHD M (S.D)** | **Control M (S.D)** | **Degrees of freedom** | $t$ | **Sig.** | **Effect size (Cohen's d)**[c] |
|---|---|---|---|---|---|---|
| SNR | 26.22 (2.67) | 26.53 (2.77) | 653 | 1.45 | 0.148 | - |
| nbLog | **4.94 (2.40)** | **3.79 (1.75)** | **653** | **-7.05** | **< 0.0001**[a] | **0.55** |
| SNR/nbLog | **6.55 (3.15)** | **8.09 (2.90)** | **653** | **6.46** | **< 0.0001**[a] | **0.51** |
| $t_0$ | **0.69 (0.77)** | **0.41 (0.69)** | **2854** | **-10.07** | **< 0.0001**[a] | **0.38** |
| D | **62.83 (74.70)** | **73.29 (84.36)** | **2854** | **3.51** | **< 0.0001**[a] | **0.13** |
| $\mu$ | -1.55 (0.76) | -1.64 (0.74) | 2854 | -2.27 | 0.023[b] | - |
| $\sigma$ | **0.34 (0.28)** | **0.31 (0.25)** | **2854** | **-3.17** | **0.002**[a] | **0.11** |
| $\theta_s$ | -0.23 (6.51) | -0.86 (5.99) | 2854 | -2.65 | 0.008[b] | - |
| $\theta_e$ | -0.51 (13.04) | -0.88 (9.23) | 2854 | -0.84 | 0.399 | - |
| M | **0.94 (0.75)** | **0.64 (0.68)** | **2854** | **-10.81** | **0.000**[a] | **0.41** |
| m | **0.96 (0.75)** | **0.66 (0.68)** | **2854** | **-10.96** | **0.000**[a] | **0.42** |
| $\bar{t}$ | **0.95 (0.75)** | **0.65 (0.68)** | **2854** | **-10.87** | **0.000**[a] | **0.41** |
| s | **0.053 (0.04)** | **0.045 (0.03)** | **2854** | **-6.15** | **0.000**[a] | **0.24** |
| Ac | **0.15 (0.18)** | **0.13 (0.16)** | **2854** | **-3.30** | **0.001**[a] | **0.13** |

[a]Significant with Bonferroni corrections for multiple comparisons ($p < 0.00625$).
[b]Significant at $p < 0.05$.
[c]Effect size is considered large when Cohen's d is at about 0.80, medium at about 0.50, and small at 0.20 [Cohen, 1988].

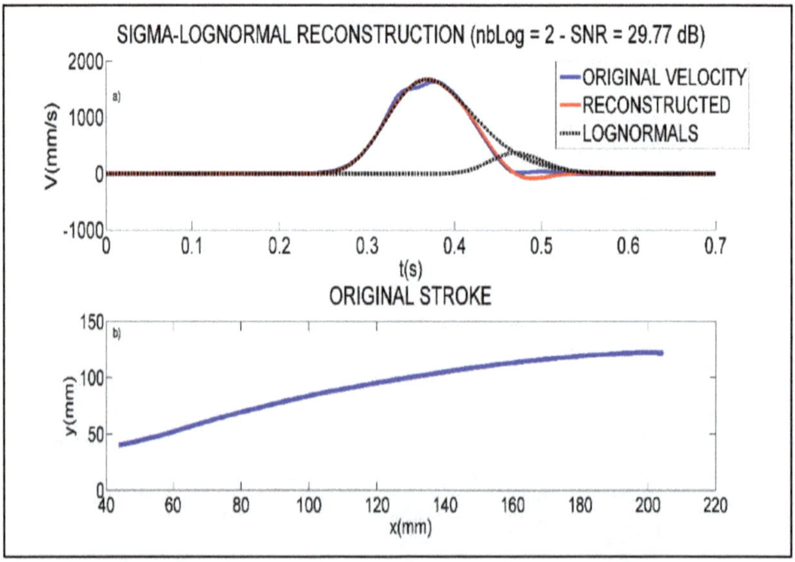


Fig. 1. Sigma-Lognormal Reconstruction of the Stroke velocity executed by a control subject with an SNR of 29.77 dB.

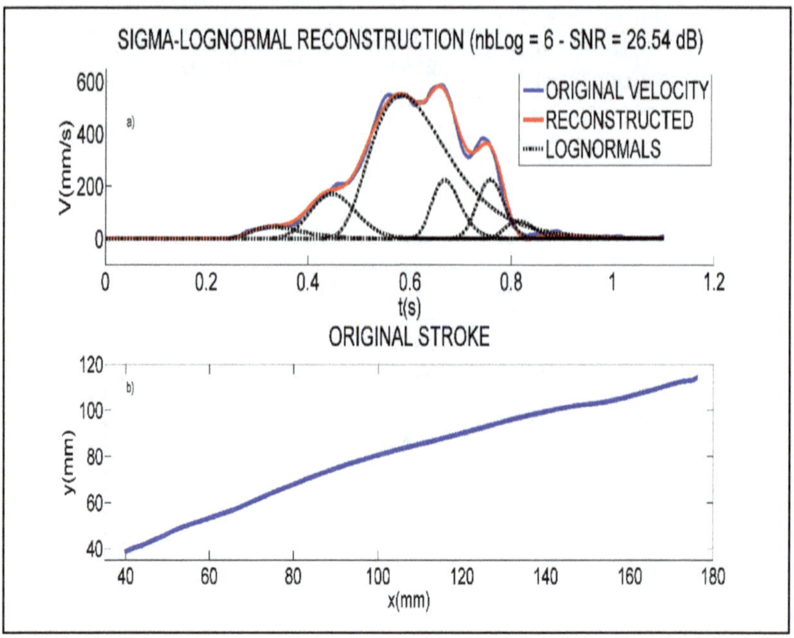


Fig. 2. Sigma-Lognormal Reconstruction of the Stroke velocity executed by an ADHD subject with an SNR of 26.54 dB.

## 4. Discussion of ADHD – Control Group Differences

As expected, a significant between-group difference was found on the SNR/nbLog parameter, which was significantly lower in the ADHD group than the controls. This suggests that children with ADHD have poorer motor control of single pen strokes. The sigma-lognormal model applied in the context of the kinematic analysis of fast pen strokes thus seems sensitive to the motor and graphomotor skill problems typical of children with ADHD. This is consistent with previous studies showing that the SNR and nbLog parameters derived from the sigma-lognormal model are useful in describing motor control skills [Plamondon and Djioua, 2006; Plamondon *et al.*, 2013; Duval *et al.*, 2015; Remi *et al.*, 2007, 2016, 2017].

A significant between-group difference was also observed on nbLog, but not on SNR. This can be explained by the fact that jerky movements tend to require more lognormals for reconstruction, whereas it is possible to reconstruct them with a good SNR [Duval *et al.*, 2013]. More lognormals were required to reconstruct the pen strokes of children with ADHD versus controls. The SNR parameter could not discriminate between groups as the required value to be reached by the software was fixed (> 15). The number of lognormals required to reconstruct the signal thus seems to best discriminate between groups.

The amplitude of movement D was significantly smaller in the ADHD group; that is, more motor commands were needed to produce a stroke, but of a smaller amplitude, to cover the same distance as typically developing children. This might reflect a problem with programming movements in children with ADHD [Langmaid *et al.*, 2014]. However, more research is needed to better understand the mechanisms underlying this between-group difference in the D parameter. This could be done by studying more complex movements.

The time required for the neuromuscular system to emit a motor command, $t_0$, was significantly higher in the ADHD group. This may reflect a lag in the perception of the start cue or diminished motor planning [Woch *et al.*, 2011] among children with ADHD. The latter hypothesis is supported by previous studies, in which children with ADHD needed more time to plan their actions [Kaiser *et al.*, 2015; van der Meere *et al.*, 1992].

It appears that in addition to prolonged movement planning in children with ADHD compared to control children, the condition is also associated with increased movement time. As such, Sigma ($\sigma$) and Response time (s) were significantly greater for children with ADHD than for controls. This shows that children with ADHD tend to take more time to execute their movements. The Time delay ($\bar{t}$) was also greater in the ADHD group than in the control group, indicating that the neuromuscular system of children with ADHD may be slower to react to a command than that of control children. These results are consistent with other studies that show slower handwriting in children with ADHD compared to controls [Shen *et al.*, 2012; Yan and Thomas, 2002].

This exploratory study was limited by its very small sample size (24 children of which 12 have ADHD) and revealed no direct relationship between level of inattention and hyperactivity, and motor control skills represented by the lognormal parameters. Moreover, as suggested in Eliasson *et al.* (2004), inattention and hyperactivity symptoms might not be entirely responsible for the motor impairments observed in children with ADHD. Therefore, the hypothesis of a comorbid DCD (Developmental Coordination Disorder) underlying motor difficulties in ADHD seems more plausible than the hypothesis that motor deficits are secondary to the central symptoms of ADHD (i.e., inattention, hyperactivity or impulsivity) [Laniel *et al.*, 2019]. Since the group was small, comorbidities associated with ADHD were not controlled (e.g., DCD). However, the lognormal parameters are interesting in their ability to discriminate between children with and without ADHD. This could be useful in helping establish an ADHD objective diagnosis. Futures studies should include larger samples of children with ADHD and control for comorbidities to ensure the robustness and generalization of the results.

## 5. ADHD Classification

In the second part of this study, we explored the possibility of making an automatic classification of the subjects based on their graphomotor skills as described by the Sigma-Lognormal representation using different techniques (see Methods below). In this two class problem, there are handwriting samples from 24 children (12 ADHD and 12 control children). Since there was only a small number of samples, we repeated the following process $T$ times to generate the training-testing sets (ratio of 10:2). For a training set, 10 children with ADHD were randomly selected from the ADHD group and 10 control children from the control group. After selecting the samples for the training set, the rest of the samples were assigned to the testing set. The following machine learning models were trained by the training sets and evaluated on the testing sets.

According to the t-test results from Table 1 (results of t-test for independent measures between ADHD group and control group on sigma-lognormal parameters), four features (number of lognormal (nbLog), ratio of SNR/nbLog, $t_0$, and D) were statistically significant different features between the children with ADHD and the control group. Thus, we used these four features to classify children from the ADHD and control groups. For each child, a four-dimensional feature vector was computed by averaging the above four features from all strokes.

### 5.1. *Methods*

Principal Component Analysis (PCA) is used to detect the combination of features ranking these according to the highest variances, which are named as principal components [Hotelling, 1933]. Linear Discriminant Analysis (LDA) also computes the features that represent the highest variances between classes [Fisher, 1936]. However, LDA is a supervised method which requires labels of training samples while PCA is an unsupervised method which does not require any labels.

k-Nearest Neighbor (kNN) identifies the testing samples based on k closest training samples in feature space; it is a nonparametric method [Altman, 1992]. The problem of this method is to choose the correct value of *k*, since it depends on the data. In particular, larger values of *k* reduce the effect of noise on the classification but make boundaries between classes less distinct. Even though a good *k* can be selected by various heuristic techniques, we tried with *k* values from 1 to 10 since there were only 10 children belonging to the training set in this study.

Support Vector Machine (SVM) is a well-known method based on separating hyperplane to solve the binary classification task [Cortes and Vapnik, 1995; Dreyfus, 1990]. For SVM, there are two trade-offs: getting a larger margin and lowering misclassification rate. In general, if the margin is increased, the misclassification rate becomes higher, and vice-versa. For archiving more generalized result (larger margin), a smaller value of regularization parameter (parameter *C*) is employed. For more accurate results, (smaller margin) on training set, a larger value of the *C* parameter is used. Moreover, the *Gamma* parameter is used to adjust the variance and bias values in the SVM model. In the case of a small Gamma value, the SVM model has low bias and high variance; and vice-versa. Thus, we employed a grid search approach for searching for the appropriate values of the *C* and *Gamma* parameters, since there is no rule of thumb to select their best values.

Artificial Neural Network (ANN) consists of multiple connected layers of non-linear units. The ANN model is trained by a momentum gradient descent algorithm. In this research, we tried both of the two following structures: Feedforward Neural Network (FNN) and Long Short-Term Memory (LSTM) [Hochreiter and Schmidhuber, 1997]. We employed a FNN model with three hidden layers of 20 hidden units. Also, we employed 3 different one-hidden layer LSTM networks with the number of hidden units being 8, 16, 32. Both structures of ANN did not achieve a good performance since there was only 12 samples for each group (ADHD and control groups), which is insufficient.

## 5.2. *Results*

In our experiments, T equals 100 for SVM, LDA, PCA, kNN methods and T equals 10 for the ANN-based methods. Despite using one hidden layer neural network (shallow network), ANN achieves a performance around 50%. The ANN did not converge due to lack of samples. Other methods including SVM, LDA, PCA, kNN performed better than ANN. The results of PCA, LDA, kNN, and SVM on the ADHD dataset are presented in the next figures. All of these following methods were implemented in Python using scikit-learn package. Figure 3 illustrates the representation/projection of the ADHD dataset with only the first two principal components, that is the two feature combinations with the highest variance using PCA.

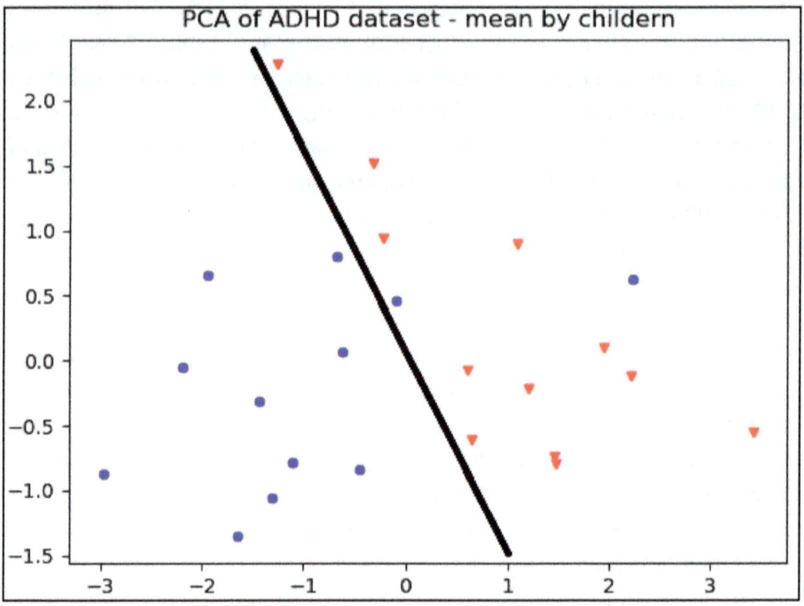


Fig. 3. The representation of samples from the ADHD dataset by the two features with the most PCA variance. Blue circles represent the ADHD group, and red triangles represent the Control group.

The black line in Fig. 3 represents the boundary that separates the two groups. In that way, two samples of ADHD group are misclassified. Thus, the best PCA result was 91.67% (22/24).

Figure 4 shows the results when LDA is used. As in Fig. 3, the blue dots are for the samples from the ADHD group and the red triangles are for the samples from the control group. The black line in Fig. 4 represents a threshold of LDA, which showed two misclassified samples among the 24 samples. The best result obtained by LDA was 91.67% which is comparable with the best result from PCA. These results suggest that non-linear classifiers such as kNN, SVM or ANN should be employed for this dataset, since the data points are not linearly separable in both PCA and LDA results.

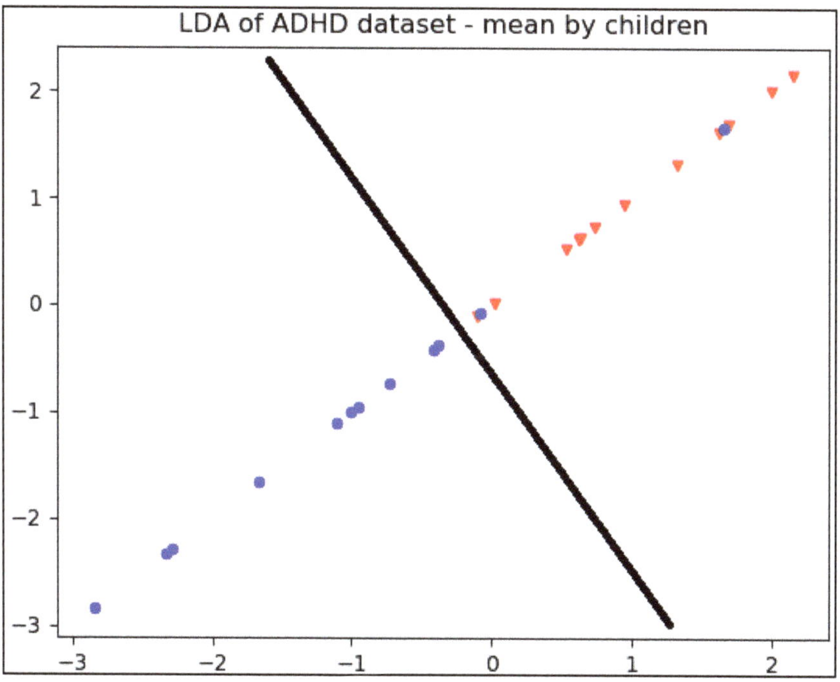


Fig. 4. The representation of samples from the ADHD dataset by LDA.

Figure 5 shows our experiments on the ADHD dataset using the kNN method with different values of $k$. Each blue dot is a result of an individual test and each red square marker is an average result for a specific $k$ value. The best performance (around 85%) was obtained for $k = 3$. Even for the larger number of $k$ (up to 10), the kNN performance remained at 80%.

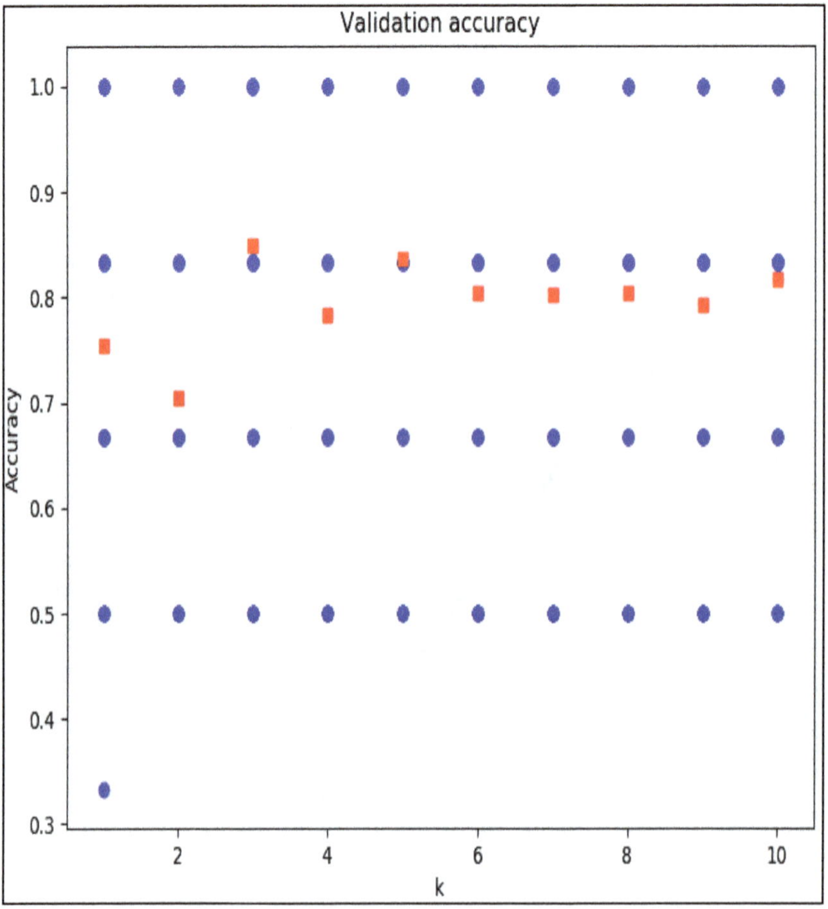


Fig. 5. Results on the ADHD dataset using the kNN method.

The SVM models with Radial Basis Function (RBF) kernel achieved the best average accuracy, around 85%, with the pairs of $C$ and *gamma* as $\{10^{l}, 10^{-(l+2)}\}$ and $l$ from 1 to 5. Figure 6 shows the results of SVM on the ADHD dataset. Inspired by the promising SVM results, we tried to compute the four-dimensional feature vector for each child by the mean from a set of $n$ random strokes.

Figure 7 shows the performance of SVM on randomly selected n strokes, where the value of $n$ is 3 in Fig. 7(a) (~10% of all strokes) and is 10 in Fig. 7(b) (~33% of all strokes). The best results with 3-strokes were over 75% with the pairs of $C$ and *gamma* at $\{10^{l},10^{-(l+1)}\}$ and where $l$ is from 0 to 7. For 10-strokes, the best results were also over 75% with the pairs of $C$ and *gamma* at $\{10^{l},10^{-(l+2)}\}$ and where $l$ is set from 0 to 7.

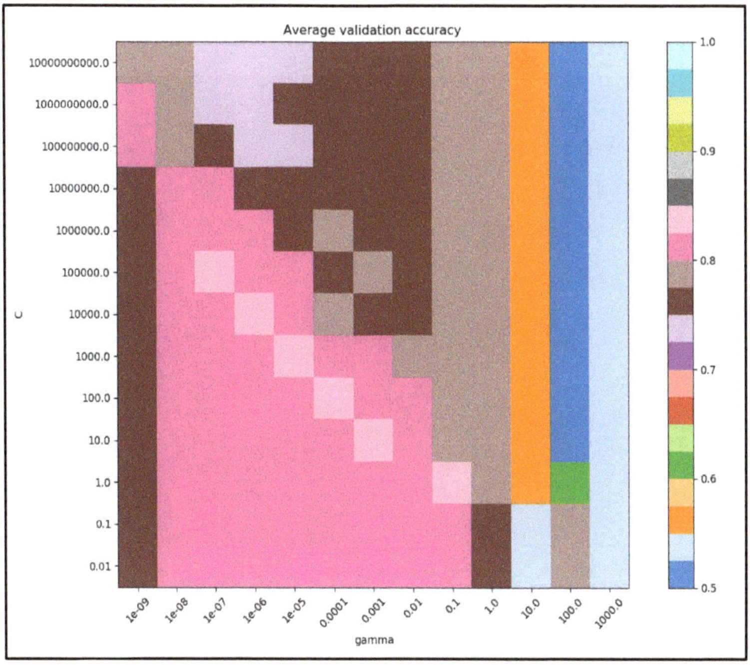


Fig. 6. SVM results on the ADHD dataset. The pink area is where a performance of more than 80% could be achieved.

a. 3 random strokes

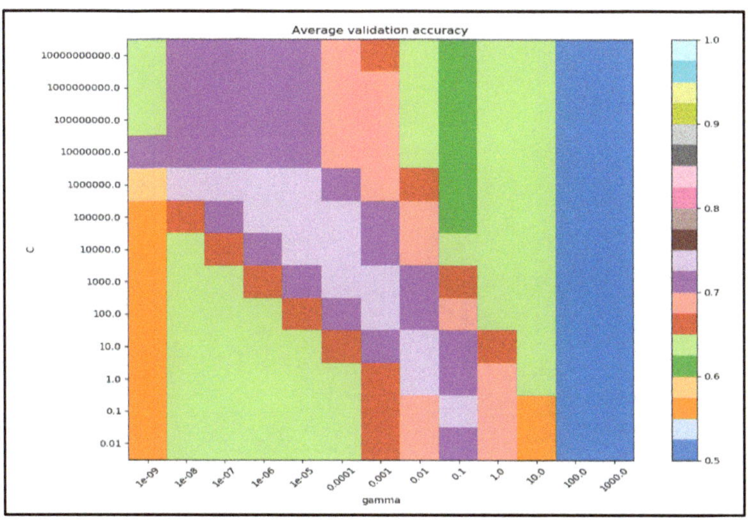


Fig. 7. SVM results by randomly selecting strokes.

b. 10 random strokes

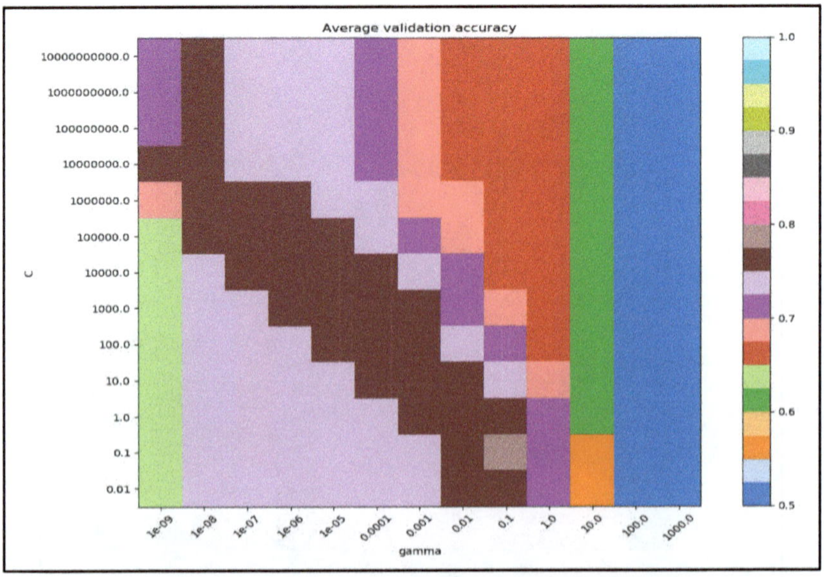


Fig. 7. SVM results by randomly selecting strokes.

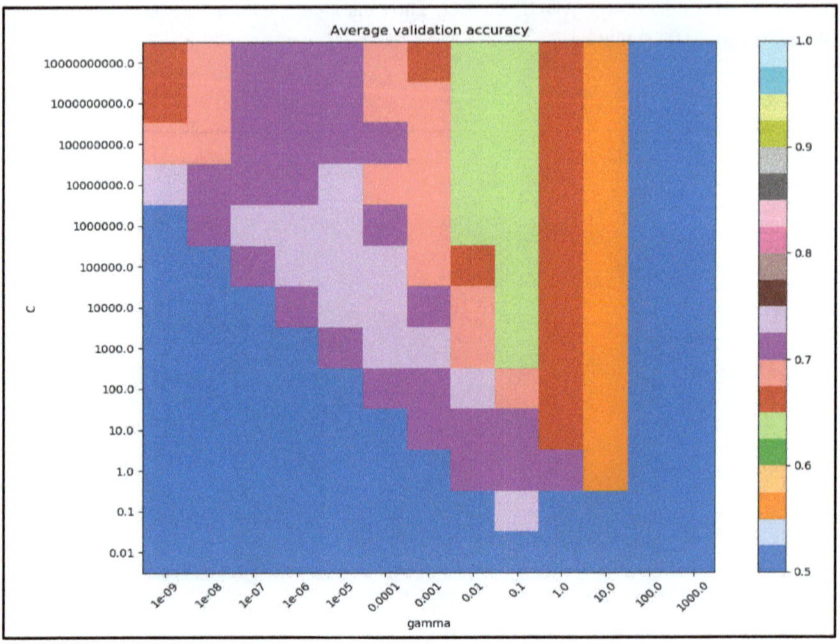


Fig. 8. SVM results by grouping strokes.

Instead of randomly selecting strokes, we investigated another selecting scheme, which is to group every ten strokes and compute the feature vector by averaging them. This scheme comes from the assumption that every child changes from a learning state at the beginning of the task to a stable state in the middle and then to a more fatigued state at the end of the data acquisition protocol. Figure 8 reports the results of SVM by grouping strokes. The best results were around 75% with the pairs of $C$ and *gamma* at $\{10^{l},10^{-(l+2)}\}$, where $l$ is from 0 to 7.

### 5.3. *Further analyses*

For further analyses, we computed the following measures: Type I and II errors, sensitivity and specificity as shown in Table 2, which measure the efficiency of our SVM classifiers on the ADHD classification task. Type I errors happen when "A child x belongs to the control group but is identified as a child with ADHD" (false positives). Type II errors happen when "A child x belongs to the ADHD group but is identified as a control child" (false negatives). The sensitivity expresses the ability of SVM to classify correctly those who have ADHD while the specificity shows the ability of SVM to classify correctly those who do not have ADHD (in Control group).

Table 2. The statistical measures to evaluate the SVM classification.

| | | Reality | |
|---|---|---|---|
| | | ADHD group | Control group |
| Prediction | ADHD | True Positive (TP) | False Positive (FP) |
| | Control | False Negative (FN) | True Negative (TN) |
| statistical measures | Errors type | II = FN/(TP+FN) | I = FP/(TN+FP) |
| | Sens. & Spec. | Sensitivity = TP/(TP+FN) | Specificity = TN/(TN+FP) |

Figure 9 shows the average type I and II errors on the ADHD dataset by SVM. The lowest false positive values were around 5% where $C$ is from $10^{-2}$ to $10^{10}$ and *gamma* is $10^{3}$, while the lowest false negative values were around 10% where $C$ is from $10^{-2}$ to $10^{8}$ and *gamma* is $10^{-9}$. According to these subfigures, type I error is higher than type II error in general. Since the SVM models achieve low error of type II, fewer children with ADHD are misidentified as control children, which is an important distinction in actual clinical scenarios. Although the type I error is higher than the type II, a medical doctor or a neuropsychologist could confirm and help correct the type I error cases.

a. Type I errors (false positive)

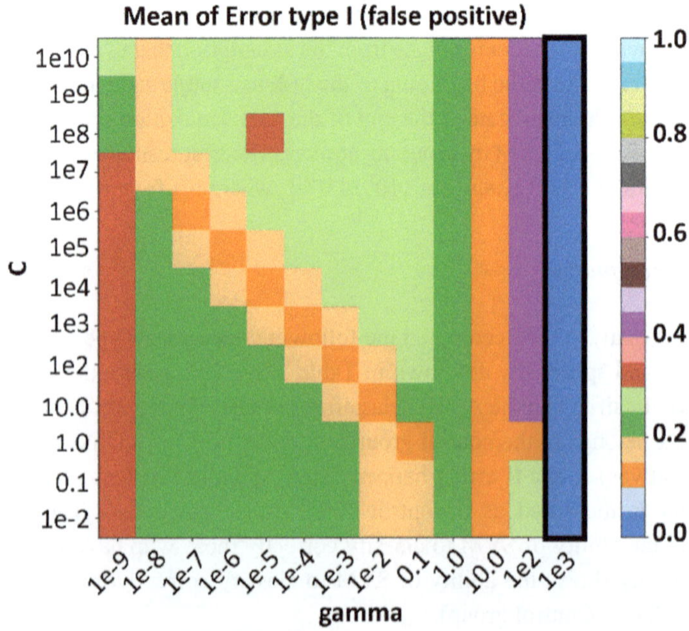


b. Type II errors (false negative)

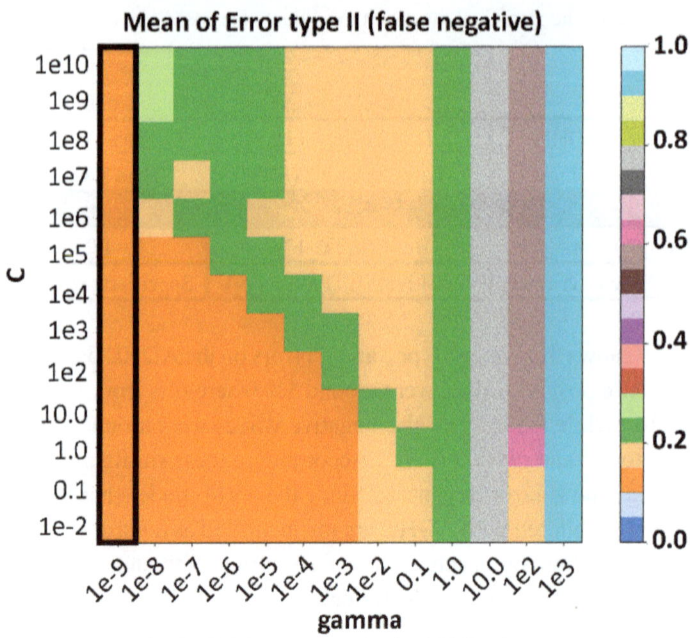


Fig. 9. SVM error types in the ADHD dataset.

a. Specificity (true negative)

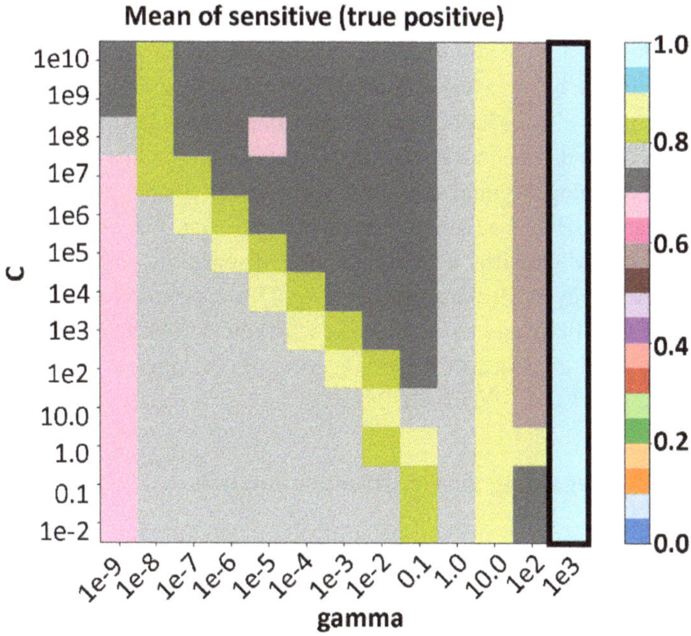


b. Sensitivity (true positive)

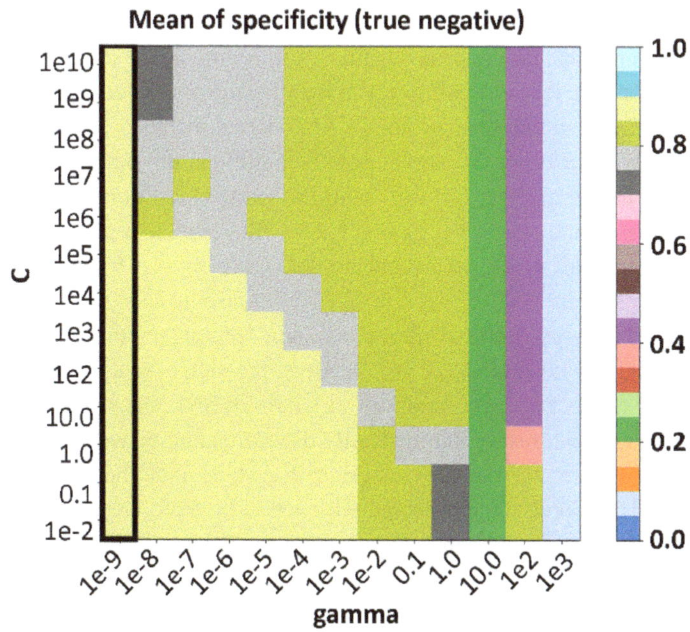


Fig. 10. Specificity and Sensitivity of SVM on the ADHD dataset.

The "sensitivity" and "specificity" present the true positive and true negative results of SVM. Thus, they are inversely related as shown in Fig. 10 which expresses the sensitivity and specificity by SVM in the ADHD dataset. The highest specificity values were around 95% when $C$ was from $10^{-2}$ to $10^{10}$ and *gamma* was $10^3$, while the highest sensitivity values were nearly 90% when $C$ was from $10^{-2}$ to $10^8$ and *gamma* was $10^{-9}$. Here again assuming that a medical doctor or a neuropsychologist could point out more easily a non ADHD child than diagnosing an ADHD, focussing on the sensitivity, that is maximizing the true positives, the SVM classifier achieves its highest sensitivity values, nearly 90%, while the specificity values are from 55% to 75%. The maximum compromise between the specificity and the sensitivity, where both values vary between 80% and 85%, is when the value of the pair (*gamma*, $C$) belongs to {(0.1, 0.1); (0.1, 0.01); (0.01, 1.0); (0.01, 100.0); (1e-7; 1e7)}.

## 6. Discussion Regarding the Classification Analysis

As shown in the classification analysis presented in Sec. 5, the traditional machine learning methods such as PCA, LDA, and kNN achieved the highest results using only a few samples while SVM achieved slightly lower results due to a lack of samples. According to the results of PCA and LDA, two children belonging to the ADHD group have similar features compared to those of children in the Control group, which means that the false negative ratio is around 16%. Due to the small size of this dataset, it is difficult to claim that the above methods could maintain their performance in practice. Although ANN-based methods generally achieve state-of-the-art performance in image recognition or online handwriting recognition problems, the lack of samples caused the non-convergence or underfitting of neural networks in this study. However, it is expected that the ANN-based methods would perform better with more samples from a larger group of children.

The performance of these above machine learning methods supports the statistical analysis in Table 1 of Sec. 4. Thus, the sigma-lognormal parameters consisting of number of lognormal (nbLog), ratio of SNR/nbLog, $t_0$, and D seem to be the critical features for automatically discriminating between children with ADHD and children in the control group based on their handwritten stroke production. These preliminary results will need to be reproduced and confirmed not only using more data from a larger number of children, but also in studies using more varied presentations of ADHD that take into consideration comorbidities and other diagnostic complexities. Nevertheless, the results of the sensitivity and specificity are promising considering the size of the data set.

## 7. Conclusion

In this study, we aimed to examine whether the sigma-lognormal model derived from the Kinematic Theory of rapid human movements could discriminate between the handwriting strokes produced by children with and without ADHD. As we predicted, a significant statistical difference was found between children with and without ADHD on several lognormal parameters: SNR/nbLog, nbLog, $t_0$, D, $\sigma$, Time delay ($\bar{t}$) and Response time (s). These results suggest that ADHD may be associated with deficits in graphomotor skills. The sigma-lognormal analysis shows promise as it offers the possibility to study the internal functioning of the neuromotor system of children with ADHD.

According to the preliminary classification results on the ADHD dataset, the machine learning methods are also promising for identifying children with ADHD based on their handwriting samples. To deploy and further apply these machine learning methods, the handwriting samples from a large number of children are required. This is a prerequisite for any follow up of the present proof of concept study.

## Acknowledgements

This work was partly supported by the NSERC-Canada Grant RGPIN-2015-06409 to R. Plamondon.

## References

Adi-Japha, E., Landau, Y. E., Frenkel, L., Teicher, M., Gross-Tsur, V. and Shalev, R. S. (2007). ADHD and dysgraphia: underlying mechanisms. *Cortex*, *43*(6), 700–709.

Altman, N. S. (1992). An introduction to kernel and nearest-neighbor nonparametric regression. *The American Statistician*, *46*(3), 175–185.

American Psychiatric Association (2013). *Diagnostic and statistical manual of mental disorders (DSM-5®)*. American Psychiatric Pub.

Brossard-Racine, M., Majnemer, A., Shevell, M., Snider, L. and Ageranioti Bélanger, S. (2011). Handwriting capacity in children newly diagnosed with attention deficit hyperactivity disorder. *Research in Developmental Disabilities*, *32*(6), 2927–2934.

Cohen, J. (1988). *Statistical power analysis for behavioral sciences* (2 ed.). New Jersey: Lawrence Erlbaum Associates.

Cortes, C. and Vapnik, V. N. (1995). Support-vector networks. *Machine Learning*, *20*(3), 273–297.

Dreyfus, S. E. (1990). Artificial neural networks, back propagation, and the Kelley-Bryson gradient procedure. *Journal of Guidance, Control, and Dynamics*, *13*(5), 926–928.

Duval, T., Rémi, C., Plamondon, R. and O'Reilly, C. (2013). On the use of the sigma-lognormal model to study children handwriting. In M. Nakagawa, M. Liwicki and B. Zhu (Eds.). *Recent progress in graphonomics: Learn from the past*, (pp. 26–30). Nara, Japan.

Duval, T., Rémi, C., Plamondon, R., Vaillant, J. and O'Reilly, C. (2015). Combining sigma-lognormal modeling and classical features for analyzing graphomotor performances in kindergarten children. *Human Movement Science*, *43*, 183–200.

Eliasson, A.-C., Rosblad, B. and Forssberg, H. (2004). Disturbances in programming goal-directed arm movements in children with ADHD. *Developmental Medicine and Child Neurology*, *46*(1), 19–27.

Farone, S. V. (2015). Attention deficit hyperactivity disorder and premature death. *The Lancet*, *385*(9983), 2132–2133.

Fisher, R. A. (1936). The Use of Multiple Measurements in Taxonomic Problems. *Annals of Eugenics*, *7*(2), 179–188.

Goulardins, J. B., Marques, J. C. and Casella, E. B. (2011). Quality of life and psychomotor profile of children with attention deficit hyperactivity disorder (ADHD). *Arquivos de neuro-psiquiatria*, *69*(4), 630–635.

Graham, S., Fishman, E. J., Reid, R. and Hebert, M. (2016). Writing characteristics of students with attention deficit hyperactive disorder: A meta-analysis. *Learning Disabilities Research & Practice*, *31*(2), 75–89.

Hochreiter, S. and Schmidhuber, J. (1997). Long short-term memory. *Neural computation*, *9*(8), 1735-1780.

Hotelling, H. (1933). Analysis of a complex of statistical variables into principal components. *Journal of Educational Psychology*, *24*, 417–441 and 498–521.

Kaiser, M. L., Schoemaker, M. M., Albaret, J. M. and Geuze, R. H. (2014). What is the evidence of impaired motor skills and motor control among children with attention deficit hyperactivity disorder (ADHD)? Systematic review of the literature. *Research in Developmental Disabilities*, *36C*, 338–357.

Kaiser, M. L., Schoemaker, M. M., Albaret, J. M. and Geuze, R. H. (2015). What is the evidence of impaired motor skills and motor control among children with attention deficit hyperactivity disorder (ADHD)? Systematic review of the literature. *Research in developmental disabilities*, *36C*, 338–357. doi:10.1016/j.ridd.2014.09.023.

Langmaid, R. A., Papadopoulos, N., Johnson, B. P., Phillips, J. G. and Rinehart, N. J. (2014). Handwriting in Children with ADHD. *Journal of Attention Disorders*, *18*(6), 504–510.

Langmaid, R. A., Papadoupolos, N., Johnson, B. P., Phillips, J. and Rinehart, N. (2016). Movement scaling in children with ADHD-combined type. *Journal of Attention Disorders*, *20*(2), 131–137.

Laniel, P., Faci, N., Plamondon, R., Beauchamp, M. H. and Gauthier, B. (2018). Kinematic analysis of fast pen strokes in children with ADHD using the Sigma-lognormal model. *Proceedings of ICPRAI 2018. First International Workshop on the Lognormality Principle and its Applications*, Montréal, Canada pp. 784–789.

Laniel, P., Faci, N., Plamondon, R., Beauchamp, M. H. and Gauthier, B. (2019). Kinematic analysis of fast pen strokes in children with ADHD. *Applied Neuropsychology: Child*, 1–16.

Mcleod, K. R., Langevin, L. M., Goodyear, B. G. and Dewey, D. (2014). Functional connectivity of neural motor networks is disrupted in children with developmental coordination disorder and attention-deficit/hyperactivity disorder. *NeuroImage: Clinical, 4*, 566–575.

O'Reilly, C., Plamondon, R. and Lebrun, L.-H. (2014). Linking brain stroke risk factors to human movement features for the development of preventive tools. *Frontiers in Aging Neuroscience, 6.*

O'Reilly, C., Plamondon, R., Landou, M. K. and Stemmer, B. (2013). Using kinematic analysis of movement to predict the time occurrence of an evoked potential associated with a motor command. *European Journal of Neuroscience, 32*(2), 173–180.

Pitcher, T. M., Piek, J. P. and Barrett, N. C. (2002). Timing and force control in boys with attention deficit hyperactivity disorder: Subtype differences and the effect of comorbid developmental coordination disorder, 21(5–6), *Human Movement Science, 21*(5–9), 919–945.

Plamondon, R. (1995a). A kinematic theory of rapid human movements: Part I: Movement representation and generation. *Biological Cybernetics, 72*(4), 295–307. Retrieved 11 12, 2017, from https://link.springer.com/article/10.1007/bf00199059.

Plamondon, R. (1995b). A kinematic theory of rapid human movements. Part II. Movement time and control. *Biological Cybernetics, 72*(4), 309–320. Retrieved 11 12, 2017, from https://ncbi.nlm.nih.gov/pubmed/7748960.

Plamondon, R. and Djioua, M. (2006). A multi-level representation paradigm for handwriting stroke generation. *Human Movement Science, 25*(4), 586–607.

Plamondon, R., Feng, C. and Woch, A. (2003). A kinematic theory of rapid human movement. Part IV: a formal mathematical proof and new insights. *Biological Cybernetics, 89*(2), 126–138. Retrieved 11 12, 2017, from https://link.springer.com/article/10.1007/s00422-003-0407-9.

Plamondon, R., O'Reilly, C., Rémi, C. and Duval, T. (2013). The lognormal handwriter: learning, performing, and declining. *Frontiers in Psychology, 4.*

Plamondon, R., Yu, L.-d., Stelmach, G. E. and Clément, B. (1991). On the automatic extraction of biomechanical information from handwriting signals. *IEEE Transactions on Systems, Man, and Cybernetics, 21*(1), 90–101.

Remi, C., Nagau, J., Vaillant, J. and Plamondon, R. (2016). Multimodal study of the impact of graphomotor training. In Y. Mazabraud, T. Forissier, A. Pietrus, O. Gros and C. Anjou (Eds.). *Caribbean Academy of Sciences 20th General Assembly and Biennial Meeting, 37.*

Remi, C., Nagau, J., Vaillant, J. and Plamondon, R. (2017). Preliminary study of t0, a sigma-lognormal parameter extracted from young children's controlled scribbles.

*Cultural heritage and education. Symposium conducted at the 18th Biennial Conference of the International Graphonomics Society*. Gaeta.

Remi, C., Vaillant, J. and Plamondon, R. (2007). Etude multimodale de l'effet de l'entrainement graphomoteur chez des enfants d'âge préscolaire [Multimodal study of the effects of graphomotor training on preschoolers]. *International Symposium on Educational Literacy*. Ajaccio, France.

Rosa Neto, F., Goulardins, J. B., Rigoli, D., Piek, J. P. and De Oliveira, J. A. (2015). Motor development of children with attention deficit hyperactivity disorder. *Revista Brasileira de Psiquiatria, 37*(3), 228–234.

Rosenblum, S., Epsztein, L. and Josman, N. (2008). Handwriting performance of children with attention deficit hyperactive disorders: A pilot study. *Physical & Occupational Therapy in Pediatrics, 28*(3), 219–234.

Schoemaker, M. M., Ketelaars, C. E., Van Zonneveld, M., Minderaa, R. B. and Mulder, T. (2005). Deficits in motor control processes involved in production of graphic movements of children with attention-deficit–hyperactivity disorder. *Developmental Medicine and Child Neurology, 47*(6), 390–395.

Shaw, P., Eckstrand, K., Sharp, W., Blumenthal, J., Lerch, J., Greenstein, D. and Rapoport, J. (2007). Attention-deficit/hyperactivity disorder is characterized by a delay in cortical maturation. *Proceedings of the National Academy of Sciences, 104*(49), 19649–19654.

Shen, I.-H., Lee, T.-Y. and Chen, C.-L. (2012). Handwriting performance and underlying factors in children with Attention Deficit Hyperactivity Disorder. *Research in Developmental Disabilities, 33*(4), 1301–1309.

Tucha, O., Tucha, L. and Lange, K. W. (2008). Graphonomics, automaticity and handwriting assessment. *Literacy, 42*(3), 145–155.

van der Meere, J., Vreeling, H. J. and Sergeant, J. (1992). A motor presetting study in hyperactive, learning disabled and control children. *Journal of Child Psychology and Psychiatry, 33*(8), 1347–1351.

Van Gemmert, A., Plamondon, R. and O'Reilly, C. (2013). Using the Sigmalognormal model to investigate handwriting of individuals with Parkinson's disease. *Proceedings 16th Biennial Conf. of the Graphonomics Society*, (pp. 119–122). Nara, Japan.

Woch, A., Plamondon, R. and O'Reilly, C. (2011). Kinematic characteristics of bidirectional delta-lognormal primitives in young and older subjects. *Human Movement Science, 30*(1), 1–17.

Yan, J. H. and Thomas, J. R. (2002). Arm movement control: differences between children with and without attention deficit hyperactivity disorder. *Research Quarterly for Exercise and Sport, 73*(1), 10–18.

https://doi.org/10.1142/9789811226830_0006

# Chapter 6

# Analysing the Evolution of Children's Neuromotor System Lognormality after Mild Traumatic Bain Injury

Nadir Faci[1,*], Naddley Désiré[2,†], Miriam H. Beauchamp[3,4,‡], Isabelle Gagnon[5,§] and Réjean Plamondon[1,ǂ]

[1]*Laboratoire Scribens, Département de génie électrique, Polytechnique Montréal*

[2]*Alberta Children's Hospital, University of Calgary*

[3]*Département de neuropsychologie, Université de Montréal*

[4]*Centre de recherche de l'hôpital Ste-Justine, Montréal*

[5]*School of Physical and Occupational Therapy/Pediatrics; McGill University, Montréal*

[*]*nadir.faci@polymtl.ca,* [†]*naddley.desire@sickkids.ca,* [‡]*miriam.beauchamp@umontreal.ca,* [§]*isabelle.gagnon8@mcgill.ca,* [ǂ]*rejean.plamondon@polymtl.ca*

This pilot study investigates whether the Kinematic Theory of rapid human movements through the Sigma-Lognormal model can help to detect differences between handwriting strokes produced by youth with mild traumatic brain injuries (mTBI) 1 and 3 months after their injury. Thirty-two (32) participants, aged between 6 and 17 years, who sustained mTBI were asked to produce handwriting strokes on a digitizing tablet. Strokes were reconstructed and analyzed using the Sigma-Lognormal model. Some parameters of this model showed significant differences between the two sets of strokes produced 1- and 3-months' post-injury, with the quality of strokes improving over time for ten participants while decreasing for seventeen participants. Four participants did not show any significant changes. These results were eventually compared to the results of the Post-Concussion Symptom Inventory questionnaire. The findings point out the potential benefits of using the Sigma-Lognormal model to assess and monitor the integrity of neuromotor systems following pediatric mTBI.

# 1. Introduction

## 1.1. *Background*

Pediatric traumatic brain injury (TBI) is a public health burden and a leading cause of disability worldwide [Neurological disorders: public health challenges: World Health Organization, 2006]. Each year, millions of children sustain TBI, with mild traumatic brain injuries (mTBI, including concussion) accounting for more than 90% of all TBI cases [National Center for Injury Prevention and Control, 2003; World Health Organization, 2006]. Mild TBI can result in detrimental consequences for children and potentially interfere with development, impacting children's academic, cognitive, psychological, social and behavioural functioning. Children with mTBI often report post-concussive symptoms (PCS) which include somatic (e.g., headache, dizziness), cognitive (e.g., inattention, forgetfulness), and affective (e.g., irritability, dysphoria) complaints. Previous studies have shown that 15–30% of children with mTBI continue to experience PCS for several months following injury, which in turn can result in functional deficits and declines in quality of life [Barlow, 2016; Moran *et al.*, 2011; Zemek *et al.*, 2013]. These challenges have led government organizations, such as the World Health Organization and the Centers for Disease Control and Prevention, to call for additional research to inform clinical care and improve outcomes in youth with mTBI [Report to Congress on mild traumatic brain injury in the United States: Steps to prevent a serious public health problem, 2003; Brain Traumatic, 1999].

In light of the need for advancing knowledge in mTBI, extensive research is also required to study brain-behaviour disturbances following mTBI and to examine how these changes may be associated with functional impairments. Consequences of pediatric mTBI may differ from those observed in the adult population, as there is evidence that childhood brain injuries can result in poorer outcomes in some children due to the vulnerability of the developing brain [Giza *et al.*, 2007]. The cognitive, motor, and social functions that develop during childhood and adolescence strongly depend on the integrity of particular brain structures and networks at critical developmental stages. Research is therefore imperative to identify children who may be at risk of developing persistent problems.

Although the majority of children and adolescent with mTBI are expected to have a favourable long-term outcome, a subset may also experience cognitive and motor difficulties in addition to presenting with PCS. Specifically, problems with attention, processing speed, and executive functioning have been reported within the first few months post-injury [Babikian and Aasarnow, 2009; Niogi *et al.*, 2008]. The investigation of motor functioning in children following mTBI has been scarcer. A few studies have reported difficulties with complex and dynamic gross motor skills that would go undetected with standard neurological testing. Studies to date show that although some motor deficits (e.g., balance problems) tend to diminish over time after injury, others remain several months after mTBI, even after PCS have resolved [Gagnon *et al.*, 2001, 2004; Brosseau-Lachaine *et al.*, 2008; Rhine *et al.*, 2017]. This is an important finding to consider as children are often recommended to gradually return to daily activities (i.e., school, physical activities) after being asymptomatic without further assessment or intervention [McCrory *et al.*, 2017]. The investigation of fine motor skills following mTBI has also been limited. Fine motor difficulties may impact a child's visuo-motor processing speed which can result in problems with producing written work efficiently [Gagnon *et al.*, 2004]. Children with fine motor difficulties may also require additional time to complete school assignments as they continue to recover from mTBI. However, there is currently a lack of accurate objective and developmentally appropriate tools to sensitively assess fine motor skills after mTBI.

In the current study, we aimed to investigate fine motor control in children and adolescents with mTBI using a handwriting stroke task that can produce objective parameters to obtain information about the integrity of the neuromuscular system after pediatric mTBI. We hypothesized that children and adolescents with mTBI would display atypical fine motor control as reflected by poorer performance at an earlier stage (1 month post-injury) of recovery than at a later stage (3 months post-injury). We also sought to examine change in reported post-concussive symptoms (PCS) over time in comparison to changes in neuromuscular parameters at 1-month and 3-months post-injury.

### 1.2. *The Kinematic Theory and Lognormality*

The Kinematic Theory stipulates that to make a simple rapid pointing movement referred to as a stroke, a synergy made up of two neuromuscular systems has to plan and execute the movement, an agonist and an antagonist neuromuscular system. The first is responsible of initiating and accelerating the movement. The second is in charge of deceleration and ending the movement. According to this theory, when a person is in perfect control of their movement, the velocity of the movement will tend to be closer to lognormality. This means that each neuromuscular system involved in the movement will generate a lognormal velocity curve. In the case of an ideal rapid pointing movement, it will consist of two lognormals that can be analyzed by the Delta-Lognormal model [Plamondon R., 1995a, 1995b, 1998; Plamondon *et al.*, 2003, 1991]. As shown in [Plamondon *et al.*, 2013], typically developing children tend to get closer to the Delta-Lognormal model as they get older, but tend to use more than two lognormals in their movements. To address this issue, the Kinematic Theory proposes the Sigma-Lognormal model to fit these complex movements.

The Sigma-Lognormal model provides a methodology to extract parameters that can describe the state of the neuromuscular systems involved. As described in Sec. 2.4.2, three parameters describe the general state of a neuromotor system. Four parameters are referred to as central, they describe the planning of the movement, that is the action plan and the input command of the system. The other two parameters are referred to as peripheral, they describe the execution of the movement. Some of those parameters are more likely to be significant, for example, the global parameters might have more impact because they reflect an individual's global Lognormality. As shown in previous studies [Plamondon *et al.*, 2013; Duval *et al.*, 2015; Van Gemmert *et al.*, 2013], those parameters are most likely to discriminate between a clinical group and a control group.

In this research project, the Sigma-Lognormal model was used to fit the velocity profile of the strokes produced by adolescent participants. Although typically developing adolescents tend towards lognormality, they have not yet reached it. In addition, the participants in the current study sustained mTBI, therefore they have hypothetically moved away from lognormality, and may tend to make more than two lognormals per stroke. Using the Sigma-Lognormal model, the 1-month post-injury results for each participant were compared to their 3-month post-injury findings. It was expected that the quality of handwriting movements would improve over time with recovery.

## 2. Methods and Analyses

### 2.1. *Design*

This is a sub-study of a larger prospective, longitudinal, multicenter cohort study [Gagnon *et al.*, 2018] designed to collect a standard set of common data element measures for children and adolescents in pediatric mTBI follow-up programs across Canada and explore outcomes (PCS, motor, cognitive, social) and predictive factors.

## 2.2. *Participants*

Participants included children and adolescents who presented to the emergency department (ED) of two tertiary care pediatric hospitals (i.e., Montreal Children's Hospital and CHU Sainte-Justine). Ninety children and adolescents were initially recruited for the sub-study. Then, 56 children were excluded because they did not execute the test properly. From the 34 remaining children two of them were rejected due to lack of data because they did not produce more than 15 strokes. Finally, valid handwriting stroke data were available for 32 participants (M = 13.99, SD = 2.03 years; 19 females). Inclusion criteria were being 6.00–17.99 years of age, having sustained a mTBI as defined by the WHO task force [Neurological disorders: public health challenges: World Health Organization, 2006], presenting to the ED within 28 days of having suffered the initial injury, and being proficient in either French or English. Exclusion criteria included presenting with a Glasgow Coma Scale score below 13, multi-system injury requiring hospitalization, operating room or procedural sedation, neurosurgical intervention, intubation, or intensive care unit admission required, intoxication at time of ED presentation as per clinical judgment, significant developmental delays resulting in communication difficulties that could impact a child's ability to complete testing.

## 2.3. *Procedure*

Recruitment was done on a voluntary basis. Children were enrolled either in the ED or from mTBI follow-up programs. Families were introduced to the study by the mTBI clinic coordinator after confirmation of their diagnosis. Families who agreed to participate were contacted by a study coordinator who explained procedures, screened for eligibility, and obtained written consent. Information and details regarding participant's demographics, past medical and developmental history, injury characteristics, and acute symptoms were then collected by a research assistant or nurse. Participants who were recruited as part of the larger study underwent motor and cognitive assessments at 2-weeks/1-month, 3-months and 6-months post-injury. However, the stroke task was administered only at the 1-month and 3-month time points.

## 2.4. *Data extraction and analyses*

### 2.4.1. *Data extraction*

Each participant was asked to produce 30 valid, fast single strokes on a tablet digitizer (Wacom Intuos2), one at a time. The trials were recorded at 100 Hz. Every stroke had to begin from a starting point located at the middle of a guide sheet, and to end at one of the sides of the sheet. The direction side was set depending on the laterality of the participant. The guide sheet is shown in Fig. 1. Participants start drawing the stroke when they see a green light signal displayed by a stimulator. The stimulator displays the cue stimulus after a random time set by a homemade software called Sign@medic. This software manages and synchronizes the digitizer and the stimulator. Of the 90 children who took the test, 56 of them did not execute the test properly due to several factors, including violation of the study instructions, as well as methodological errors in protocol administration. Therefore, these pen strokes were not analyzed. For the 34 children included in the analyses, 2351 strokes were collected (1150 for participants at 1 month and 1201 for those at 3 months post-injury), and approximately 34 strokes per participant were performed to obtain 30 valid strokes. For example, every stroke that began before the cue was displayed were automatically removed by the software. Data rejected in this way were counted as Behavioural rejects. (For more details on the test, see [O'Reilly *et al.*, 2014]).

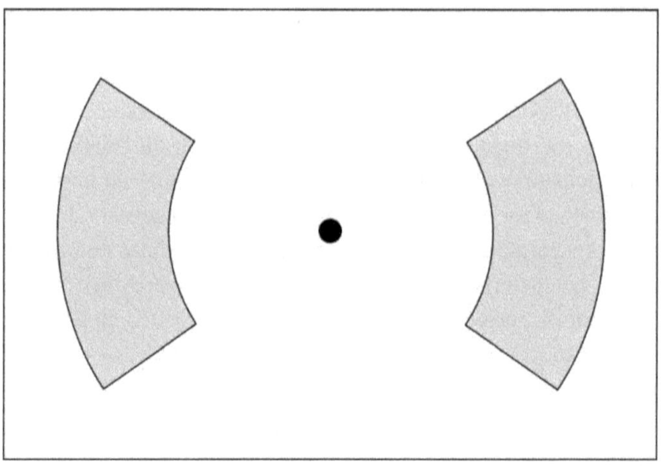

Fig. 1. The guide sheet used in this study.

Furthermore, the original data were filtered to segment each stroke. The time when the cue was displayed indicates the beginning of the trial, and the moment the pressure returns to 0 fixes the end of the trial (the pen could then be lifted from the digitizer). If the velocity of the movement was not null at the end of the trial, this means that the pen was lifted before the stroke was finished. In this case, the stroke was not executed properly and it was rejected. Those rejected strokes were also counted as Behavioural rejects. There was a high number of Behavioural rejects at both 1 and 3 months post-injury. 645 strokes (27.44%) were rejected in total [307 (26.70%) at 1 month and 338 (28.14%) at 3 months]. The average Behavioural rejects per participant at 1 month was M = 9.03, SD = 5.24, and M = 9.94, SD = 5.34 at 3 months post-injury. There was no difference between the two timestamps (starting too early or ending at non null velocity) ($p = 0.48$).

The velocity profiles were extracted using the Script Studio software [O'Reilly *et al.*, 2009; O'Reilly and Plamondon, 2009a; O'Reilly *et al.*, 2013; Duval *et al.*, 2014]. Script Studio is a software that extracts the parameters of the Sigma-Lognormal model. Eight parameters can be extracted. The first two are the number of Lognormal per stroke (nbLog) and SNR. These parameters are the global parameters for each stroke. The following six parameters are related to each Lognormal composing the movement: $t_0$, D, $\theta_s$, $\theta_e$, $\mu$, and $\sigma$. Figure 2 shows an example of a reconstruction using the Sigma-Lognormal model with 4 lognormals and a SNR of 26.55 dB. Table 1 gives the values of each parameter for each lognormal.

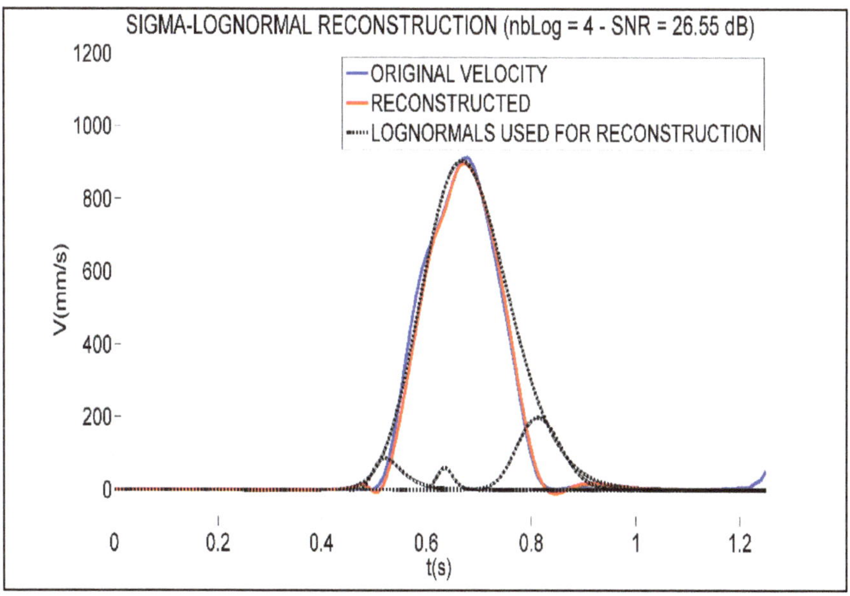


Fig. 2. Example of a reconstruction using the Sigma-Lognormal model.

Table 1. The Sigma-Lognormal parameters extracted by the software for the velocity profile presented in Fig. 2.

| GLOBAL PARAMETERS | | | | | |
|---|---|---|---|---|---|
| | nbLog 4 | | | SNR (dB) 26.55 | |
| LOGNORMAL PARAMETERS | | | | | |
| $t_0$(s) | D (mm) | $\mu$ | $\sigma$ | $\theta_s$ | $\theta_e$ |
| 0.037 | 185.02 | -0.446 | 0.128 | 0.143 | -0.156 |
| 0.266 | 20.81 | -0.597 | 0.076 | -3.050 | -3.424 |
| 0.449 | 6.21 | -2.488 | 0.367 | 2.710 | 5.024 |
| 0.362 | 2.30 | -1.294 | 0.056 | -2.931 | -2.889 |

After the extraction, 67 strokes (2.85%) were rejected due to software processing errors. Those rejected strokes are referred to as Software rejects. Two criteria were used to reject strokes. First, a SNR lower than 15 dB means that the stroke was not correctly reconstructed by the software. Following this criterion 37 trials (1.57%) were rejected [23 (2.00%) at 1 month and 14 (1.17%) at 3 months post-injury]. Poor reconstruction can be due to a software problem or a lognormality problem. It was not possible with such a small number of samples to clarify the source of the problem. Second, if a $t_0$ is negative and a $\mu$ is positive, the stroke is rejected. 31 trials (1.32%) were rejected according to this criterion [20 (1.74%) at 1 month and 11 (0.92%) at 3 months]. One stroke was rejected for both criteria. There was no significant difference between the two timepoints on the number of strokes rejected for this reason ($p = 0.17$). Two participants had fewer than 15 valid strokes for their 1 month trial block, therefore both participants were excluded from the analyse. Following rejection, 1567 strokes were finally used for the statistical analyse. $D_{tot}$ was calculated as the total amplitude of the movement for each stroke. $D_{tot}$ was calculated as a vector addition of the D of each lognormal. Strokes considered to be too short or too long were counted. This parameter was called $D_{tot\text{-}reject}$.

### 2.4.2. *Sigma-Lognormal parameters*

For each participant, the mean value of each parameter was calculated. The following three parameters describe the global state of the neuromotor system [Plamondon R., 1995a, 1995b; Plamondon *et al.*, 2003; O'Reilly *et al.*, 2013]:

- SNR: The signal-to-noise ratio between the original speed profile and the reconstructed speed profile, computed in decibels (dB). This is a measure of the quality of the sigma-lognormal reconstruction. Higher SNR indicates better reconstructions and a better fitting of the velocity.

- nbLog: The number of lognormal functions required to reconstruct the signal. This parameter represents fluidity of movement by the participant. Higher nbLog, indicates less fluid movements.
- SNR/nbLog: This ratio is used as a performance criterion. The capacity to reconstruct a movement's speed profile with lognormals can be interpreted as an indicator of the quality of motor control since the lognormal speed profile corresponds to perfect motor control [Plamondon *et al.*, 2013]. Higher SNR/nbLog indicates that the movement is closer to ideal lognormal behavior.

Other parameters are an index of the internal functioning of the neuromotor system. The four following central parameters represent the neuromotor action plan:

- $t_0$: The time required (in seconds) for the brain to produce a motor command. $t_0$ should not to be confused with the moment a sound cue is emitted ($t = 0$) nor with the response time (RT) measured by moments of movement onset. $t_0$ is the moment when a population of neurons sends a motor command. In other words, $t_0$ occurs after the start cue, it represents the amount of time that the central nervous system needs to perceive the light cue and prepare the appropriate response command. This time occurrence happens before RT, given that a delay is necessary for the motor command to be transmitted and for enough muscle cells to be recruited to begin the movement effectively [O'Reilly *et al.*, 2013]. Longer $t_0$ are indicative of longer delays before a motor command is emitted and, therefore, longer time to perceive the stimulus and prepare the command. $t_{0init}$ is the first motor command produced by the brain.
- D: The amplitude of the movement associated with each motor command, in millimeters. Greater D indicates greater amplitude of the movement associated with a command.
- $\theta s$: The starting angular direction of each stroke in radians.
- $\theta e$: The ending angular direction of each stroke in radians.

The following parameters are intrinsic characteristics of each lognormal and related to motor program execution [Plamondon *et al.*, 2003; Plamondon and Djioua, 2006]:

- $\mu$: Logtime delay, the time required to reach the middle of movement distance on a logarithmic scale. It can be associated with the general response speed of the neuromuscular system to a command.

- σ: Logresponse time, the response time on a logarithmic scale of the neuromuscular system's response to a command. Sigma (σ) is used to compute an estimate of movement duration.

The following figure explains the impact of each parameter on the lognormal curve shape.

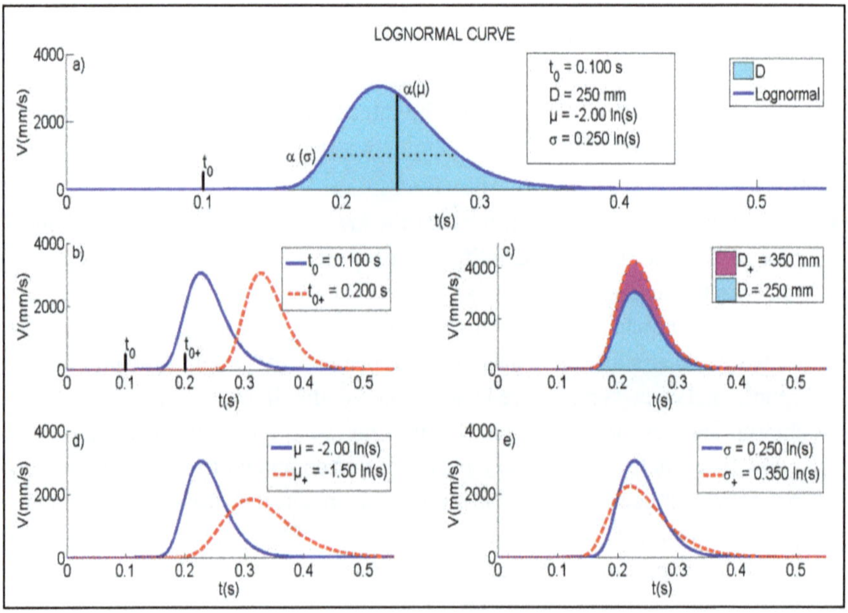


Fig. 3. The impact of each parameter on the lognormal curve shape.

## 2.5. *Statistical analysis*

The analysis was performed for every participant separately. The results at 1 month post-injury were compared to those at 3 months post-injury for each participant. In order to compare the evolution of each individual over time, a t-test for independent groups was applied. Bonferroni correction was calculated using the 8 original parameters. Thus, to be considered significant, the p-value had to be less than 0.00625 ($\alpha = 0.05/8 = 0.00625$).

## 2.6. *Analysis of the evolution of the neuromotor system state*

### 2.6.1. *Evolution of the neuromotor system state according to the sigma-lognormal model*

After statistical analyses, the results were examined to determine if the state of the neuromotor system improved or deteriorated. The most discriminating parameters of the Sigma-Lognormal model were used to classify the subjects according to the evolution of their state. To this end, the participants were classified into four categories: "better, worse, stable, and undetermined". In order to improve classification, another parameter was added. This parameter is the number of strokes that did not respect the distance restriction.

The classification was done using the evolution of each parameter over the 3 month time laps for each subject. To determine if the change in parameters is evolving positively, there has to be a significant statistical difference between the 1- and 3- month mean post-injury. Participants were considered as having an improved neuromotor system state when there was a decrease in the mean value of the following parameters: $t_0$, $t_{0init}$, $\sigma$, rejectD and nbLog. Similarly, participants had a better neuromotor system state after 3 months post-injury if there was an increase of the mean value of the following parameters: $\mu$, SNR and SNR/nbLog. In this classification, more weight is attributed to the parameters related to the global state of the neuromotor system. Based on previous studies, those parameters are most likely to discriminate between clinical and control groups [Plamondon *et al.*, 2013; Duval *et al.*, 2015; Van Gemmert *et al.*, 2013].

### 2.6.2. *Evolution of the neuromotor system state according to the PCSI*

To measure the somatic, cognitive, emotional, and fatigue/sleep-related symptoms present in each participant, parents were asked to complete the Post-Concussion Symptom Inventory [PCSI, Sady, Vaughan, and Gioia, 2014] to document their child's symptoms. The parent-report version of the PCSI consists of 26 items, where responses are rated for severity on a 7-point Likert scale (i.e., from 0 to 6; 0 = Not a problem, 6 = Severe problem).

As part of the analyses, we examined how the number of reported symptoms, as endorsed on the parent PCSI, progressed over time. Specifically, the total score obtained on the parent PCSI (i.e., total number of symptoms) for each participant was compared at 1-month and 3-months post-injury. Second, change in reports of post-concussive symptoms (PCS) over time were examined in comparison to changes in neuromuscular parameters at 1-month and 3-months post-injury. To this end, the participants were classified into four categories: "better, worse, stable, and undetermined".

## 3. Results

### 3.1. *Results of the analysis of the sigma-lognormal parameters*

Of the 32 valid participants, 29 had significant differences for at least one parameter when comparing their result at 1- and 3-months post-injury. The other 3 children did not show any significant differences between their parameters. Table 2 presents the number of participants having a significant difference by the most discriminant parameters between the 1-month and 3-months results post-injury. There were seventeen (53%) participants who had a significant difference, with the Bonferroni correction, for the SNR/nbLog parameter. As shown in Table 2, eleven participants (34%) had a significant difference between their 1- and 3-months results for the total amplitude of movement ($D_{tot}$). With regards to the number of lognormals (nbLog), twelve participants (38%) displayed a significant difference between their 1- and 3-months results post-injury. These findings suggest that the quality of the performance on the stroke task evolves over time. That said, further analysis was needed to determine if participants tend to get closer to or further from Lognormality; these results are reported below.

Table 2. Number of participants with a significant difference for the most discriminant parameters between the 1-month and 3-months results post-injury.

| **Parameters** | **$D_{tot}$** | **nbLog** | **SNR/nbLog** |
|---|---|---|---|
| **Number of participants** | 11 | 12 | 17 |
| **ratio** | **34%** | **38%** | **53%** |

### 3.2. *Results of the analysis of the evolution of the neuromotor system state according to Sigma-Lognormal model*

Of the 32 participants, 10 (31%) exhibited improvement of their neuromotor system state after 3 months. 17 (53%) participants showed a deterioration of their neuromotor system state during that laps of time. Four (13%) revealed a stabilization of the neuromotor system state over the first three months. For one (3%) participant, it was not possible to make a decision and he was classified as undetermined regarding the evolution of the neuromotor system state, because improvement was found for some parameters and deterioration for others. The results are presented in Table 3.

Table 3. Distribution of the evolution of the neuromotor system state according to the results of the Sigma-Lognormal analysis.

| **Evolution of the state** | **Number of subjects** | **% subjects** |
|---|---|---|
| Better | 10 | 31% |
| Worse | 17 | 53% |
| Stable | 4 | 13% |
| Undetermined | 1 | 3% |

### 3.3. *Results of the analysis of the evolution of the neuromotor system state in relation to the PCSI questionnaire*

Of the 32 parents who completed the PCSI questionnaire, 14 (44%) reported improvement in post-concussive symptoms (PCS). Conversely, three (9%) reported a deterioration of symptoms over time. There was no change in the number of symptoms seen over time for 11 (34%) participants. Finally, there was no data for four (13%) participants. Analysis of the evolution of post-concussive symptoms as measured on the parent-rated PCSI are presented in Table 4.

Table 4. Distribution of the evolution of the neuromotor system state according to the PCSI questionnaire.

| | **PCSI results** | |
|---|---|---|
| **Evolution of the state** | Number of subjects | % subjects |
| Better | 14 | 44% |
| Worse | 3 | 9% |
| Stable | 11 | 34% |
| No data | 4 | 13% |

To compare the PCSI total score results to the results of the Sigma-Lognormal analysis, only the 28 participants who had data for both 1- and 3-months timepoints were used. From those 28 individuals:

- There was complete concordance between the results of the Sigma-Lognormal analysis and the results of the PCSI questionnaire for nine (32%) subjects. From those 9 participants:

  - The results of the questionnaire and the Sigma-Lognormal analysis suggest that there was an improvement in the neuromotor system state and PCSI for six participants.
  - The results of the questionnaire and the Sigma-Lognormal analysis indicate that there was a deterioration of the neuromotor system state and PCSI for two participants.
  - For the remaining participant, both PCSI and Sigma-Lognormal results remained stable over time.
- From the nineteen (68%) remaining participants, there was a discordance between Sigma-Lognormal results and PCSI total scores for seven (37%) participants.
  - According to the results of the PCSI questionnaire, there was a stabilization of PCSI over time for nine (47%) participants, but the results of the Sigma-Lognormal analysis stipulated a change (improvement or deterioration) of the neuromotor state during the same lapse of time.
  - According to the results of the PCSI questionnaire, there was a change (improvement or deterioration) of total number of PCSI over time for two (11%) participants, but the results of the Sigma-Lognormal analysis demonstrated a stabilization of the neuromotor state during the same lapse of time for that participant.
  - No decision could be made according to the results of the Sigma-Lognormal analysis for one (5%) participant, but the PCSI questionnaire stipulates a stabilization of PCSI over time.

## 4. Discussion

The primary aim of this preliminary pilot study was to examine fine motor control parameters in children and adolescents who sustained mTBI. Consistent with initial study expectations, the majority of children in the sample showed reduced fine motor control between 1-month and 3-months post-injury, as reflected by changes in the Sigma-Lognormal parameters. These findings are consistent with previous studies that have shown that pediatric mTBI may be associated with fine motor deficits several months post-injury, such as reduced fine motor speed and dexterity and grapho-motor processing speed. Future research should seek to examine the evolution of the state of the neuromotor system according to the Sigma-Lognormal model in relation to other fine motor tests used by the health care professionals in clinical settings.

By comparing the results of the Sigma-Lognormal analysis and PCSI answers, a concordance was observed between the two tests for only nine (32%)

of the 28 participants. This is not surprising as the PCSI questionnaire and Sigma-Lognormal parameters assess different aspects of functioning, and thus, are likely to yield different patterns of results. First, post-concussive symptoms were documented using subjective parents' reports on questionnaire, and it has previously been suggested that parents of children with mTBI may tend to over-report symptoms on questionnaires [Bernard *et al.*, 2016]. Second, the sigma-lognormal analysis is an objective methodology based on the lognormality principle. It could be used to compare with subjective reports and provide unbiased data to confirm or infirm the evolution of the mTBI. Nevertheless, the majority of participants showed stable or improved PCS over time, which is consistent with previous research studies suggesting that most children following a mTBI tend to have favourable outcomes. However, for a minority of individuals, there was a worsening of PCS reported, which may also suggest that a smaller subset of participants may actually have a more protracted recovery following mTBI. As this was a preliminary study, additional research is required in order to examine associations between objective neurocognitive measures and Sigma-Lognormal parameters with regards to recovery following mTBI.

The major limitation of this preliminary study is the rejection of the 56 children due to the lack of compliance to instructions. Specifically, as many participants did not respect the instructions (i.e., stroke should end and be held on the tablet for 1 second), the final velocity was not null and the reconstruction with the Sigma-Lognormal model was not possible. From a qualitative perspective, a large number of participants with mTBI appeared to display a tendency to incorrectly perform the task despite the instructions provided. Although speculative, this may be suggestive of executive dysfunction, such as impulsivity. Problems with executive control are not uncommon after mTBI [Niogi *et al.*, 2008], and should be taken into account when interpreting the results in future studies, as they may be indicative of subtle neurocognitive and neuromuscular deficits.

From a technology point of view, future studies should add a detector to the homemade software Sign@medic, to monitor if the pen is lifted at the end of the stroke and the velocity is null. This solution would serve to counter the lack of compliance to instruction observed in this study. The second solution is to ensure that research assistants are extensively trained to detect incorrect execution of the stroke and follow a strict standardized protocol. Moreover, to better understand and examine the clinical value of the sigma-lognormal parameters, it will be essential to include a matched control group and explore group differences. This control group could consist of typically developing children or children with orthopedic injuries. The pilot data are promising enough to justify future research using stronger methodologies (e.g., control group, additional cognitive and motor evaluation measures, larger sample, narrower age bands, and more advanced statistical analyses will be carried out) to more clearly characterize the process by which fine motor changes occur. Lastly, the preliminary findings provide interesting results regarding the integrity of neuromotor systems following mTBI, and it may become useful to add the objective Sigma-Lognormal analysis of the fast stroke task to screening assessments in order to investigate neuromotor and neuropsychological outcomes after mTBI.

## Acknowledgements

Portions of this work were funded by the Canadian Institutes of Health Research and the Fonds de Recherche du Québec – Santé. I.G. is supported by a clinical investigator award from the Fonds de Recherche du Québec – Santé. MB is supported by a Canada Research Chair from the Canadian Institutes of Health Research. This work was partly supported by the NSERC-Canada Grant RGPIN-2015-06409 to R. Plamondon.

## References

Babikian, T. and Aasarnow, R. (2009). Neurocognitive outcomes and recovery after pediatric TBI: meta-analytic review of the literature. *Neuropsychology*, *23*(3), 283.

Barlow, K. M. (2016). Postconcussion syndrome: a review. *Journal of Child Neurology*, *31*(1), 57–67.

Bernard, C. O., Ponsford, J. A., McKinlay, A., McKenzie, D. and Krieser, D. (2016). Predictors of post-concussive symptoms in young children: injury versus non-injury related factors. *Journal of the International Neuropsychological Society*, *22*(8), 793–803.

Brain Traumatic (1999). Rehabilitation of persons with traumatic brain injury. *JAMA*, *282*(10), 974–983.

Brosseau-Lachaine, O., Gagnon, I., Forget, R. and Faubert, J. (2008). Mild traumatic brain injury induces prolonged visual processing deficits in children. *Brain Injury*, *22*(9), 657–668.

Duval, T., Rémi, C., Plamondon, R., Vaillant, J. and O'Reilly, C. (2015). Combining sigma-lognormal modeling and classical features for analyzing graphomotor performances in kindergarten children. *Human Movement Science*, *43*, 183–200.

Duval, T., Rémi, C., Prevost, L., Dorville, A. and Plamondon, R. (2014). Etude de la faisabilité de l'évaluation de l'efficacité des mouvements de tracé du jeune apprenti-scripteur. *IHM'14, 26e conférence francophone sur l'Interaction Homme-Machine*. Lile, France.

Gagnon, I., Friedman, D., Beauchamp, M. H., Christie, B., DeMatteo, C., Macartney, G. and Zemek, R. (2018). The Canadian Pediatric Mild Traumatic Brain Injury Common Data Elements Project: Harmonizing Outcomes to Increase Understanding of Pediatric Concussion. *Journal of Neurotrauma, 35*(16), 1849–1857.

Gagnon, I., Friedman, D., Swaine, B. and Forget, R. (2001). Balance findings in a child before and after a mild head injury. *Journal of Head Trauma Rehabilitation, 16*(6), 595–602.

Gagnon, I., Friedman, D., Swaine, B. and Forget, R. (2004). Children show decreased dynamic balance after mild traumatic brain injury1. *Archives of Physical Medicine and Rehabilitation, 85*(3), 444–452.

Gagnon, I., Swaine, B., Friedman, D. and Forget, R. (2004). Visuomotor response time in children with a mild traumatic brain injury. *Journal of Head Trauma Rehabilitation, 19*(5), 391–404.

Giza, C. C., Mink, R. B. and Madikians, A. (2007). Pediatric traumatic brain injury: not just little adults. *Current Opinion in Critical Care, 13*(2), 143–152.

McCrory, P. M., Dvorak, J., Aubry, M., Bailes, J., Broglio, S. and Davis, G. A. (2017). Consensus statement on concussion in sport. *The 5th International Conference on Concussion in Sport, 51*, pp. 838–847. Berlin.

Mccrory, P., Johnston, K., Meewisse, W. *et al.* (2005). Summary and agreement statement of the 2nd International Conference on Concussion in Sport, Prague 2004. *British Journal of Sports Medicine, 39*(1), i78–i86.

Moran, L. M., Taylor, H. G., Rusin, J., Bangert, B., Dietrich, A., Nuss, K. and Yeates, K. (2011). Quality of life in pediatric mild traumatic brain injury and its relationship to postconcussive symptoms. *Journal of Pediatric Psychology, 37*(7), 736–744.

National Center for Injury Prevention and Control (2003). *Report to Congress on mild traumatic brain injury in the United States: steps to prevent a serious public health problem*. Atlanta, GA, USA: Centers for Disease Control and Prevention.

Neurological disorders: public health challenges: World Health Organization (2006).

Niogi, S., Mukherjee, P., Ghajar, J., Johnson, C., Kolster, R., Sarkar, R. and McCandliss, B. (2008). Extent of microstructural white matter injury in postconcussive syndrome correlates with impaired cognitive reaction time: a 3T diffusion tensor imaging study of mild traumatic brain injury. *American Journal of Neuroradiology, 29*(5), 967–973.

O'Reilly, C., Plamondon, R. and Lebrun, L.-H. (2014). Linking brain stroke risk factors to human movement features for the development of preventive tools. *Frontiers in Aging Neuroscience, 6*(3), 150. doi:10.3389/fnagi.2014.00150.

O'Reilly, Plamondon, R., Landou, M. K. and Stemmer, B. (2013). Using kinematic analysis of movement to predict the time occurrence of an evoked potential associated with a motor command. *European Journal of Neuroscience, 32*(2), 173–180.

O'Reilly, Plamondon, R., Lebrun, L.-H., Clément, B. and Mathieu, P. (2009). Sigma-lognormal analysis of a complex movements neuromuscular test. *14th Conference of the International Graphonomics Society*. Dijon, France.

O'Reilly, C. and Plamondon, R. (2009a). Development of a Sigma-Lognormal representation for on-line signatures. *Pattern Recognition*, *42*(12), 3324–3337.

Plamondon, R. (1995a). A kinematic theory of rapid human movements: Part I: Movement representation and generation. *Biological Cybernetics*, *72*(4), 295–307. Retrieved 11 12, 2017, from https://link.springer.com/article/10.1007/bf00199059.

Plamondon, R. (1995b). A kinematic theory of rapid human movements. Part II. Movement time and control. *Biological Cybernetics*, *72*(4), 309–320. Retrieved 11 12, 2017, from https://ncbi.nlm.nih.gov/pubmed/7748960.

Plamondon, R. (1998). A kinematic theory of rapid human movements: Part III. Kinetic outcomes. *Biological Cybernetics*, *78*(2), 133–145.

Plamondon, R. and Djioua, M. (2006). A multi-level representation paradigm for handwriting stroke generation. *Human Movement Science*, *25*(4), 586–607.

Plamondon, R., Feng, C. and Woch, A. (2003). A kinematic theory of rapid human movement. Part IV: a formal mathematical proof and new insights. *Biological Cybernetics*, *89*(2), 126–138. Retrieved 11 12, 2017, from https://link.springer.com/article/10.1007/s00422-003-0407-9.

Plamondon, R., O'Reilly, C., Rémi, C. and Duval, T. (2013). The lognormal handwriter: learning, performing, and declining. *Frontiers in Psychology*, *4*.

Plamondon, R., Yu, L.-d., Stelmach, G. E. and Clément, B. (1991). On the automatic extraction of biomechanical information from handwriting signals. *IEEE Transactions on Systems, Man, and Cybernetics*, *21*(1), 90–101.

Report to Congress on mild traumatic brain injury in the United States: Steps to prevent a serious public health problem (2003).

Rhine, T., Quatman-Yates, C. and Clark, R. A. (2017). A longitudinal examination of postural impairments in children with mild traumatic brain injury: implications for acute testing. *Journal of Head Trauma Rehabilitation*, *32*(2), E18–E23.

Van Gemmert, A., Plamondon, R. and O'Reilly, C. (2013). Using the Sigmalognormal model to investigate handwriting of individuals with Parkinson's disease. *Proceedings 16th Biennial Conf. of the Graphonomics Society*, (pp. 119–122). Nara, Japan.

World Health Organization (2006). *Neurological disorders: public health challenges*. Geneva, Switzerland: World Health Organization.

Zemek, R., Osmond, M. H. and Barrowman, N. (2013). Predicting and preventing postconcussive problems in paediatrics (5P) study: protocol for a prospective multicentre clinical prediction rule derivation study in children with concussion. *BMJ open*, *3*(8), p. e003550.

https://doi.org/10.1142/9789811226830_0007

# Chapter 7

# Evaluating Parkinson's Disease in Voice and Handwriting Using the Same Methodology

Cristina Carmona-Duarte[1,*], Miguel A. Ferrer[1,†], Pedro Gómez-Vilda[2,‡], Arend W. A. Van Gemmert[3,§] and Réjean Plamondon[4,**]

[1]*Instituto Universitario para el Desarrollo Tecnológico y la Innovación en Comunicaciones, University of Las Palmas de G.C. Las Palmas de G.C., Las Palmas, Spain*

[2]*Neuromorphic Speech Processing Lab, Center for Biomedical Engineering, Universidad Politécnica de Madrid, Madrid, Spain*

[3]*Fine Motor Control & Learning Laboratory School of Kinesiology, Louisiana State University, Baton Rouge, LA 70803, USA*

[4]*Laboratoire Scribens, Département de génie électriquee, Polytechnique Montréal, Montréal, Canada*

[*]*ccarmona@idetic.eu*, [†]*miguelangel.ferrer@ulpgc.es*, [‡]*pedro@fi.upm.es*, [§]*gemmert@lsu.ede*, [**]*rejean.plamondon@polymtl.ca*

Parkinson's disease is manifested as well in handwriting as in voice. Different procedures have been used in previous research to estimate the dysfunctions of the illness in voice and handwriting. This paper proposes one parameter to evaluate the influence of the illness on both voice and handwriting as the symptoms affecting these two behaviors have a common origin. Specifically, the parameter proposed is based on the Kinematic Theory of rapid human movements. It allows to quantify the deficits caused by Parkinson's disease in handwriting and voice. The velocity profile obtained to characterize voice between the first and second formant is computed by a spatio-temporal approximation. In handwriting, the velocity profile is obtained from the sampled positions of the pen on a digitizer tablet. Once the velocity profile is derived, it is transformed to fit into the lognormal model in which similarities between voice and handwriting have been found for the performance of these tasks by Parkinson's patients. The experiments with different databases of voice and handwriting recorded from different patients in different labs display positive results pointing out a common parameter that is affected by the two types of communication modalities.

## 1. Introduction

Parkinson's disease (PD) is a neurodegenerative disease which has symptoms that manifest in deficiencies affecting both handwriting and voice. PD symptoms are the result of a dopaminergic deficiency characterized by the presence of two or more cardinal motor symptoms (i.e., bradykinesia, rest tremor, rigidity, and postural disturbances) [Tykalova *et al.*, 2014]. PD is a slow progressive disease with a long duration where clinical treatment and rehabilitation can help to improve the quality of life. Therefore, an early diagnosis and a continuous monitoring of the effects of treatments are important. PD has been monitored by recording handwriting and processing its signals and recording the voice and processing its signals.

In handwriting changes of the kinematic aspects of movements and the analysis of in-air movements have been proposed as useful methodologies to monitor and diagnose early the disease [Eichhorn and Gasser, 1996; Walton, 1997; Broderick *et al.*, 2009; Drotár *et al.*, 2014].

On the other hand, in voice processing there have been recent studies about the evaluation of the voice of individuals with Parkinson's disease using the variability of the pitch, the voice rate and pausing [Skodda and Schlegel, 2008; Rusz, Cmejla *et al.*, 2011; Rusz, Čmejla *et al.*, 2011; Galaz *et al.*, 2015].

In the current paper, we propose using a common methodology to analyze handwriting and voice. As both voice and handwriting are complex tasks involving the neurological and muscular systems, in which the muscular system is synchronized to communicate an idea through sentences, words, and letters. When a person writes on a Wacom or any tablet device that allows capturing the temporal position of the pen during handwriting, the velocity information can be obtained and analyzed. To this end, the Kinematic Theory of rapid human movements [Plamondon, 1995] is applied to divide a complex movement into simple movements (strokes), each one is modelled by a sigma-lognormal function and the complex movement is then the vector summation of all the parametrized sigma-lognormals. This theory has been applied in different fields to assess movements as it pertains to handwriting variations across time [Gomez-barrero *et al.*, 2013; Plamondon *et al.*, 2013]. Thus this model has allowed to specify screening systems for neuromuscular disorders [O'Reilly and Plamondon, 2012; Van Gemmert, Plamondon and O'Reilly, 2013] and the assessment of risk factors for stroke risk [Plamondon, O'Reilly and Ouellet-Plamondon, 2014].

In voice production, the resonating cavities modifiable by the articulatory organs allow the energy of the voice signal to be concentrated at certain frequencies (formants), due to oropharyngeal tract resonators. It is well known that the formants are related to the tongue-jaw reference center (TJRC) [Gómez-Vilda *et al.*, 2015]. Also, the TJRC is related to the first and second formant [Peterson and Barney, 1952]. In recent studies [Carmona-Duarte, Alonso *et al.*, 2016; Carmona-Duarte, Plamondon *et al.*, 2016; Carmona-Duarte *et al.*, 2017] a relationship between the formants and the lognormal model have been demonstrated.

The present work is intended to compare voice and handwriting production and the derived velocity signals as both tasks can be captured by a common parameter, which can be validated to detect Parkinson disease. This validation of a common parameter in both domains reinforces earlier findings that could assist, in the long term, to develop a more reliable diagnosis.

## 2. From Voice or Handwriting to Velocity

In order to get the kinematic signal from voice, formants are calculated as it is shown in Fig. 1. The formant estimation is obtained by adaptive inverse filtering using the software BioMetroLing presented in ['http://www.glottex.com/'; Gómez-Vilda *et al.*, 2013] (Fig. 1). The first (F1) and second (F2) formants correspond to the two first maximum values in the LPC spectrum.

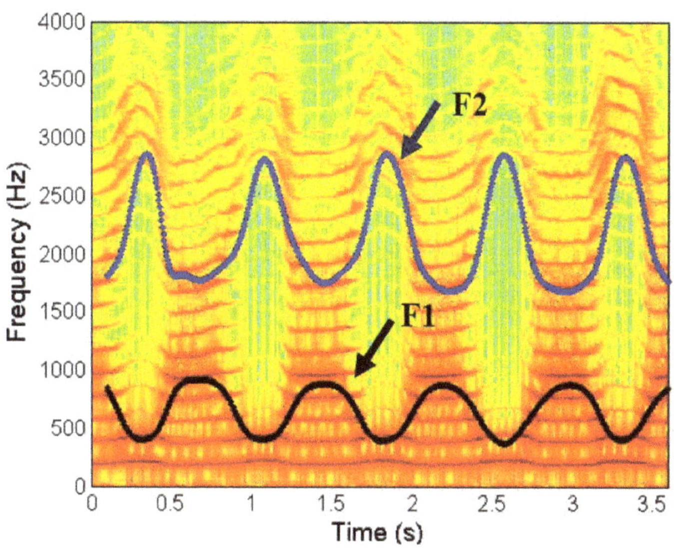


Fig. 1. From formants to velocity. Spectrogram and formants.

As it is well known [Peterson and Barney, 1952], the first formant $F1$ is related to the vertical movement of the TJRC and the second formant $F2$ is related to its horizontal movement as it is shown in Fig. 2 and Fig. 3. These movements can be correlated with the formants positions in the plane $F1$ vs $F2$ [Gómez-Vilda *et al.*, 2015] as:

$$\begin{bmatrix} \Delta y \\ \Delta x \end{bmatrix} = \begin{bmatrix} c_{11} & c_{12} \\ c_{21} & c_{22} \end{bmatrix} \begin{bmatrix} F1 \\ F2 \end{bmatrix} \tag{1}$$

where $\Delta x$ and $\Delta y$ are the relative displacements from the previous position of the TJRC. $c_{ij}$ are the weights of the combination matrix. If we assume that first and second formant are perpendicular movements [Peterson and Barney, 1952], as we can see in Fig. 1 and 2, c21 and c12 can be approximated to zero.

Once the displacement is calculated (Fig. 2), the velocity signal $v(t)$ (Fig. 4) is estimated as:

$$v_{ov}(t) = \sqrt{\left(\frac{c_{11}\Delta F_1)}{\Delta t}\right)^2 + \left(\frac{c_{22}\Delta F_2}{\Delta t}\right)^2} \tag{2}$$

In the case of handwriting, the patient is asked to write down on a tablet, in our case a WACOM digitizer that records the position $x(t)$ and $y(t)$ of the pen on the tablet 200 times per second. As a result, the pen velocity can be worked out as:

$$v_{oh}(t) = \sqrt{\left(\frac{\mathrm{x(t)-x(t-1))}}{\Delta t}\right)^2 + \left(\frac{\mathrm{y(t)-y(t-1)}}{\Delta t}\right)^2} \tag{3}$$

Both voice signals are expressed in mm/s.

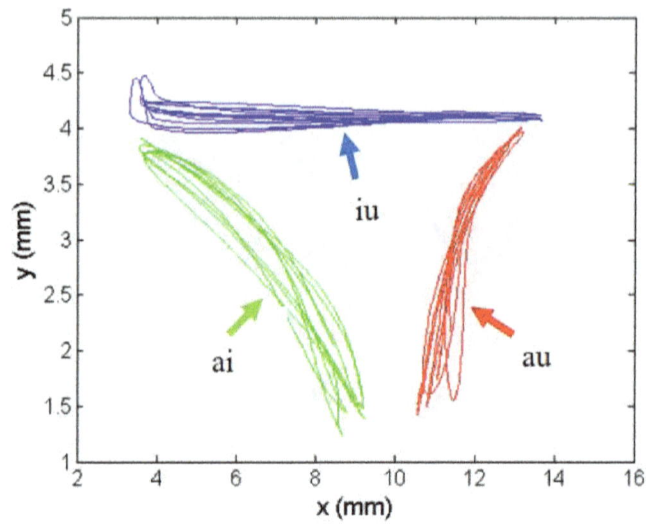


Fig. 2. Space representation.

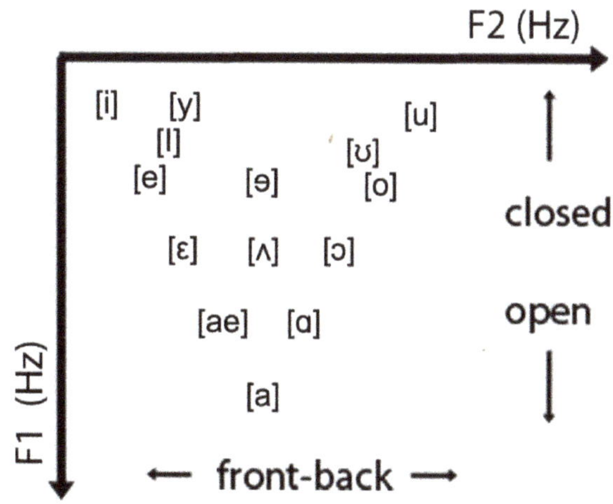


Fig. 3. Vowel representation space.

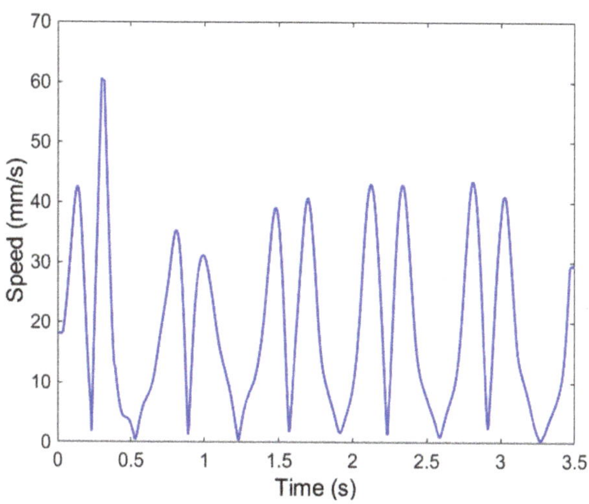


Fig. 4. Velocity profile.

## 3. From Velocity to Sigma-Lognormal Model

The velocity profile of both voice and handwriting signal can be parameterized using the Kinematic Theory of rapid movements framework. In this way, the amplitude of the velocity profile $v_{ov}(t)$ or $v_{oh}(t)$ can be modeled as a linear combination of *Nblog* lognormals $v_r(t)$ [O'Reilly and Plamondon, 2009] as follows:

$$v_r(t) = \sum_{j=1}^{Nblog} v_{rj}(t; D_j, \tau_{0j}, \mu_j, \sigma_j^2) \quad (4)$$

being $v_r(t) = v_{rv}(t)$ for voice and $v_r(t) = v_{rh}(t)$ for handwriting, where the velocity profile of each stroke $v_{rj}(t)$ is defined as:

$$v_{rj}(t; \tau_j, \mu_j, \sigma_j^2) = \frac{D_j}{\sigma_j\sqrt{2\pi}(t-\tau_{0j})} exp\left\{-\frac{[ln(t-\tau_j)-\mu_j]^2}{2\sigma_j^2}\right\} \quad (5)$$

with $t$ the time basis, $\tau_{0j}$ the time of stroke occurrence, $D_j$ the amplitude of each stroke, $\mu_j$ the stroke time delay and $\sigma_j$ the stroke response time, both on a logarithmic time scale.

The error between $v_{rv}(t)$ and $v_{oh}(t)$ or $v_{rh}(t)$ and $v_{ov}(t)$ is then evaluated using the Signal-to-Noise-Ratio (SNR) between both. Specifically, the SNR is defined as in O'Reilly and Plamondon [2012]:

$$SNR = 10 \log \left(\frac{\sum_1^{Nblog} v_o^2(t)\Delta t}{\sum_1^{Nlog} |v_o(t) - v_r(t)|^2 \Delta t}\right) \quad (6)$$

being $v_o$ the original velocity: $v_{oh}(t)$ or $v_{ov}(t)$.

Based on the facts that Parkinson patients perform shorter movements and have some difficulties to plan next movements, among all of the expected effect of the disease on the lognormal decomposition a lower stroke logresponse time $\sigma_j$ has been observed [Broderick *et al.*, 2009; Carmona-Duarte, Plamondon *et al.*, 2016]. Therefore, in this study, we propose as a common parameter to characterize Parkinson disease in voice and handwriting computed from $\sigma_j$: the averaged value of the mean stroke logresponse time sequential differences $\overline{\Delta\sigma}$ as defined in Eq. (7):

$$\Delta\bar{\sigma} = \frac{\sum_{j=1}^{Nlog} |\sigma_j - \sigma_{j-1}|}{\text{Nblog}} \quad (7)$$

It is expected that people with some degree of Parkinson's disease will show a lower $\overline{\Delta\sigma}$ than healthy people both in speech and handwritten signals.

## 4. Method

### 4.1. *Voice database*

A database comprising phonations from five PD patients selected by neurologists and five control subjects was used in this study. The subjects, aged from 52 to 78 years old, were diagnosed with PD as grade 2 or 3. Each subject was asked to utter the vowel /a/ at the same normal loudness and their most natural way (modal phonation). Each sample of the database comprises the recording of the vowel /a/

from the Spanish vowel set ([a] from the International Phonetic Alphabet) uttered in a sustained way, each utterance lasting approximately two seconds, separated by silences from repeated utterances. All patients were tested on medication. The voice records were taken by a hand recorder at 16 KHz and 16 bits, in the neurologists' office. They were segmented automatically by an energy-based method and the central part of the phonation selected, avoiding the initial and final transients.

### 4.2. *Handwriting database*

A database comprising of five individuals with Parkinson's disease and seven age-matched controls were used in this study. All patients were tested on medication. All participants had written six loops that progressed to the right (i.e., cursive connected '*llllll*') with an electronic pen. The participants were instructed to match the size of the cursive 'l's which were displayed before each trial. After each trial the performance of the participant was displayed between two lines, which were 25 mm (i.e., 2.5 cm) apart. So the participants could see if she/he matched the required size. Each participant did repeat the writing task 8 times, in which they tried to match the 25 mm size requirement.

The data were automatically segmented with a custom made segmentation procedure that searched for the first zero crossing in vertical velocity after the first full loop. Thereafter searched for the first zero crossing in vertical velocity after the fourth full loop, i.e., each segment consisted of three loops per repetition (i.e., the connected second, third, and fourth loops of each trial).

### 4.3. *Signal processing*

To process the voice production samples, first the central part of the signal was isolated to avoid the transients, where the speed is higher in Parkinson's patients [Carmona-Duarte *et al.*, 2017]. Second, the signal was down-sampled, to get a sampling frequency of 8 KHz. Third, to obtain the better SNR in the reconstructed signal, the computation processing has been carried out with a window length of 15 Hz with a displacement of 0.001 seconds and using an 8-order prediction-error lattice-ladder filter [Deller, Hansen and J. G. Proakis, 2000]. Finally, the first and second formants were identified, after which the velocity was calculated as explained in Sec. 2.

Once the velocity signals of the voice and the handwriting samples were obtained, the procedure to estimate the velocities was the same, i.e., $v_{ov}(t)$ and $v_{oh}(t)$. The position signal data were fed into the sigma-lognomal estimator, after the signals were filtered with a cutoff frequency of 7 Hz. The lognormal parameters were calculated using the low-pass filtered velocities as explained in Sec. 3. A new analytical signal was reconstructed from the estimated parameters as it is shown in Fig. 3 and Fig. 4.

## 5. Results

### 5.1. *Qualitative comparison of Parkinson and normal voice and handwriting profiles*

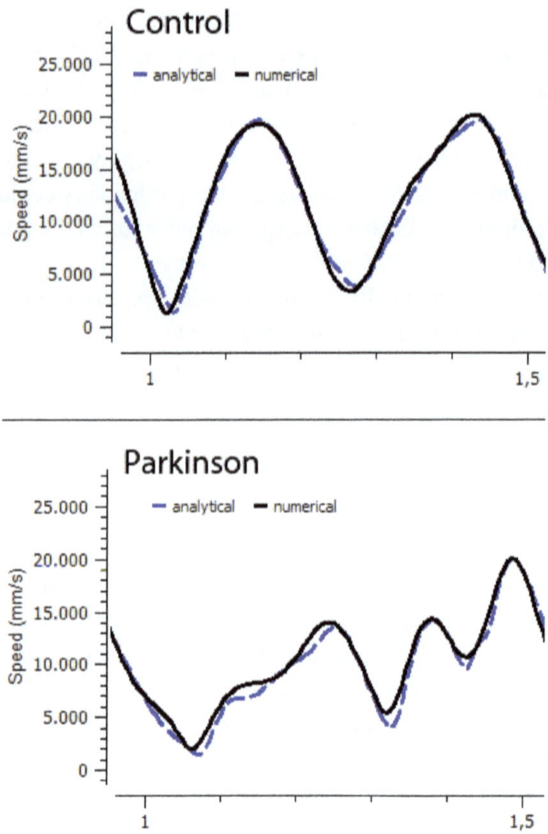


Fig. 5. Speed profile of the handwriting of a healthy control (upper panel) and a PD patient (lower panel). Original speed profile (continuous black line) and the reconstructed (discontinuous blue line).

As a first step, we compared the low-pass filtered velocity profile (original) with the analytical one. As can be observed in Fig. 5, the analytical and the original look similar. The SNR between the original and analytical signals is 21.7 dB in PD and 25.38 dB in control, which reflects a good reconstruction quality in the sigma-lognormal domain [O'Reilly and Plamondon, 2009].

If we compare the handwriting signal (Fig. 5) with the voice signal (Fig. 6), we can observe that they are similar but the time between peaks is longer in handwriting (i.e. 0.35 s) than in voice (0.1 s). This could be due to the requirement of keeping the tongue position at a constant target to utter the vowel /a/. Instead, in handwriting, where simple movements are joined, the movement is fast and longer. The overlapping of two consecutives lognormals depends on the initial time ($t_0$) of each lognomal and on the stroke logresponse time ($\sigma$), therefore, the longer is the movement, the longer the stroke logresponse time will be, and there will be more overlapping between lognormals, as it is explained in [Cristina Carmona-Duarte *et al.*, 2017].

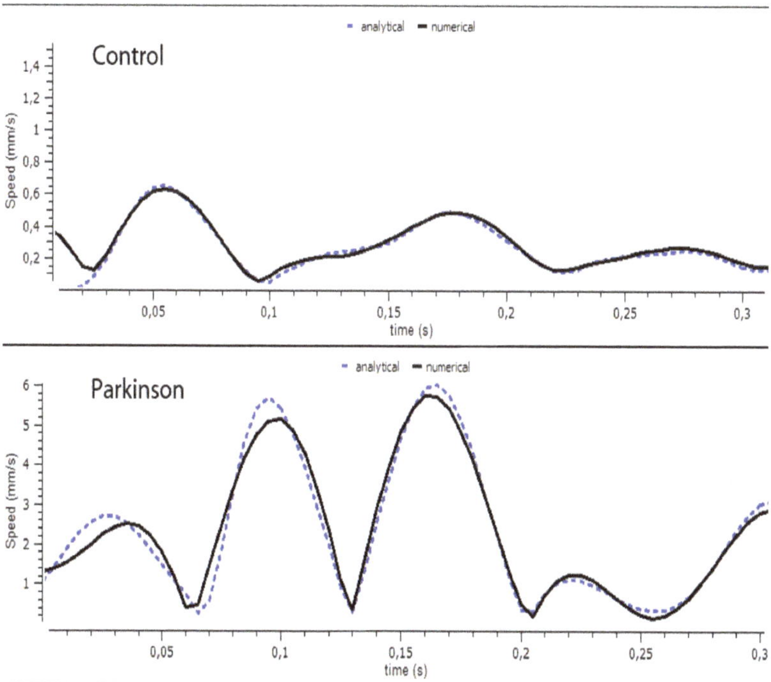


Fig. 6. Speed profile of the voice of a healthy control (upper panel) and a PD patient (lower panel). Original speed profile (continuous black line) and the reconstructed (discontinuous blue line).

Comparing speed profiles of a control (Fig. 5 and Fig. 6 upper panels) with a PD (Fig. 5 and Fig. 6 lower panels), one can observe how the velocity signal of the PD has more and shorter peaks, while it is also noticeable that the speed profiles are more variable in the PD. When inspecting the voice signal it can also be seen that the speed is higher for the PD. This higher speed could be due to the disability of PDs to keep the tongue in the same position.

When comparing the spatial trace of the control and the PD (Fig. 7), one can observe that the healthy control draws continuous lines that are more curved, while the writing of the Parkinson's patient is less curved and has some points where the pen is paused. A similar effect is observed in the trajectory of the TJRC for voice (Fig. 8). The movement is wider in the case of PD than in control due to the difficulties to keep the same position.

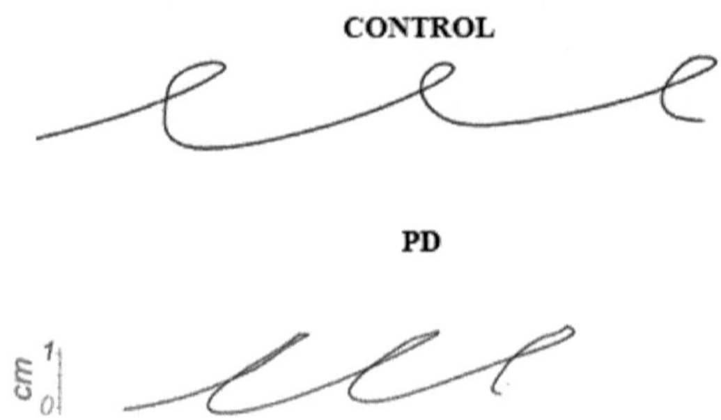


Fig. 7. Handwriting of a representative healthy control (upper panel) and a PD patient (lower panel).

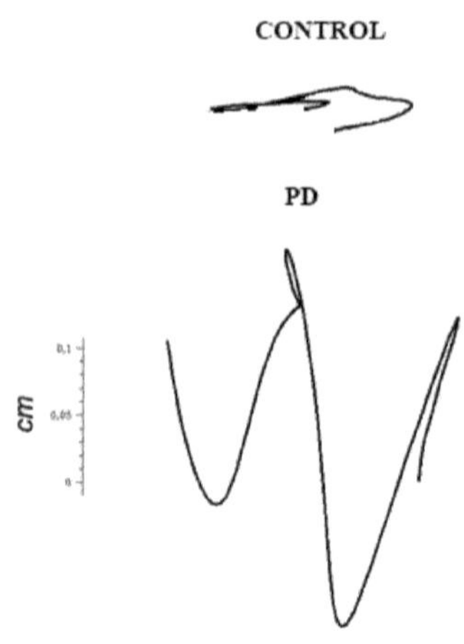


Fig. 8. Voice articulation trajectory representation from formants of a healthy control (upper panel) and a PD patient (lower panel).

### 5.2. *Quantitative comparison of voice and handwriting*

To evaluate whether a statistical difference exists between the $\overline{\Delta\sigma}$ values of controls and PDs, an ANOVA (Analysis of Variance) was performed (using the statistical toolbox of Matlab). The two groups were considered different when the residual *p-value* is close to 0 and statistically similar if the *p-value* is greater than 0.05 [Hogg and Ledolter, 1987]. The parameters obtained from voice and handwriting across the two databases and their respective *p-value* are given in Table 1, Table 2.

Table 1. Average Parameters: Normal vs. Parkinson Voices.

| | Control | PD | p-value |
|---|---|---|---|
| $\overline{\Delta\sigma}$ | **0.2** | **0.08** | **0.005** |
| Nblog/s | 18 | 20 | 0.6 |
| SNR (dB) | 24 | 20 | 0.53 |
| $\overline{D}$ | 0.03 | 0.57 | 0.22 |
| $\overline{\mu}$ | -1.79 | -1.89 | 0.6456 |
| $\overline{\sigma}$ | 0.2 | 0.15 | 0.19 |

Table 2. Average Parameters: Normal vs. Parkinson handwriting.

| | Control | PD | p-value |
|---|---|---|---|
| $\overline{\Delta\sigma}$ | **0.32** | **0.23** | **0.04** |
| **Nblog/s** | **6** | **8** | **0.015** |
| SNR (dB) | 24 | 25 | 0.7 |
| $\overline{D}$ | 13 | 10 | 0.16 |
| $\overline{\mu}$ | -1.44 | -1.6 | 0.4 |
| $\overline{\sigma}$ | 0.32 | 0.28 | 0.16 |

As observed, the number of lognormals in both voice and handwriting are smaller for healthy controls than PD patients. The difference is greater and statistically significant in handwriting, where the *p-value* estimated is lower than 0.05. As seen in chapter 1 of this book, this is consistent with the lognormality principle [Plamondon *et al.*, 2013] which predicts that the production of complex movements is accomplished by time superimposing and, summing up lognormal vectors, with the goal of minimizing their number in a given task, to produce efficient, fluent gestures optimizing the energy required to generate these.

Finally, the SNR is greater for healthy controls than for PD patients for voice but not for handwriting, but these differences are not significant. Similarly, for D, $\mu$ and $\sigma$ there are no statistical differences either.

The only parameter with *p-values* lower than 0.05 for handwriting and voice is $\overline{\Delta\sigma}$. The average of the values obtained for $\overline{\Delta\sigma}$ for the voice and handwriting data for healthy controls and PD patients are presented in Fig. 9 and Fig. 10. These figures show clearly that the mean of the parameter $\overline{\boldsymbol{\Delta\sigma}}$ is lower for handwriting and voice production in PDs. These differences in $\overline{\boldsymbol{\Delta\sigma}}$ values between controls and PD patients proved to be significant in both handwriting and voice (*p-values* lower than 0.05). However, the voice shows a larger and statistically more significant difference between healthy and control samples than that found in handwriting, suggesting that voice production is might be more discriminative than handwriting.

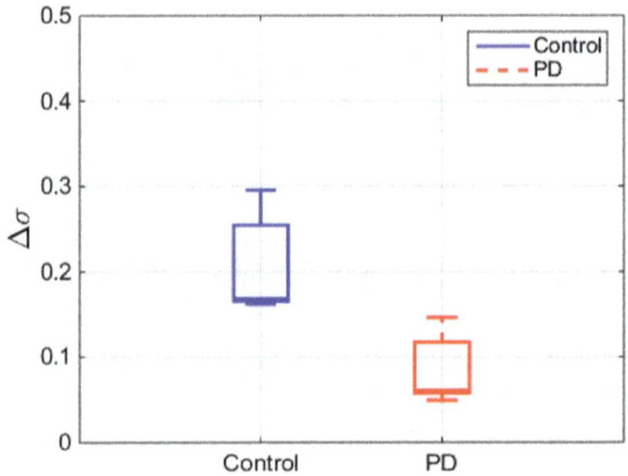


Fig. 9. Box-plot of the averaged stroke response time sequential difference ($\overline{\Delta\sigma}$) for Controls and PD patients in voice.

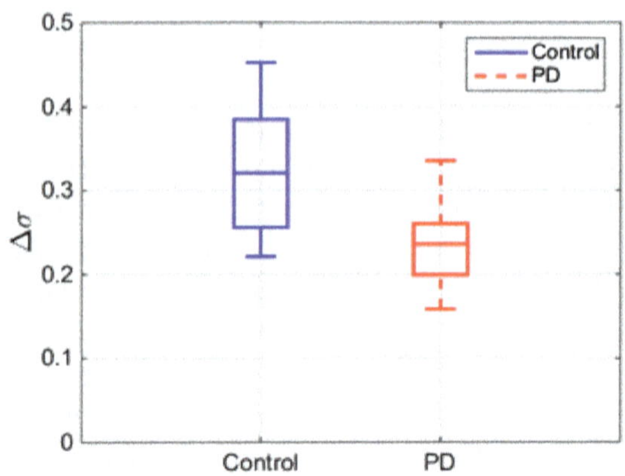


Fig. 10. Box-plot of the averaged stroke response time sequential difference ($\overline{\Delta\sigma}$) for Controls and PD patients in handwriting.

## 6. Conclusion

These results support the suggestion that one can use a common parameter to assess voice production and handwriting generation. Furthermore, the current study shows that Parkinson's disease affects the velocity profile of both handwriting and voice behavior.

Comparing handwriting and voice productions of PD patients, seem to indicate that parameters obtained from voice production could better discriminate performances of PDs from controls. This could be due to fact that voice requires smaller movements and less inertia because the tongue is used as opposed to the hand, a more massive system which is used to produce handwriting movements. The difference in characteristics between these two fine motor tasks could affect the width of the lognormal and the separation between them. However, to confirm this latter suggestion, this study should be repeated using voice and handwriting samples of the same participants. This will be the subject of forthcoming studies.

## Acknowledgment

This study was funded by the Spanish government's MIMECO TEC2016-77791-C4-1-R and TEC2016-77791-C4-4-R research project and European Union FEDER program/funds and the NSERC-Canada Grant RGPIN-2015-06409 to R. Plamondon. C. Carmona-Duarte was supported by a Juan de la Cierva contract (IJCI-2016-27682) and the "José Castillejo" mobility grant from Spanish government CAS18/00315.

## References

Broderick, M. P., Van Gemmert, A. W. A., Shill, H. A. and Stelmach, G. E. (2009). Hypometria and bradykinesia during drawing movements in individuals with Parkinson's disease, *Experimental Brain Research*, *197*(3), 223–233.

Carmona-Duarte, C., Alonso, J. B., Diaz, M., Ferrer, M. A., Gómez-Vilda, P. and Plamondon, R. (2016). Kinematic modelling of dipthong articulation. In A. Esposito, M. Faundez-Zanuy, A. M. Esposito, G. Cordasco, T. Drugman, J. Solé-Casals and F. C. Morabito (Eds.), *Recent Advances in Nonlinear Speech Processing* (Vol. 48, pp. 53–60).

Carmona-Duarte, C., Ferrer, M. A., Parziale, A. and Marcelli, A. (2017). Temporal evolution in synthetic handwriting *Pattern Recognition*, *68*, 233–244.

Carmona-Duarte, C., Gómez-Vilda, P., Plamondon, R., Rodellar-Biarge, V., Palacio-Alonso, D., Ferrer-Ballester, M. A. and Alonso-Hernández, J. B. (2017). Phonatory and articulatory correlates in kinematic neuromotor degeneration In P. J. Ibáñez J.,

González-Vargas J., Azorín J., Akay M. (Eds.), *Converging Clinical and Engineering Research on Neurorehabilitation II.* (Biosystems, Vol. 15, pp. 203–208).

Carmona-Duarte, C., Plamondon, R., Gómez-Vilda, P., Ferrer, M. A., Alonso, J. B. and Londral, A. R. M. (2016). Application of the lognormal model to the vocal tract movement to detect neurological diseases in voice In J. L. Chen YW., Tanaka S., Howlett R. (Eds.), *Innovation in Medicine and Healthcare 2016. Smart Innovation, Systems and Technologies* (Vol. 60, pp. 25–35).

Deller, J. R., Hansen, J. H. L. and J. G. Proakis (2000). *Discrete-Time Processing of Speech Signals*, Wiley.

Drotár, P., Mekyska, J., Rektorová, I., Masarová, L., Smékal, Z. and Faundez-Zanuy, M. (2014). Analysis of in-air movement in handwriting: A novel marker for Parkinson's disease, *Computer Methods and Programs in Biomedicine*, *117*(3), 405–411.

Eichhorn, T. E. and Gasser, T. (1996). Computational Analysis of Open Loop Handwriting Movements in Parkinson's Disease : A Rapid Method to Detect Dopamime tic Effects, *Brain*, *11*(3), 289–297.

Galaz, Z., Mekyska, J., Mzourek, Z., Smekal, Z., Rektorova, I., Eliasova, I., Berankova, D. (2015). Prosodic analysis of neutral, stress-modified and rhymed speech in patients with Parkinson's disease, *Computer Methods and Programs in Biomedicine*, *127*, 301–317.

Gomez-Barrero, M., Galbally, J., Fierrez, J., Ortega-garcia, J. and Plamondon, R. (2013). Variations of Handwritten Signatures with Time: A Sigma-Lognormal Analysis. *International Conference on Biometrics (ICB)*.

Gómez-Vilda, P., Londral, A. R. M., Rodellar-Biarge, V., Ferrández-Vicente, J. M. and de Carvalho, M. (2015). Monitoring amyotrophic lateral sclerosis by biomechanical modeling of speech production. *Neurocomputing*, *151*(P1), 130–138.

Gómez-Vilda, P., Rodellar-Biarge, V., Nieto-Lluis, V., Muñoz-Mulas, C., Mazaira-Fernández, L. M., Martínez-Olalla, R., Fernández-Fernández, M. (2013). Characterizing Neurological Disease from Voice Quality Biomechanical Analysis, *Cognitive Computation*, *5*(4), 399–425.

Hogg, R. V. and Ledolter, J. (1987). Engineering Statistics (MacMillan). New York. http://www.glottex.com/. (n.d.).

O'Reilly, C. and Plamondon, R. (2009). Development of a Sigma-Lognormal representation for on-line signatures, *Pattern Recognition*, *42*(12), 3324–3337.

O'Reilly, C. and Plamondon, R. (2012). Design of a Neuromuscular Disorders Diagnostic System Using Human Movement Analysis, *The 11th International Conference on Information Sciences, Signal Processing and Their Applications*, 787–792.

Peterson, G. E. and Barney, H. L. (1952). Control Methods Used in a Study of the Vowels, *The Journal of the Acoustical Society of America*, *24*(2), 175–184.

Plamondon, R. (1995). A kinematic theory of rapid human movements - Part II. Movement time and control, *Biological Cybernetics*, *72*(4), 309–320.

Plamondon, R., O'Reilly, C. and Ouellet-Plamondon, C. (2014). Strokes against stroke - Strokes for strides, *Pattern Recognition*, *47*(3), 929–944.

Plamondon, R., O'Reilly, C., Rémi, C. and Duval, T. (2013). The lognormal handwriter: Learning, performing, and declining, *Frontiers in Psychology*, *4*(DEC), 1–14.

Rusz, J., Čmejla, R., Růžičková, H., Klempíř, J., Majerová, V., Picmausová, J., Růžička, E. (2011). Acoustic assessment of voice and speech disorders in Parkinson's disease through quick vocal test, *Movement Disorders*, *26*(10).

Rusz, J., Cmejla, R., Ruzickova, H. and Ruzicka, E. (2011). Quantitative acoustic measurements for characterization of speech and voice disorders in early untreated Parkinson's disease, *The Journal of the Acoustical Society of America*, *129*(1), 350–367.

Skodda, S. and Schlegel, U. (2008). Speech rate and rhythm in Parkinson's disease, *Movement Disorders*, *23*(7), 985–992.

Tykalova, T., Rusz, J., Cmejla, R., Ruzickova, H. and Ruzicka, E. (2014). Acoustic investigation of stress patterns in Parkinson's disease, *Journal of Voice*, *28*(1), 129.e1–129.e8.

Van Gemmert, A. W. A., Plamondon, R. and O'Reilly, C. (2013). Using the Sigma-lognormal model to investigate handwriting of individuals with Parkinson's disease, *16th International Graphonomics Society Conference*, 119–122.

Walton, J. (1997). Handwriting changes due to aging and Parkinson's syndrome, *Forensic Science International*, *88*(3), 197–214.

https://doi.org/10.1142/9789811226830_0008

# Chapter 8

# Sigma-Lognormal Modelling to Assess Mobility Deficits: The Case of Turn Signature Assessed with Inertial Measurement Units (IMUs)

Karina Lebel[1,*], Christian Duval[2,†], Réjean Plamondon[3,‡], Nadir Faci[3,§] and Patrick Boissy[1,**]

[1]*Centre de recherche sur le vieillissement, Université de Sherbrooke, QC, Canada*

[2]*Département des Sciences de l'activité Physique, Université du Québec à Montréal, Canada*

[3]*Laboratoire Scribens, Département de génie électrique, Polytechnique Montréal*

[*]*Karina.Lebel@usherbrooke.ca*, [†]*duval.christian@uqam.ca*, [‡]*rejean.plamondon@polymtl.ca*, [§]*nadir.faci@polymtl.ca*, [**]*patrick.Boissy@USherbrooke.ca*,

When assessing mobility performance in humans, the specific way by which a task is performed is often referred to as its movement signature. If one can determine a *normal* signature for a task, deviations from that reference signature can potentially be used to identify and quantify mobility deficits. This chapter presents a conceptual framework for assessing turn mobility deficits based on movement signatures derived from the sigma-lognormal model, using orientation data captured at multiple segments with inertial measurement units (IMUs). The concept, tested on a small sample of older adults and participants with early Parkinson's disease, suggests that even at early stages of the disease, the complexity and fluidity of the turn signature are impacted. Specific parameters of the main lognormal also appear to have the potential of capturing the effect of medication on turning deficits. As such, the framework proposed herein, based on the sigma-lognormal model, has a true potential for mobility deficits assessment.

## 1. Introduction

We traditionally refer to a signature as the handwritten pattern executed to identify ourselves. However, the signature concept can be generalized far beyond the hand scripted signature. Indeed, one can define a movement signature as the specific way (timing, amplitude, velocity, direction and force) a task is performed. If one can deduce a normal pattern, or patterns, in a task signature, significant deviations from this reference signature could potentially identify mobility deficits or enhance understanding of adaptive strategies. Alternatively, specific motor deficits might also have their own signature. Furthermore, a task may involve series of actions that have to be coordinated in order to perform the task in a safe and efficient manner. A good example of this concept is turning while walking. Turning while walking is defined as a change in direction while standing upright and walking. Turning while walking is a task often performed in everyday life and is also recognized as a demanding functional task requiring planning, coordination, and postural stability to process and execute the central nervous system instructions in a safe and efficient manner [Patla *et al.*, 1999; Mancini *et al.*, 2015; Mellone *et al.*, 2016). Healthy individual typically exhibit a specific cranio-caudal sequence during a turn while walking, where the head initiates the motion, followed by the trunk, and then the pelvis to complete the body's reorientation process towards the new direction to pursue [Fuller *et al.*, 2007; Hong *et al.*, 2009]. In other words, turning while walking can be seen as a pointing task in the orientation domain, which requires coordination between the different segments to be performed adequately. The specificities of the sequence followed to accomplish the task can be defined as a turn signature.

This chapter first covers an overview of traditional methods used to study and characterize the turn signature to assess mobility deficits. Then, the conceptual approach of using the sigma-lognormal model for cranio-caudal turn signature assessment is introduced. Finally, the usefulness of the concept is demonstrated and discussed for mobility deficit assessment in Parkinson's disease.

## 2. Turn Signature Evaluation: The Current Status

The complexity of the system involved in human movement makes the movement signature analysis complicated. Indeed, it requires not only an appropriate measurement system to acquire the movement data, but also advanced computational methods to process the data and compute metrics. So far, turn signature has been primarily studied in typical biomechanical laboratory, using advanced equipment such as optoelectronic motion capture systems [Hong *et al.*,

2009; Spildooren *et al.*, 2013; Wright *et al.*, 2012; Crenna *et al.*, 2007; Ferrarin *et al.*, 2006]. In such setting, the sequence of motion between the head, trunk and pelvis involved in turning while walking is captured and characterized with spatiotemporal characteristics on healthy individuals [Hong *et al.*, 2009; Fuller *et al.*, 2007]. It has also been demonstrated that people with neurodegenerative disease and recurrent fallers exhibit an increased coupling in these segments while turning [Hong *et al.*, 2009; Spildooren *et al.*, 2013; Wright *et al.*, 2012; Crenna *et al.*, 2007; Ferrarin *et al.*, 2006]. In other words, they appear to have a limited degree of freedom in the axial rotation, leading to a turn typically called "*en-bloc*". More recently, metrics based on wearable systems like inertial measurement systems have been considered to assess turn deficits [Sheehan *et al.*, 2014; Vervoort *et al.*, 2016; Mancini *et al.*, 2016; Zakaria *et al.*, 2015; Greene *et al.*, 2010; Mancini *et al.*, 2015; El-Gohary *et al.*, 2013; Zampieri *et al.*, 2010; Salarian *et al.*, 2010; Salarian *et al.*, 2009; Galán-Mercant and Cuesta-Vargas, 2014]. Most of the metrics derived from inertial systems are currently based on information provided by the raw sensors (gyroscopes and accelerometers). For example, metrics like turn duration, turn speed, or the number of steps required to complete a turn have shown a potential to characterize age-related changes, and to identify recurrent fallers, patients with Parkinson's disease as well as frails. However, the variability reported in those metrics require the turning while walking task to be performed in standardized conditions, somehow limiting their potential to be used in a natural context.

### 2.1. *Laboratory-based movement signature assessment*

Biomechanics laboratory are well equipped to study movement signature. Optoelectronic systems are amongst the most popular equipment to study kinematics. These systems require positioning of markers on specific anatomical landmarks. High speed cameras positioned in the environment capture the 3D position of those markers and reconstruct the body segments to deduce segments' position and orientation. Precision of such system is typically reported to be around 0.1 mm, depending on the system used and its configuration within the calibrated volume of acquisition. Based on the assumption that such precision ensures good quality of data, the accuracy of the position and orientation data is rarely verified [Windolf *et al.*, 2008]. Occasionally, a Monte Carlo approach has been used to verify the potential impact of markers' position inaccuracies on the orientation data [Lebel *et al.*, 2017a, 2014]. Regardless, this type of system is recognized as the gold standard in biomechanics. Unfortunately, the available volume of acquisition with this accuracy depends on the number of cameras, their positioning and their resolution, which in turn limits the type of movement or task one can measure in this environment. Another major limitation of camera-based system relates to the possibility of occlusions [Zhou and Hu, 2008]. Indeed, these systems require a clear line-of-sight between the cameras and the markers in order to ensure visibility, and hence measurement. This may be problematic when objects are required to be placed in the environment for a person to perform a specific task (e.g. a chair to test sit-to-stand transfers), or when more than one person is involved in the task.

Using a laboratory setting allows to standardize the environment and the task performed. For example, the exact moment where a turn is initiated as well as the importance of that turn can be controlled to a certain extent. This control over the environment and the testing conditions somehow facilitate the analysis, increasing internal validity of the experiment. However, this same control affects the external validity of the results: are those findings representative of the way this person performs the task in real-life?

Nevertheless, movement signature assessment in laboratory settings is useful and will certainly remain, at least for cases where accuracy is of a major importance (e.g. an evaluation to determine which surgical approach is appropriate) or in situations where a control over the experimental conditions is crucial, such as for initial evaluation of a new assessment approach or to validate the impact of specific factors on an assessment technic (e.g. movement velocity).

### 2.2. *Movement signature assessment using wearables*

A sensor package to be worn by an individual to measure movement characteristics is often referred to as a wearable. As its name suggests, the main advantage of this kind of equipment is the fact that they are not limited to laboratory settings; they can be used in a variety of environments. Inertial measurement unit (IMU) is a good example of wearables. An IMU is a platform which contains one or more inertial sensors (accelerometer, gyroscope, magnetometer) to assess linear acceleration, angular velocity and magnetic North. IMUs are sometimes referred to as magnetic and inertial measurement unit (MIMU), magnetic angular rate and gravity sensor (MARG), inertial and magnetic measurement unit (IMMU), or attitude and heading reference system (AHRS), depending on its exact configuration and domain of application. Nowadays, most IMUs sold for biomechanics also include a fusion algorithm which uses all the information provided by the inertial sensors to estimate the orientation of the platform, expressed in an Inertial reference frame defined by the magnetic North and the gravity vector. Due to the global nature of the reference frame, positioning an IMU on the head and one on the trunk allows to derive the relative orientation of the head to the trunk. The type of fusion algorithm may vary from one system to the other, and so can the properties. But evaluation of the accuracy of orientation issued from IMU is generally good in conditions of regular motion [Lebel *et al.*, 2017a, 2013, 2015]. However, users should be aware that velocity and magnetic perturbations may affect the accuracy of the orientation data. In regular everyday movement, these problematic situations mainly occur in extremities [Lebel *et al.*, 2017a; Palermo *et al.*, 2014]. Nevertheless, inertial systems give us the opportunity to capture body kinematics in situations where other systems cannot (e.g. turning while going through a doorway) and have the potential to be used in natural environments, increasing the external validity of the assessment.

Inertial systems' popularity has been constantly increasing over the past few decades. However, metrics used to assess movement signature are mainly based on the raw inertial signals, that is the accelerometers and the gyroscopes. For example, accelerometers are typically used to derive the number of steps required to perform a turn, while gyroscopes enables to derive metrics regarding the speed of the turn (mean or max) as well as the angle of the turn. Yet, orientation data offer the possibility to link segments' information thus offering complementary information to enable movement signature assessment to go one step further.

## 3. Sigma-lognormal for Movement Signature Assessment

Movement is produced when a command is given to the neuromuscular system by the central nervous system. However, to fulfill its purpose, this movement must follow a specific trajectory, and coordination between the different segments involved shall be ensured. This specific way of accomplishing a task represents a movement signature. Gait is a good example of a movement signature. Indeed, gait has been studied extensively to understand the relationship between distance, timing, left/right asymmetry, and orientation of the different lower-limb segments. From these observations, a so-called *normal* gait pattern can be identified, and deviations form this *normal* pattern are considered as impairments. In other words, one could argue that there is a *normal* gait signature. However, the approach taken to develop gait signature can be considered as a bottom-up method since it is largely based on parameters derived from specific timing events, such as hill strike and toe-off.

By opposition, signal modelling is a top-down approach where a mathematical model is used to fit the data, reducing the complexity of the analysis. Indeed, signal modelling is not a new approach to biomechanics. Different paradigms can be used to model human movements, among which the equilibrium point models, the models based on kinematics, the minimization-based models, and the neural networks [Plamondon *et al.*, 2014]. The sigma-lognormal model is based on the Kinematics Theory which stipulates that a human movement is the end-effector of a cumulative response of numerous biological systems [Plamondon, 1995a, 1995b, 1998; Plamondon *et al.*, 2003]. In other words, each system produces a velocity vector, and the cumulative sum of these vectors will result in the motion of the segment [Lebel *et al.*, 2017b]. As movement is planned and controlled at the central nervous system level, assessment of human motion can be seen as a window onto the fundamentals of the motor control systems [Wolpert *et al.*, 1995].

Specifically, the sigma-lognormal model corresponds to a mathematical representation of the neuromuscular system (NMS) impulse response for rapid pointing movements. The NMS, as intended in this model, covers the motor cortex down to the muscles. The NMS is composed of multiple motor units. This system, once linearized, is characterized by an impulse response that follows an asymmetric bell-shaped curve, which can be optimally described using the lognormal function. As such, the parameters of the lognormal function are directly linked to the individual's NMS characteristics, as demonstrated using EEG [O'Reilly *et al.*, 2013] and EMG [Plamondon *et al.*, 2013a] techniques. Equation (1) below illustrates the relationship between the lognormal function and the physiological interpretation of the parameters.

$$\underline{v}(t) = \frac{D}{\sigma(t-t_0)\sqrt{2\pi}} e^{\left(\frac{[\ln(t-t_0)-\mu]^2}{-2\sigma^2}\right)}, \tag{1}$$

where $\underline{v}$ is the column matrix of velocity vector components
$t_0$ time of occurrence of the input command
$\mu$ logtime delay, the time delay of the NMS on a logarithmic scale
$\sigma$ logresponse time, the response time of the NMS on a logarithmic scale
$D$ the amplitude of the command sent to the NMS

The previous equation clearly demonstrates that for a specific task, a particular set of commands is sent to the system which responds according to its individual characteristics (delay, response time). According to the lognormality principle (Chapter 1), a lognormal velocity profile can be seen as a basic unit or primitive of human movement. The production of complex movements is accomplished through the exploitation of this convergence principle, by time superimposing and summing up lognormal vectors, with the goal of minimizing their number in a given task, to produce efficient, fluent gestures minimizing the energy required to generate these. In other words, for a similar task performed by multiple individuals, the variations observed in the signal can be interpreted as changes in the process or the functioning of the NMS.

So far, this model has been used extensively in the linear domain, from signature verification [Woch and Plamondon, 2010; Woch *et al.*, 2011; Plamondon *et al.*, 2013b; Djioua and Plamondon, 2009; Diaz *et al.*, 2016] to fine motor control investigation [O'Reilly *et al.*, 2014; O'Reilly and Plamondon, 2011; Duval *et al.*, 2015; Rémi *et al.*, 2015]. But the fundamentals on the movement characteristics can also be seen in the orientation domain, such as during a turn. As already mentioned in the introduction to this chapter, turning while walking involves a specific cranio-caudal sequence to be executed in order to safely reorient the body in a new direction to pursue. As such, turning can be seen as a multiple-segments pointing task performed in the orientation domain and its execution should be guided by the inherent lognormality of successful tasks.

### 3.1. *Conceptual framework for turn signature assessment using the sigma-lognormal*

Cranio-caudal signature involves a sequence of movement led by the head and followed by the trunk and pelvis. As such, it was hypothesized that each of these motions would require a specific command to be sent by the central nervous system, which impulse response could be captured through the analysis of the velocity profile (Fig. 1).

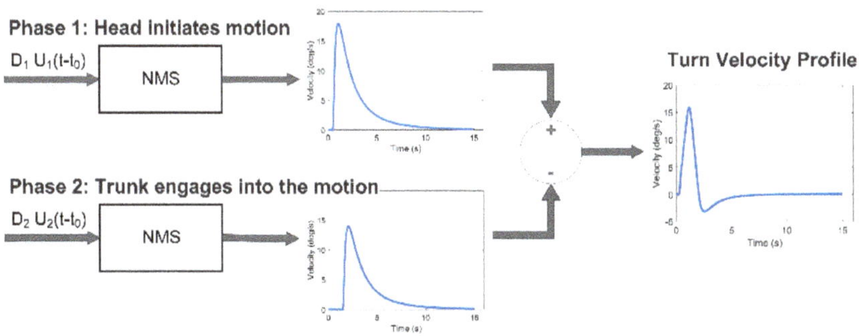


Fig. 1. Turn velocity profile generation.

The sigma-lognormal model is based on the premises that the velocity of a segment represents the cumulative impulse response of the neuromuscular system. Transposing this concept into the orientation domain requires to identify an equivalence to that velocity signal, namely the angular velocity. However, the axial control involved in turning also requires coordination between different segments, and it was therefore determined that the relative angular velocity between the head and the trunk should be considered as the main input to the sigma-lognormal model. Figure 2 illustrates the conceptual framework for the turn signature assessment based on the sigma-lognormal model. The initial step consists in identifying the turn task. This step can either be achieved manually using specific guidelines [Lebel *et al.*, 2017b] or through the use of automatic segmentation algorithms [Nguyen *et al.*, 2015; Hung *et al.*, 2017; Ayachi *et al.*, 2016], depending on the context and the available information. Once segmented, relative angle between segments of interest can be computed (e.g. head to trunk) and low-pass filtered to enable computation of the relative angular velocity signal.

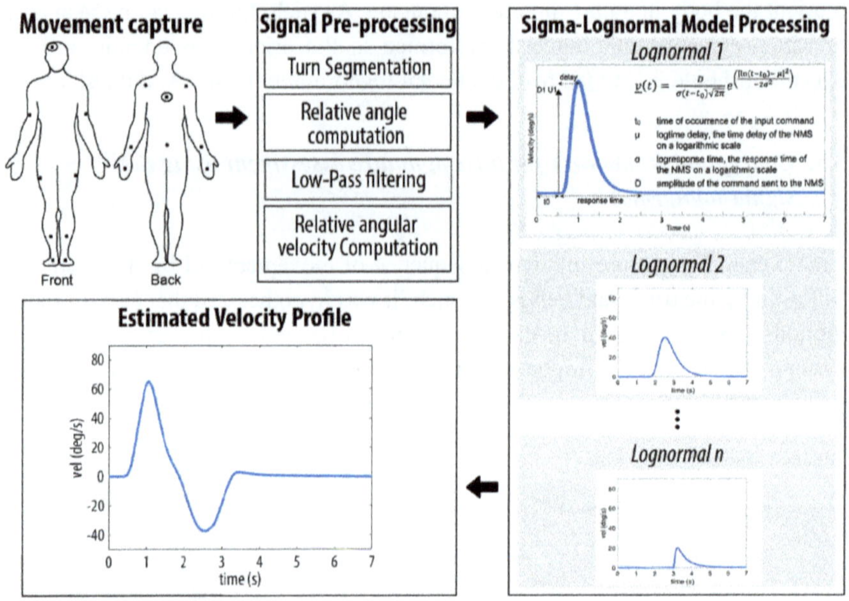


Fig. 2. Turn Signature Assessment Conceptual Framework.

Indeed, relative angular velocity can be derived from the relative angular profile using Eq. (2) [Lebel *et al.*, 2017b].

$$\boldsymbol{q} = cos\left(\frac{\theta}{2}\right) + \left(u_x\boldsymbol{i} + u_y\boldsymbol{j} + u_z\boldsymbol{k}\right) sin\left(\frac{\theta}{2}\right)$$

$$\boldsymbol{\omega} = \dot{\theta}(t)\hat{\mathbf{u}}(t) + \dot{\hat{\mathbf{u}}}\sin(\theta(t)) + \hat{\mathbf{u}}(t) \times \dot{\hat{\mathbf{u}}}\left(1 - \cos(\theta(t))\right), \quad (2)$$

where $\theta$ is the rotation angle

$\vec{u} \triangleq \boldsymbol{u} = (u_x\boldsymbol{i} + u_y\boldsymbol{j} + u_z\boldsymbol{k})$ is the unit vector, expressed with the Cartesian axes i, j, k

**q** is the quaternion representing the relative orientation, and

**ω** is the relative angular velocity vector

From the previous set of equations, one can then extract the axial component of the angular velocity profile to be used for the turn signature analysis. Further details on the pre-processing steps can be found in Lebel *et al.* [2017b]. Turn signature can then be assessed using the sigma-lognormal approach.

Outputs generated by this analysis include global information such as the number of lognormals required to reconstruct the signal (NbLogs) as well as the quality of the reconstruction, assessed using a signal to noise ratio (SNR). Derived from those two parameters is a ratio (SNR/NbLogs) which is meant to qualify the fluidity of the motion. Specific NMS parameters can also be output for each lognormal (i.e. the time of occurrence of the input command, the time delay of the NMS, the NMS response time and the amplitude of the command). From these data, further characteristics such as the time delay ($\bar{t}$), defined as the rapidity at which the system responds to the command, and the time response (s), corresponding to the time it takes the system to react and execute the movement, can be derived using Eqs. (3) and (4) [Plamondon *et al.*, 2003].

$$\bar{t} = \int_{t_0}^{+\infty} t\,\Lambda(t, t_0, \mu, \sigma^2)dt = t_0 + e^{\mu+0.5\sigma^2} \tag{3}$$

$$s = \sqrt{\int_{t_0}^{+\infty} t\,\Lambda(t, t_0, \mu, \sigma^2)dt} = \sqrt{e^{2\mu+\sigma^2}\left(e^{\sigma^2} - 1\right)}$$

$$= (\bar{t} - t_0)\sqrt{\left(e^{\sigma^2} - 1\right)} \tag{4}$$

Analysis of the signature can therefore be made at two levels. Indeed, the number of lognormals required to reconstruct the signal and the ratio of the SNR over NbLogs can be considered as proxies for the level of complexity involved in the task and the fluidity achieved doing so, while the parameters extracted from the lognormals give information on the NMS response for each phase of the turn.

## 4. Case Study: Turn Signature to Assess Mobility Deficits in Parkinson's disease

The concept of turn signature assessment for mobility deficits identification was first verified using a simplified approach of the sigma-lognormal model where only the two major lognormals were extracted, per turn [Lebel *et al.*, 2017b,

2018]. To illustrate the true potential of the sigma-lognormal model for mobility deficits identification, the concept is deepened and herein experienced using an optimized approach on two different samples: a group of 11 healthy older adults (mean age = 69.3 ± 6.0 years old; 55% male, 45% female) and a group of 11 patients diagnosed with early Parkinson's disease (mean age = 67.1 ± 6.7 years old; 64% male, 36% female; Hoehn and Yahr between 1 and 2, i.e. early Parkinson) [Lebel *et al.*, 2017b, 2018]. Both groups were asked to perform multiple trials of a Timed-Up and Go where the participants initiated the trials sitting down, got up upon signal, walked straight forward for 10 m, turned around, and came back to their initial seating position. Specifically, older adults were asked to perform two trials at normal pace, and participants with PD performed trials OFF medication and ON medication. While doing so, participants wore a motion capture suit (IGS-180, Synertial Ltd, UK) integrating 17 IMUs (f = 60Hz) positioned on multiple segments of the body and allowing the capture of the entire body's kinematics. The turn portion of each trial was first manually identified with the help of the avatar available in IGS-Bio, a software application provided with the IGS-180. Specifically, the segmentation steps included:

(i) Visual assessment of the initial misalignment in axial rotation (i.e. initial misalignment between the head, trunk and pelvis);
(ii) Identification of the beginning of preceding gait cycle, marking the beginning of the turn;
(iii) Assessment of the moment where the head, trunk and pelvis are realigned;
(iv) Establishment of the following gait cycle initiation point, corresponding to the end of the turn.

Using Matlab (v2015a MathWorks, US), relative orientation of the head to the trunk ($\underline{q}_{H2T}$) during the turn was then calculated and low-pass filtered (Butterworth filter, fc = 1.5Hz). A residual-based analysis of the relative orientation over repeated trial was used to determine the optimal cut-off frequency [Carbonneau *et al.*, 2013]. The resulting signal was then derived using Eq. (2) to determine the head to trunk relative angular velocity profile ($\underline{\omega}_{H2T}$). Further processing of the sigma-lognormal model was done using a proprietary software referred to as script studio [O'Reilly and Plamondon, 2009]. Non-parametric statistical analysis (i.e. Mann-Whitney U test for differences between healthy older adults (OA) and patients living with PD OFF medication (PDoff), Wilcoxon signed rank test for differences between ON and OFF states of patients living with PD) were performed using SPSS (v23, IBM), with a predetermined level of significance of 0.05.

Figure 3 illustrates typical results for the angular velocity profiles captured during a turn for an older individual, a participant with early Parkinson's disease OFF medication, and that same participant ON medication. As shown in this figure, the sigma-lognormal model appears to be a good fit for the turn cranio-caudal signature. However, it can be observed that the participant with Parkinson's disease seems to have a more complex signature with diminished amplitude in both phases of the turn which reflects departures from lognormality.

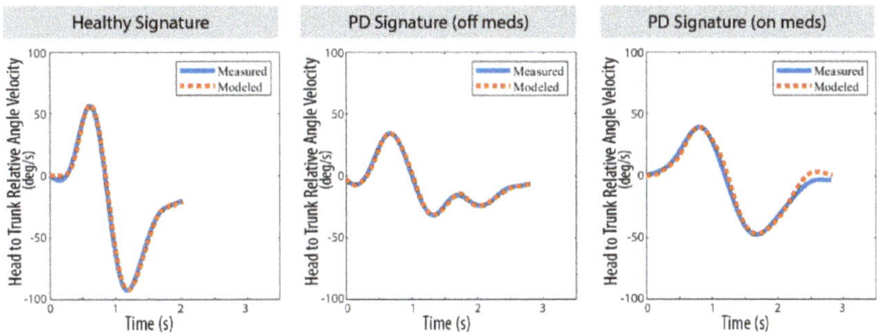


Fig. 3. Typical Results for Turn Signature Assessment Using Sigma-Lognormal Modelling.

Table 1. Sigma-Lognormal Model Parameters. Sigma-Lognormal Model Parameters for Older Adults (OA), Patients with Parkinson's disease ON (PDon) and OFF (PDoff) medication, calculated for Head to Trunk relative angular velocity. Values reported are medians along with the $1^{st}$ and $3^{rd}$ quartiles.

| Parameters | OA (n = 11) | PDoff (n = 12) | PDon (n = 12) | $p_{OA-PDoff}$ | $p_{PDoff-on}$ |
|---|---|---|---|---|---|
| ***Global*** | | | | | |
| Nb Logs | 3 [3, 3.5] | 4.6 [4, 5.4] | 4.5 [4, 5] | **0.005** | 0.837 |
| SNR | 29.5 [27.5, 29.6] | 28.0 [27.1, 29.5] | 28.4 [26.8, 28.7] | -- | -- |
| SNR/NbLogs | 9.6 [8.1, 10.6] | 6.3 [5.2, 7.8] | 6.4 [6.0, 6.8] | **0.003** | 0.722 |
| ***Specific Phase1*** | | | | | |
| D1 | 21.7 [15.1, 29.0] | 11.0 [8.4, 20.4] | 15.6 [8.9, 22.5] | **0.039** | 0.929 |
| $\bar{t}1$ | 0.66 [0.52, 0.74] | 0.57 [0.34, 0.88] | 0.66 [0.41, 0.79] | 0.684 | 0.722 |
| s1 | 0.16 [0.18, 0.26] | 0.19 [0.16, 0.26] | 0.18 [0.14, 0.21] | 1.000 | 0.286 |
| ***Specific Phase2*** | | | | | |
| D2 | 26.7 [18.5, 39.3] | 15.4 [12.6, 19.4] | 23.2 [15.8, 31.9] | **0.039** | **0.026** |
| $\bar{t}2$ | 1.40 [1.21, 1.53] | 1.83 [1.79, 2.27] | 1.65 [1.16, 2.20] | **0.001** | 0.859 |
| s2 | 0.23 [0.20, 0.34] | 0.36 [0.24, 0.46] | 0.33 [0.32, 0.59] | 0.100 | **0.021** |

Results compiled over the two groups of participants, reported in Table 1, also tend to demonstrate the higher complexity involved in turning for participants with Parkinson's disease, as shown by the significantly higher number of lognormals required to model the turn cranio-caudal signature adequately ($p = 0.005$). Similarly, the lower ratio (SNR/NbLogs) support the idea that fluidity in the motion is affected, even early in the disease ($p = 0.003$). But changes in those global parameters appears to be insignificant with medication ($p = 0.837$ and $p = 0.722$ for NbLogs and the ratio of SNR/NbLogs, respectively).

A look at the specific parameters suggests that amplitude of the commands and time delay of the NMS for the second phase of the turn are all affected by the disease, although the clinical importance of those changes still has to be assessed. Interestingly, these results also suggest that medication increases the amplitude of the main commands sent to the NMS, specifically during the second phase of the turn ($p = 0.026$).

## 5. Discussion & Conclusion

This chapter intended to demonstrate the potential of the sigma-lognormal model and its underlying lognormality principle to assess mobility deficits. Specifically, a conceptual framework based on the sigma-lognormal model was developed for the model to be used in the orientation domain, to characterize relative motion. This framework, applied on motion data acquired with wearable sensors, opens possibilities for the sigma-lognormal model approach to be used in a less constraint context, such as to characterize the cranio-caudal turn signature. As demonstrated in the case study, the high level of signal-to-noise ratio obtained for both the older adults and the participants with early Parkinson's disease confirms the ability of the sigma-lognormal model to characterize the turn cranio-caudal signature. At a global level, the impact of the disease on the complexity and fluidity of the motion appears to be captured using the number of lognormals required to model the signal as well as the ratio of the SNR over the number of lognormals. However, these preliminary results suggest that medication have a limited impact on these parameters. On the other hand, medication appears to affect specificities of the main lognormals, both at the command (D) and peripheral levels ($\bar{t}$ and s) [Lebel *et al.*, 2018]. To maximize the potential impact of the sigma-lognormal model on mobility deficit assessment, future work should probably include the development of a modified signal-to-noise ratio including the noise attributed to the disease. Such outcome could potentially be helpful to assess disease progression or be used to evaluate the effect of medication. Furthermore, the consideration of the pelvis rotation in the turn signature assessment may reveal further details.

In conclusion, the sigma-lognormal model can be used to capture movement signature both in the linear and orientation domains. The demonstrated capability of the model to be used on relative motion, on signals acquired from inertial sensors, opens-up possibilities to assess movement coordination at different levels. The framework proposed herein to assess mobility deficits during a turn and the results presented in the case study support the potential of the sigma-lognormal model to be used for mobility deficits assessment.

## Acknowledgment

The authors would like to thank the participants who took part in this study. This work was partly supported by a Canadian Institute of Health Research (CIHR) team in Mobility in Aging grant (200907AMG-208147-AML-CFEB-56596): Quantifying, characterizing, and modeling the whole-body mobility of individuals in their natural environment; from normal aging to Parkinson's disease, as well as by the NSERC-Canada Grant RGPIN-2015-06409 to R. Plamondon.

## References

Ayachi, F. S., Nguyen, H. P., Lavigne-Pelletier, C., Goubault, E., Boissy, P. and Duval, C. 2016. Wavelet-based algorithm for auto-detection of daily living activities of older adults captured by multiple inertial measurement units (IMUs). *Physiol Meas*, 37, 442–61.

Carbonneau, E., Fontaine, R. and Smeesters, C. A practical approach to determine apropriate cutoff frequencies for motion analysis data. 37th Annual Meeting of the American Society of Biomechanics, September 4–7 2013 2013 Omaha, NE.

Crenna, P., Carpinella, I., Rabuffetti, M., Calabrese, E., Mazzoleni, P., Nemni, R. and Ferrarin, M. 2007. SIAMOC Best Clinical Paper Award 2006: The association between impaired turning and normal straight walking in Parkinson's disease. *Gait & Posture*, 26, 172–178.

Diaz, M., Fischer, A., Ferrer, M. A. and Plamondon, R. 2016. Dynamic Signature Verification System Based on One Real Signature. *IEEE Trans Cybern.*

Djioua, M. and Plamondon, R. 2009. A New Algorithm and System for the Characterization of Handwriting Strokes with Delta-Lognormal Parameters. *IEEE Transactions on Pattern Analysis & Machine Intelligence*, 31, 2060–2072.

Duval, T., Remi, C., Plamondon, R., Vaillant, J. and O'Reilly, C. 2015. Combining sigma-lognormal modeling and classical features for analyzing graphomotor performances in kindergarten children. *Hum Mov Sci*, 43, 183–200.

El-Gohary, M., Pearson, S., McNames, J., Mancini, M., Horak, F., Mellone, S. and Chiari, L. 2013. Continuous monitoring of turning in patients with movement disability. *Sensors (Basel)*, 14, 356–69.

Ferrarin, M., Carpinella, I., Rabuffetti, M., Calabrese, E., Mazzoleni, P. and Nemni, R. 2006. Locomotor disorders in patients at early stages of Parkinson's disease: A quantitative analysis. *Conference Proceedings: Annual International Conference of the IEEE Engineering in Medicine & Biology Society*, 1, 1224–7.

Fuller, J. R., Adkin, A. L. and Vallis, L. A. 2007. Strategies used by older adults to change travel direction. *Gait & Posture*, 25, 393–400.

Galán-Mercant, A. and Cuesta-Vargas, A. I. 2014. Differences in trunk accelerometry between frail and non-frail elderly persons in functional tasks. *BMC research notes*, 7, 100–100.

Greene, B. R., McGrath, D., O'Neill, R., O'Donovan, K. J., Burns, A. and Caulfield, B. 2010. An adaptive gyroscope-based algorithm for temporal gait analysis. *Med. Biol. Eng. Comput.*, 48, 1251–60.

Hong, M., Perlmutter, J. S. and Earhart, G. M. 2009. A kinematic and electromyographic analysis of turning in people with Parkinson disease. *Neurorehabil Neural Repair*, 23, 166–76.

Hung, N., Lebel, K., Boissy, P., Bogard, S., Goubault, E., Duval, C. and Nguyen, H. 2017. Auto detection and segmentation of daily living activities during a Timed Up and Go task in people with Parkinson's disease using multiple inertial sensors. *Journal of NeuroEngineering & Rehabilitation (JNER)*, 14, 1–13.

Lebel, K., Boissy, P., Duval, C., Jog, M., Speechley, M., Karelis, A., Vincent, C. and Edwards, R. Assessing the Validity of Attitude and Heading Reference Systems for Biomechanical Evaluation of Motions - A Methodological Proposal. International Conference on Biomedical Electronics and Devices, 2014 Angers, France. 230–237.

Lebel, K., Boissy, P., Hamel, M. and Duval, C. 2013. Inertial measures of motion for clinical biomechanics: comparative assessment of accuracy under controlled conditions—effect of velocity. *PLoS ONE*, 8.

Lebel, K., Boissy, P., Hamel, M. and Duval, C. 2015. Inertial Measures of Motion for Clinical Biomechanics: Comparative Assessment of Accuracy under Controlled Conditions – Changes in Accuracy over Time. *PLoS ONE*, 10, e0118361.

Lebel, K., Boissy, P., Nguyen, H. and Duval, C. 2017a. Inertial measurement systems for segments and joints kinematics assessment: towards an understanding of the variations in sensors accuracy. *BioMed. Eng. OnLine*, 16, 56.

Lebel, K., Duval, C., Nguyen, H. P., Plamondon, R. and Boissy, P. 2018. Cranio-Caudal Kinematic Turn Signature Assessed with Inertial Systems As a Marker of Mobility Deficits in Parkinson's Disease. *Frontiers in Neurology*, 9.

Lebel, K., Nguyen, H., Duval, C., Plamondon, R. and Boissy, P. 2017b. Capturing the Cranio-Caudal Signature of a Turn with Inertial Measurement Systems: Methods, Parameters Robustness and Reliability. *Frontiers in Bioengineering and Biotechnology*, 5.

Mancini, M., El-Gohary, M., Pearson, S., McNames, J., Schlueter, H., Nutta, J. G., King, L. A. and Horak, F. B. 2015. Continuous monitoring of turning in Parkinson's disease: Rehabilitation potential. *NeuroRehabilitation*, 37, 3–10.

Mancini, M., Schlueter, H., El-Gohary, M., Mattek, N., Duncan, C., Kaye, J. and Horak, F. B. 2016. Continuous Monitoring of Turning Mobility and Its Association to Falls and Cognitive Function: A Pilot Study. *J. Gerontol. A Biol. Sci. Med. Sci.*, 71, 1102–8.

Mellone, S., Mancini, M., King, L. A., Horak, F. B. and Chiari, L. 2016. The quality of turning in Parkinson's disease: a compensatory strategy to prevent postural instability? *Journal of NeuroEngineering & Rehabilitation (JNER)*, 13, 1–9.

Nguyen, H. P., Ayachi, F., Lavigne–Pelletier, C., Blamoutier, M., Rahimi, F., Boissy, P., Jog, M. and Duval, C. 2015. Auto detection and segmentation of physical activities during a Timed-Up-and-Go (TUG) task in healthy older adults using multiple inertial sensors. *Journal of NeuroEngineering and Rehabilitation*, 12, 36.

O'Reilly, C. and Plamondon, R. 2011. Impact of the principal stroke risk factors on human movements. *Hum. Mov. Sci.*, 30, 792–806.

O'Reilly, C., Plamondon, R., Landou, M. K. and Stemmer, B. 2013. Using kinematic analysis of movement to predict the time occurrence of an evoked potential associated with a motor command. *Eur. J. Neurosci.*, 37, 173–80.

O'Reilly, C., Plamondon, R. and Lebrun, L. H. 2014. Linking brain stroke risk factors to human movement features for the development of preventive tools. *Front Aging Neurosci.*, 6, 150.

O'Reilly, C. and Plamondon, R. 2009. Development of a Sigma–Lognormal representation for on-line signatures. *Pattern Recognition*, 42, 3324–3337.

Palermo, E., Rossi, S., Patanè, F. and Cappa, P. 2014. Experimental evaluation of indoor magnetic distortion effects on gait analysis performed with wearable inertial sensors. *Physiological Measurement*, 35, 399.

Patla, A. E., Adkin, A. and Ballard, T. 1999. Online steering: coordination and control of body center of mass, head and body reorientation. *Experimental Brain Research*, 129, 629–634.

Plamondon, R. 1995a. A kinematic theory of rapid human movements - Part I. Movement representation and generation. *Biological Cybernetics*, 72, 295–307.

Plamondon, R. 1995b. A kinematic theory of rapid human movements - Part II. Movement time and control. *Biological Cybernetics*, 72, 309–320.

Plamondon, R. 1998. A kinematic theory of rapid human movements: Part III. Kinetic outcomes. *Biol. Cybern.*, 78, 133–45.

Plamondon, R., Djioua, M. and Mathieu, P. A. 2013a. Time-dependence between upper arm muscles activity during rapid movements: observation of the proportional effects predicted by the kinematic theory. *Hum. Mov. Sci.*, 32, 1026–39.

Plamondon, R., Fend, C. and Woch, A. 2003. A kinematic theory of rapid human movement. Part IV: a formal mathematical proof and new insights. *Biol. Cybern.*, 89, 126–38.

Plamondon, R., O'Reilly, C., Remi, C. and Duval, T. 2013b. The lognormal handwriter: learning, performing, and declining. *Front Psychol.*, 4, 945.

Plamondon, R., O'Reilly, C., Galbally, J., Almaksour, A. and Anquetil, É. 2014. Recent developments in the study of rapid human movements with the kinematic theory: Applications to handwriting and signature synthesis. *Pattern Recognit. Lett.*, 35, 225–235.

Rémi, C., Vaillant, J., Plamondon, R., Prevost, L. and Duval, T. Exploring the Kinematic Dimensions of Kindergarten Children's Scribbles. 17th Biennial Conference of the International Graphonomics Society, 2015. 79–82.

Salarian, A., Horak, F. B., Zampieri, C., Carlson-Kuhta, P., Nutt, J. G. and Aminian, K. 2010. iTUG, a Sensitive and Reliable Measure of Mobility. *Neural Systems and Rehabilitation Engineering, IEEE Transactions on*, 18, 303–310.

Salarian, A., Zampieri, C., Horak, F. B., Carlson-Kuhta, P., Nutt, J. G. and Aminian, K. 2009. Analyzing 180 degrees turns using an inertial system reveals early signs of progression of Parkinson's disease. *Conference Proceedings: Annual International Conference of the IEEE Engineering in Medicine & Biology Society*, 2009, 224–7.

Sheehan, K. J., Greene, B. R., Cunningham, C., Crosby, L. and Kenny, R. A. 2014. Early identification of declining balance in higher functioning older adults, an inertial sensor based method. *Gait and Posture*, 39, 1034–1039.

Spildooren, J., Vercruysse, S., Heremans, E., Galna, B., Vandenbossche, J., Desloovere, K., Vandenderghe, W. and Nieuwboer, A. 2013. Head-pelvis coupling is increased during turning in patients with Parkinson's disease and freezing of gait. *Movement Disorders*, 28, 619–25.

Vervoort, D., Vuillerme, N., Kosse, N., Hortobagyi, T. and Lamoth, C. J. 2016. Multivariate Analyses and Classification of Inertial Sensor Data to Identify Aging Effects on the Timed-Up-and-Go Test. *PLoS One*, 11, e0155984.

Windolf, M., Götzen, N. and Morlock, M. 2008. Systematic accuracy and precision analysis of video motion capturing systems-exemplified on the Vicon-460 system. *Journal of Biomechanics*, 41, 2776–2780.

Woch, A. and Plamondon, R. 2010. Characterization of bi-directional movement primitives and their agonist-antagonist synergy with the delta-lognormal model. *Motor Control*, 14, 1–25.

Woch, A., Plamondon, R. and O'Reilly, C. 2011. Kinematic characteristics of bidirectional delta-lognormal primitives in young and older subjects. *Hum. Mov. Sci.*, 30, 1–17.

Wolpert, D. M., Ghahramani, Z. and Jordan, M. I. 1995. Are arm trajectories planned in kinematic or dynamic coordinates? An adaptation study. *Exp. Brain Res.*, 103, 460–70.

Wright, R. L., Peters, D. M., Robinson, P. D., Sitch, A. J., Watt, T. N. and Hollands, M. A. 2012. Differences in axial segment reorientation during standing turns predict multiple falls in older adults. *Gait & Posture*, 36, 541–545.

Zakaria, N. A., Kuwae, Y., Tamura, T., Minato, K. and Kanaya, S. 2015. Quantitative analysis of fall risk using TUG test. *Comput. Methods Biomech. Biomed. Engin.*, 18, 426–37.

Zampieri, C., Salarian, A., Carlson-Kuhta, P., Aminian, K., Nutt, J. G. and Horak, F. B. 2010. The instrumented timed up and go test: potential outcome measure for disease modifying therapies in Parkinson's disease. *J. Neurol. Neurosurg. Psychiatry*, 81, 171–6.

Zhou, H. and Hu, H. 2008. Human motion tracking for rehabilitation-A survey. *Biomed. Signal Process. Control*, 3, 1–18.

https://doi.org/10.1142/9789811226830_0009

# Chapter 9

# Investigating the Sigma-Lognormal Model for Disease Classification by Handwriting

Donato Impedovo*, Giuseppe Pirlo†, Fabrizio Balducci, Vincenzo Dentamaro, Lucia Sarcinella and Gennaro Vessio
*Department of Computer Science, Università degli Studi di Bari Aldo Moro*
**donato.impedovo@uniba.it, †giuseppe.pirlo@uniba.it*

This chapter investigates the use of the sigma-lognormal model of the Kinematic Theory of rapid human for discriminating parkinsonian handwriting from healthy controls. Sigma-lognormal features are combined with more classic measures thus improving state-of-the-art results.

## 1. Introduction

The sigma-lognormal ($\Sigma\Lambda$) model of the Kinematic Theory of rapid hu-man movements [Plamondon, 1995a; 1995b; O'Reilly and Plamondon, 2009] has been successfully employed, so far, in many practical applications as for example automatic signature verification [Plamondon *et al.*, 2014; Pirlo *et al.*, 2015]. However, only recently, it has been investigated in characterizing and analyzing neurodegenerative disorders (NDs). For example, it has been applied with encouraging results to studying changes in handwriting due to Alzheimer's Disease (AD) [Impedovo *et al.*, 2013] and Parkinson's disease (PD) [Van Gemmert *et al.*, 2013; Carmona-Duarte *et al.*, 2018].

Neurodegeneration, in fact, results, in performance degradation of previously learned motor skills as, for instance, handwriting. Therefore, changes in handwriting can be considered as biomarker for neurodegenerative diseases assessment [Rosenblum *et al.*, 2013; Impedovo and Pirlo, 2019].

At moment, a reliable diagnosis can require months and symptoms need to be constantly monitored. In addition, in some cases, it is difficult to determine the exact type of neurodegenerative disorder, as well as its se-verity. For these reasons, identifying accurate biomarkers for early and differential diagnosis, prognosis and response to therapy is a primary goal of the research on neurodegenerative disorders. In recent years, in addition to changes in handwriting, other promising biomarkers to be used at a preclinical stage of Parkinson disease are neuroimaging techniques, such as PET, SPECT and MRI [Lotankar *et al.*, 2017]. For example, advancements have been achieved for Parkinson disease diagnosis starting from neuroimaging data [Prashanth *et al.*, 2014; Rana *et al.*, 2015]. Nevertheless, acquiring brain images is an expensive process and it cannot be used for monitoring at the patient's home. From this perspective, handwriting can cover a special role; in fact, it is a complex activity entailing cognitive, kinesthetic and perceptual-motor components [Tseng and Cermak, 1993], whose changes seem to be a prominent biomarker for the evaluation of Parkinson Disease [De Stefano *et al.*, 2019].

Micrographia, which is the abnormal reduction of the size of handwriting, can be observed in PD's patients [Lang *et al.*, 1998]. Dysgraphia, which is the progressive degeneracy and the increase in confusion of several components of handwriting, has been found in patients with moderate levels of AD [Fairhurst *et al.*, 2005; Glénat *et al.*, 2005; Onofri *et al.*, 2013].

The research challenge is to develop a computer aided diagnosis system to support the early diagnosis of the disease, as well as for monitoring its severity as the disease advances [Impedovo and Pirlo, 2019]. A computerized predictive model, in fact, is appropriate to detect subtle changes in writing behavior from a quantitative point of view. Of course, such a system is not expected to replace standard techniques or to perform a self-diagnosis, but rather to provide a complementary approach to the pathology evaluation carried out by hu-man experts, which is non-invasive and low-cost.

A typical acquisition tool is a professional tablet provided with an electronic pen. Temporal and spatial variables (named raw data) of handwriting can be recorded. Successively, features can be evaluated. In general, the most employed approach to studying the potentialities of handwriting tasks involves combining (lumping) the features coming from different tasks into a unique high dimensional feature vector and then feeding this vector into a traditional machine learning algorithm, e.g. [Drotár *et al.*, 2015]. Alternatively, some recent studies propose to combine features into ensembles of classifiers (a multi-expert system), each built on top of the feature space of every task [Moetesum *et al.*, 2018; Pereira *et al.*, 2018].

In literature, several features have been used, ranging from traditional kinematic and spatiotemporal variables of handwriting to less common measures based on entropy and signal-to-noise ratio [Drotár *et al.*, 2015] and on fractal analysis [Mucha *et al.*, 2018]. A complete review can be found in [Impedovo and Pirlo, 2019].

Features can belong to offline (static) domain or online (dynamic) domain [Zhi *et al.*, 2017]. In the latter case, the handwriting is represented as a sequence {S(n)}n = 0, 1,...,N, where S(n) is the signal captured with a sampling time of the writing process equal to n$\Delta$t with $0 \leq n \leq N$, having $\Delta$t as the sampling period (Fig. 1).

Fig. 1. Spatio temporal handwriting ("▪": pen-down; "•": pen-up).

The sigma-lognormal model represents a very interesting approach in this direction, since it considers not only the rapidity, fluency and regularity of the handwriting movement, but also physical body characteristics such the state of the neuromuscular systems responsible for the generation of the action plan.

This chapter investigates the discriminating potential of the sigma-lognormal modeling for an automatized Parkinson's disease diagnosis based on handwriting. To this end, a freely available dataset, i.e. PaHaW [Drotár *et al.*, 2016], has been considered and the impact on the classification performance of sigma-lognormal based features (alone or in combination with more classic measures) has been evaluated.

The rest of the chapter is organized as follows. Section 2 describes methods and sample description, Sec. 3 describes data processing. Section 4 focuses on the experiments and results. Section 5 concludes the chapter and sketches some future developments of the present research.

## 2. Methods & Sample Description

Data acquisition is a crucial step, which includes the recruitment of participants, the choice of the device/devices used to acquire data and the individuation of the most suitable handwriting tasks [Impedovo *et al.*, 2019].

In this work the "Parkinson's disease handwriting database" (PaHaW) has been adopted. It includes 37 PD patients and 38 age and gender-matched healthy control (HC) subjects [Drotár *et al.*, 2016]. Figure 2 depicts the distribution of the number of patients referred to the stage of the disease according to the UPDRS scale.

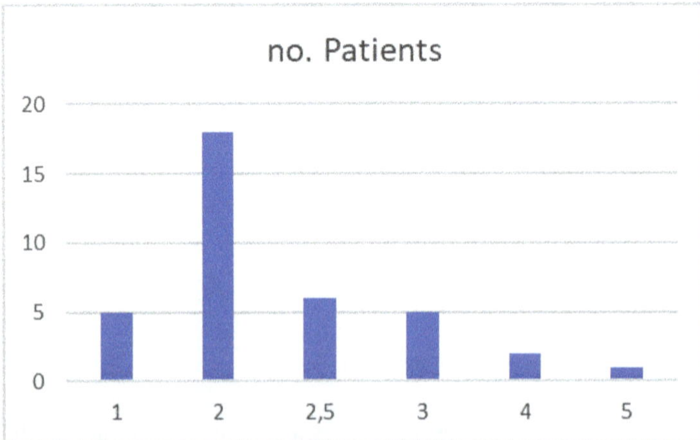


Fig. 2. No. of patients per disease stage.

Subjects were requested to complete eight handwriting tasks in accordance with a pre-filled template:

1. Drawing an Archimedes spiral;
2. Writing in cursive the letter *l*;
3. The bigram *le*;
4. The trigram *les*;
5. Writing in cursive the word *lektorka* (female teacher in Czech);
6. *porovnat* (to compare);
7. *nepopadnout* (to not catch);
8. Writing in cursive the sentence *Tramvaj dnes už nepojede* (The tram won't go today).

According to the categorization provided in [Impedovo and Pirlo, 2019], task 1 is a Simple Drawing task generally used for trajectory, tremor, dimension (size), velocity and acceleration evaluations. Tasks 2 to 8 are Simple Writing tasks. Non-sense words containing one or more characters repetitions are easy to write in a recursive and continuous way as they minimize the linguistic-comprehension processes. The "e" and "l" characters, according to the Lognormal Kinematic Theory, are constituted by just two velocity strokes. Moreover, the use of "l" and "e" involves the handwriting of the same character scaled in amplitude. The last task requires a higher degree of simultaneous processing if compared to others and it allows the capturing in-air movements between words Task 8 also allows the recording of the effect of fatigue while writing.

The signals were recorded using the Wacom Intuos 4M digitizing tablet, overlaid with an empty paper. The set of raw data acquired by the device are the x- and y-coordinates of the pen position and their timestamps, pen inclination and pressure. Finally, the button status is also available.

## 3. Data Processing

Data processing includes several steps as described in the following sub-sections.

### 3.1. *Features extraction*

A wide set of features as used in [Drotár *et al.*, 2015] are here referred to as classic features. The sigma-lognormal features, instead, are those have been additionally considered in this study. The overall set of features is summarized in Table 1. Note that the feature calculation stage resulted in either a single parameter or a vector. All features have been normalized before classification to have zero mean and unit variance, successively, for all the resulting vector features the following basic statistical measures have been also computed: mean; median; standard deviation; 1st percentile; 99th percentile; 99th percentile – 1st percentile, which is an outlier robust range.

#### 3.1.1. *Classic features*

Samples, as recorded by the tablet, have been segmented into on-surface and in-air strokes in accordance with button status.

Several features have been computed using the Cartesian coordinates of the sampled points and their time stamps. Kinematic features include: number of strokes, tangential, horizontal and vertical displacement, velocity, acceleration and jerk; number of changes of velocity/acceleration (NCV/NCA); NCA and NCV relative to writing duration. Spatiotemporal features include: stroke size and duration; speed and stroke speed; stroke height and width; on-surface and in-air time; total time; normalized on-surface and in-air time; in-air/on-surface ratio. In order to make use of the pressure signals, the following have been computed: mean pressure; number of changes of pressure (NCP); relative NCP. Furthermore, to uncover hidden complexities of handwriting, the following features have been computed for both the on-surface and in-air horizontal and vertical components of handwriting: Shannon and Rényi (second and third order) entropy; signal-to-noise ratio (SNR) and empirical mode decomposition (EMD).

### 3.1.2. *Sigma-lognormal features*

The new feature set is composed by the parameters of the sigma-lognormal reconstruction of each task. Briefly, the sigma-lognormal model considers the velocity profile of each stroke $j$ as having a lognormal shape $\Lambda$ scaled by an input command parameter $D$ and time-shifted by the time occurrence of the command $t_0$:

$$|\vec{v}_j(t)| = \frac{D_j}{\sigma_j(t - t_{0j})\sqrt{2\pi}} \exp\left( -\frac{[\ln(t - t_{0j}) - \mu_j]^2}{2\sigma_j^2} \right) \tag{1}$$

In (1), $\mu_j$ expresses the log-time delay of the neuromuscular system responsible for the generation of the action plan, while $\sigma_j$ represents the log-response time. Moreover, the model assumes each stroke as happening around a pivot, with respect to a starting angle $\theta_{sj}$ and an ending angle $\theta_{ej}$. In addition, the number of lognormal strokes has been considered as a single value feature. The complete set of features is reported in Table 1.

Table 1. Features. Abbreviations: $s$ = scalar; $v$ = vector.

| **Feature** | ***s/v*** | **Description** |
|---|---|---|
| Stroke number | *s* | Number of strokes |
| Displacement | *v* | Tangential trajectory during handwriting |
| Velocity | *v* | Rate of change of position with respect to time |
| Acceleration | *v* | Rate of change of velocity with respect to time |
| Jerk | *v* | Rate of change of acceleration with respect to time |
| Horizontal/vertical displacement | *v* | Displacement in the horizontal/vertical direction |
| Horizontal/vertical velocity | *v* | Velocity in the horizontal/vertical direction |
| Horizontal/vertical acceleration | *v* | Acceleration in the horizontal/vertical direction |
| Horizontal/vertical jerk | *v* | Jerk in the horizontal/vertical direction |
| NCV | *s* | Mean number of local extrema of velocity |
| NCA | *s* | Mean number of local extrema of acceleration |
| Relative NCV | *s* | NCV relative to writing duration |
| Relative NCA | *s* | NCA relative to writing duration |
| Stroke size | *v* | Path length of each stroke |
| Stroke duration | *v* | Movement time per stroke |
| Speed | *s* | Trajectory during handwriting divided by writing duration |

Table 1 (*continued*).

| **Feature** | *s/v* | **Description** |
|---|---|---|
| Stroke speed | *v* | Trajectory during stroke divided by stroke duration |
| Stroke height | *v* | Height of each stroke |
| Stroke width | *v* | Width of each stroke |
| On-surface time | *s* | Overall time spent on-surface |
| In-air time | *s* | Overall time spent in-air |
| Total time | *s* | On-surface time plus in-air time |
| Normalized on-surface time | *s* | On-surface time normalized by total time |
| Normalized in-air time | *s* | In-air time normalized by total time |
| In-air/on-surface ratio | *s* | Ratio of time spent in-air/on-surface |
| Mean pressure | *v* | Average pressure over all strokes |
| NCP | *s* | Mean number of local extrema of pressure |
| Relative NCP | *s* | NCP relative to writing duration |
| Horizontal/vertical Shannon entropy | *v* | Shannon entropy of the horizontal/vertical component of the pen position |
| Horizontal/vertical Rényi entropy | *v* | Second and third order Rényi entropy of the horizontal/vertical component of the pen position |
| Horizontal/vertical signal-to-noise ratio | *v* | Signal-to-noise ratio of the horizontal/vertical component of the pen position |
| Horizontal/vertical intrinsic Shannon entropy | *s* | Shannon entropy of the first and second IMF obtained by the EMD of the horizontal/vertical component of the pen position |
| Horizontal/vertical intrinsic Rényi entropy | *s* | Second and third order Rényi entropy of the first and second IMF obtained by the EMD of the horizontal/vertical component of the pen position |
| Horizontal/vertical intrinsic signal-to-noise ratio | *s* | Signal-to-noise ratio of the first and second IMF obtained by the EMD of the horizontal/vertical component of the pen position |
| **Lognormal stroke number** | *s* | **Number of lognormal strokes** |
| ***D*** **parameter** | *v* | ***D*** **parameter for all lognormal strokes** |
| $\boldsymbol{\mu}$ **parameter** | *v* | $\boldsymbol{\mu}$ **parameter for all lognormal strokes** |
| $\boldsymbol{\sigma}$ **parameter** | *v* | $\boldsymbol{\sigma}$ **parameter for all lognormal strokes** |
| $\boldsymbol{\theta_s}$ **parameter** | *v* | $\boldsymbol{\theta_s}$ **parameter for all lognormal strokes** |
| $\boldsymbol{\theta_e}$ **parameter** | *v* | $\boldsymbol{\theta_e}$ **parameter for all lognormal strokes** |

### 3.2. *Model fitting and validation*

In order to reduce the dimensionality of the feature space, a feature selection algorithm has been applied before classification. The discriminating power of each feature has been considered, evaluating its accuracy in separating PD from HC. All features have then been ranked in accordance with this score and only the *n* features providing the highest score have been retained for the final model fitting [Pedregosa *et al.*, 2011].

Support Vector Machines (SVMs) with radial basis function (RBF) kernel have been used adopting the parameter as in [Drotár *et al.*, 2015] for comparison aims.

The classification performance has been validated through a 10-fold cross-validation. With this technique, the set of examples is divided into ten folds: one-fold is treated as test set; the remaining folds form the training set. The entire procedure is iterated ten times, until each fold is used as test set once. In particular, the splitting was stratified by diagnosis, *i.e.* each fold contained roughly the same number of subjects from each diagnostic group (parkinsonian vs. healthy).

## 4. Results

In the following subsections, the results of a preliminary analysis of the features and the classification performance are reported.

### 4.1. *Features analysis*

In order to preliminary understand the predictive potential of the new set of features, compared to the classic ones, feature selection have been applied task by task. Table 2 reports the 10 most discriminating features for each task. First, it can be observed that among the most discriminating features there are those related to in-air movement and signal-to-noise ratio of handwriting. These findings confirm that in-air movements carry complementary information with the on-surface movement [Rosenblum *et al.*, 2013]. Measures of signal energy can capture the randomness of the movement during handwriting due to tremor and irregular muscle contractions.

Interestingly, some of the most relevant features have been found to be the sigma-lognormal based features, providing an initial confidence on their effectiveness in the classification task. The sigma-lognormal features are almost prevalent in the spiral task: this is important, as the task has not been used in

[Drotár *et al.*, 2015] and has been used with no significant result in [Drotár *et al.*, 2016]. Drawing a spiral is one of the most popular exercise to assess tremor [Pullman, 1998] and the velocity profile reconstruction of the task is tailored to capture velocity variations due to tremor.

Table 2. Feature importance task by task. In bold the ΣΛ features.

| **Task** | **Features** |
|---|---|
| 1 – spiral | stroke number; horizontal intrinsic Shannon entropy (2nd IMF); **$\mu$, median**; **$\theta_e$, std**; **$\theta_s$, std**; **$\theta_s$, robust range**; **$D$, 1st percentile; $D$, mean**; vertical SNR; **$D$, median** |
| 2 – *lll* | on-surface horizontal SNR, std; on-surface vertical displacement, median; **$\mu$, 1st percentile**; vertical SNR; horizontal intrinsic SNR (1st IMF); **$D$, 1st percentile**; in-air vertical Shannon entropy, median; on-surface vertical velocity, median; in-air stroke number; on-surface stroke width, 99th percentile |
| 3 – *lelele* | **$\sigma$, std**; stroke number; **$\sigma$, 99th percentile**; **$\sigma$, robust range**; on-surface vertical SNR, 1st percentile; on-surface jerk, 1st percentile; on-surface acceleration, 1st percentile; on-surface velocity; on-surface horizontal SNR, std; **$\mu$, 99th percentile** |
| 4 – *leslesles* | **$\sigma$, 1st percentile**; pressure, mean; **$D$, robust range**; **$D$, median**; horizontal SNR (2nd IMF); **$D$, 99th percentile**; vertical SNR; horizontal SNR; in-air vertical displacement, std; in-air stroke duration, median |
| 5 – *lektorka* | vertical SNR; **$D$, median**; on-surface vertical SNR, std; in-air jerk, 1st percentile; **$\sigma$, mean**; on-surface stroke width, std; in-air vertical displacement, 99th percentile; in-air vertical displacement, robust range; horizontal SNR; on-surface stroke size, 99th percentile |
| 6 – *porovnat* | on-surface vertical jerk, median; in-air stroke speed, 1st percentile; in-air vertical SNR, std; **$\mu$, std**; in-air stroke speed, median; on-surface vertical acceleration, median; vertical intrinsic SNR (1st IMF); on-surface horizontal SNR, std; **$\sigma$, median**; in-air horizontal Shannon entropy, median |
| 7 – *nepopadnout* | **$\sigma$, 1st percentile**; horizontal SNR; in-air vertical Rényi (2nd order) entropy, std; in-air vertical Rényi (3rd order) entropy, std; vertical intrinsic SNR (2nd IMF); **$D$, 1st percentile**; in-air vertical SNR, std; in-air horizontal displacement, std; horizontal Rényi entropy (2nd order); horizontal Rényi entropy (3rd order) |
| 8 – *Tramvaj dnes už nepojede* | NCP; in-air stroke speed, median; horizontal SNR; pressure, mean; **$\mu$, 99th percentile**; **$\mu$, robust range**; in-air stroke speed, 1st percentile; in-air horizontal SNR, std; on-surface stroke size, 1st percentile; on-surface stroke size, median |

The importance of the sigma-lognormal modeling emerged also in task 3 (*lelele*) and task 4 (*leslesles*). Writing non-sense words like these has been extensively studied in the literature, showing the impairment of PD patients in fine motor control during loop-like movements, *e.g.* [Van Gemmert *et al.*, 2001; Smits *et al.*, 2014]. Producing these movements offer the possibility to better evaluate the motor-planning activity between a character and the following one. Parkinsonian patients, in fact, are known to write in a more segmented fashion, showing difficulties in anticipating the forthcoming movement [Bidet-Ildei *et al.*, 2011]. A hesitation between two characters may so point out the necessity to re-plan the writing activity; a fluid writing, instead, can reveal the presence of an anticipated motor planning.

### 4.2. *Classification*

Classification results are reported in Table 3 which shows only the best mean accuracies, averaged over all the cross-validation iterations depending on the optimal number of features retained by feature selection.

Replica of [Drotár *et al.*, 2015] achieved comparable results, *i.e.* an accuracy of 88.33%, with a higher number of features, *i.e.* 233 instead of 162. These differences are marginal and may be due to implementation settings. Using only the sigma log-normal features performed 81.12%. Combining the sigma log-normal features with classic features provided an accuracy improvement up to 90.79%. For a fair comparison, the results in Table 3 do not include the spiral task, as it was not taken into account in [Drotár *et al.*, 2015].

Table 3. Accuracy performance excluding the spiral task. In bold the best result.

| **Features** | **# Features** | **Accuracy** |
|---|---|---|
| Classic | 233 | 88.33% |
| Sigma-lognormal | 51 | 81.12% |
| Classic and sigma-lognormal | 133 | **90.79%** |

Table 4. Accuracy performance including the spiral task. In bold the best result.

| **Features** | **# Features** | **Accuracy** |
|---|---|---|
| Classic | 185 | 88.33% |
| Sigma-lognormal | 57 | 87.29% |
| Classic and sigma-lognormal | 133 | **93.33%** |

Successively, results obtained considering also the spiral task (see Table 4 for details) have been computed: 93.33% of accuracy has been achieved. Similar findings, on the spiral task, have been reported in [Moetesum *et al.*, 2018], in which features automatically learned by deep learning models have been used.

Finally, a scenario where feature selection and parameter tuning are not performed on the joint set of training and test has been also considered. Under this condition, the approach here proposed achieved 79.87% of accuracy against 76.25% of classic features. Note that the *nested* feature selection enables a more reliable evaluation of the classification performance. Conversely, when assessing feature importance, feature selection has been applied on the entire set of examples for a more robust evaluation of each feature.

## 5. Conclusion

Handwriting analysis can be a powerful tool to develop a computer aided diagnosis system to support clinicians at the point of care. When designing such a system, a crucial step concerns the choice of the most appropriate features to describe the time series raw data sampled by the acquisition device. In this chapter, the sigma-lognormal model of the kinematic theory of rapid human movements has been employed with encouraging results for discriminating parkinsonian handwriting from healthy controls automatically. Combining sigma-lognormal based features to more classic measures improved state-of-the-art results.

It must be underlined that the findings reported in the present work only rely on a dynamic analysis of the handwriting tasks. An alternative approach consists in studying the static images of handwriting, eventually reconstructed from the dynamic information given by the tablet, e.g. [Moetesum *et al.*, 2018; Pereira *et al.*, 2018]. The sigma-lognormal could be also used to evaluate stability of handwriting as the disease evolves [Pirlo and Impedovo, 2010; Impedovo *et al.*, 2012; Pirlo and Impedovo, 2012]. Systems must consider the evolution of the disease by considering updating strategies [Impedovo and Pirlo, 2011; Pirlo *et al.*, 2009]. Finally, within a complete framework, the physician should be able to identify an explicit model relevant for the diagnosis based on the most informative subset of features [Senatore *et al.*, 2019].

## Acknowledgements

This work has been supported by the Italian Ministry of Education, University and Research within the PRIN2015-HAND Project under grant H96J16000820001.

## References

Bidet-Ildei, C., Pollak, P., Kandel, S., Fraix, V. and Orliaguet, J. P. (2011). Handwriting in patients with Parkinson disease: Effect of L-dopa and stimulation of the sub-thalamic nucleus on motor anticipation. *Human movement science*, 30(4), 783–791.

Carmona-Duarte, C., Ferrer, M. A., Gomez-Vilda, P., Van Gemmert, A. W. A. and Plamondon, R. (2018). A common framework to evaluate Parkinson's disease in voice and handwriting. In *International Conference on Pattern Recognition and Artificial Intelligence*, 795–798.

De Stefano, C., Fontanella, F., Impedovo, D., Pirlo, G. and Scotto di Freca, A. (2019). Handwriting analysis to support neurodegenerative diseases diagnosis: a review. *Pattern Recognition Letters, 121*, 37–45.

Drotár, P., Mekyska, J., Rektorová, I., Masarová, L., Smékal, Z. and Faundez-Zanuy, M. (2015). Decision support framework for Parkinson's disease based on novel handwriting markers. *IEEE Transactions on Neural Systems and Rehabilitation Engineering*, *23*(3), 508–516.

Drotár, P., Mekyska, J., Rektorová, I., Masarová, L., Smékal, Z. and Faundez-Zanuy, M. (2016). Evaluation of handwriting kinematics and pressure for differential diagnosis of Parkinson's disease. *Artificial intelligence in Medicine*, *67*, 39–46.

Fairhurst, M.C., Hoque, S., Boyle, T.J. (2005). Assessing behavioural characteristics of dyspraxia through on-line drawing analaysis. Proceedings of the 12th Conference of the *International Graphonomics Society* (IGS2005).

Glénat, S., Heutte, L., Paquet, T. (2005). Computer-based diagnosis of dyspraxia: The meddraw project. Proceedings of the 12th Conference of the *International Graphonomics Society* (IGS2005).

Impedovo, D., Pirlo, G. (2011). Updating knowledge in feedback-based multi-classifier systems. In proceedings of the *International Conference on Document Analysis and Recognition,* 227–231.

Impedovo, D., Pirlo, G. (2019). Dynamic handwriting analysis for the assessment of neurodegenerative diseases: a pattern recognition perspective. *IEEE Reviews in Biomedical Engineering*, 12, 209–220.

Impedovo, D., Pirlo, G., Vessio, G., Angelillo, M.T. (2019). A Handwriting-Based Protocol for Assessing Neurodegenerative Dementia. In *Cognitive Computation,* Vol. 11, issue 4, 576–586.

Impedovo, D., Pirlo, G., Mangini, F. M., Barbuzzi, D., Rollo, A., Balestrucci, A. and Plamondon, R. (2013). Writing generation model for health care neuromuscular system investigation. In *International Meeting on Computational Intelligence Methods for Bioinformatics and Biostatistics* (pp. 137–148). Springer, Cham.

Impedovo, D., Pirlo, G., Sarcinella, L., Stasolla, E., Trullo, C.A. (2012). Analysis of stability in static signatures using cosine similarity. Proceedings of the *International Workshop on Frontiers in Handwriting Recognition*, 231–235.

Lang, A.E., Lozano, A.M. (1998). Parkinson's disease. *New England Journal of Medicine*, 339(15), 1998, pp. 1044–1053.

Lotankar, S., Prabhavalkar, K.S., Bhatt, L.K. (2017). Biomarkers for Parkinson's disease: recent advancement. *Neurosci. Bull.*, 33, pp. 585–597.

Martínez-Martín, P., Gil-Nagel, A., Gracia, L.M., Gómez, J.B., Martínez-Sarriés, J., Bermejo, F. (1994). Unified Parkinson's disease rating scale characteristics and structure. *Movement disorders*, 9(1), pp.76–83.

Moetesum, M., Siddiqi, I., Vincent, N. and Cloppet, F. (2018). Assessing visual attributes of handwriting for prediction of neurological disorders—A case study on Parkinson's disease. *Pattern Recognition Letters*.

Mucha, J., Mekyska, J., Galaz, Z., Faundez-Zanuy, M., Lopez-de-Ipina, K., Zvoncak, V., Kiska, T., Smekal, Z., Brabenec, L., Rektorova, I. (2018). Identification and Monitoring of Parkinson's Disease Dysgraphia Based on Fractional-Order Derivatives of Online Handwriting. *Appl. Sci.* 2018, 8, 2566.

Onofri, E., Mercuri, M., Salesi, M., Ferrara, S., Troili, G.M. , Simeone, C., Ricciardi, M.R., Ricci, S., Archer, T. (2013). Dysgraphia in relation to cognitive performance in patients with Alzheimer's disease. *Journal of Intellectual Disability-Diagnosis and Treatment*, 1(2), 2013, pp. 113–124.

O'Reilly, C. and Plamondon, R. (2009). Development of a Sigma–Lognormal representation for on-line signatures. *Pattern Recognition*, *42*(12), 3324–3337.

Pedregosa, F., Varoquaux, G., Gramfort, A., Michel, V., Thirion, B., Grisel, O. and Vanderplas, J. (2011). Scikit-learn: Machine learning in Python. *Journal of Machine Learning Research*, *12*(Oct), 2825–2830.

Pereira, C. R., Pereira, D. R., Rosa, G. H., Albuquerque, V. H., Weber, S. A., Hook, C. and Papa, J. P. (2018). Handwritten dynamics assessment through convolutional neural networks: An application to Parkinson's disease identification. *Artificial Intelligence in Medicine*.

Pirlo, G., Diaz, M., Ferrer, M.A., Impedovo, D., Occhionero, F., Zurlo, U. (2015). Early diagnosis of neurodegenerative diseases by handwritten signature analysis. In proceedings of the *International Conference on Image Analysis and Processing*, 290–297

Pirlo, G., Impedovo, D. (2010). On the measurement of local stability of handwriting: An application to static signature verification. IEEE Workshop on Biometric Measurements and Systems for Security and Medical Applications, Taranto, 2010, pp. 41–44.

Pirlo, G., Impedovo, D. (2012). Adaptive membership functions for handwritten character recognition by Voronoi-based image zoning. *IEEE Transactions on Image Processing, 21(9)*, 3827–3837.

Pirlo, G., Trullo, C.A., Impedovo, D. (2009). A feedback-based multi-classifier system. In proceedings of the *International Conference on Document Analysis and Recognition*, 713–717.

Plamondon, R. (1995a). A kinematic theory of rapid human movements. *Biological Cybernetics*, *72*(4), 295–307.

Plamondon, R. (1995b). A kinematic theory of rapid human movements. *Biological Cybernetics*, *72*(4), 309–320.

Plamondon, R., Pirlo, G., Impedovo, D. (2014). Online signature Verification. Handbook of Document Image Processing and Recognition, pp. 917–947.

Prashanth, R., Roy, S. D., Mandal, P. K. and Ghosh, S. (2014). Automatic classification and prediction models for early Parkinson's disease diagnosis from SPECT imaging. *Expert Systems with Applications*, 41(7), 3333–3342.

Pullman, S. L. (1998). Spiral analysis: a new technique for measuring tremor with a digitizing tablet. Movement Disorders, 13(S3), 85–89.

Rana, B., Juneja, A., Saxena, M., Gudwani, S., Kumaran, S. S., Agrawal, R. K., and Behari, M. (2015). Regions-of-interest based automated diagnosis of Parkinson's disease using T1-weighted MRI. *Expert Systems with Applications*, 42(9), 4506–4516.

Rosenblum, S., Samuel, M., Zlotnik, S., Erikh, I. and Schlesinger, I. (2013). Handwriting as an objective tool for Parkinson's disease diagnosis. *Journal of neurology*, *260*(9), 2357–2361.

Senatore, R., Della Cioppa, A., Marcelli, A. (2019). Automatic Diagnosis of Neurodegenerative Diseases: An Evolutionary Approach for Facing the Interpretability Problem. In *Information*, 10, 30.

Smits, E. J., Tolonen, A. J., Cluitmans, L., van Gils, M., Conway, B. A., Zietsma, R. C. and Maurits, N. M. (2014). Standardized handwriting to assess bradykinesia, micrographia and tremor in Parkinson's disease. PloS one, 9(5), e97614.

Tseng, M. H. and Cermak, S. A. (1993). *The influence of ergonomic factors and perceptu al–motor abilities on handwriting performance.* American Journal of Occupational Therapy, 47(10), 919–926.

Van Gemmert, A. W. A., Plamondon, R. and O'Reilly, C. (2013). Using the sigma-lognormal model to investigate handwriting of individuals with Parkinson's disease. In *16th Biennal Conference of the International Graphonomics Society*, 119–122.

Van Gemmert, A. W., Teulings, H. L. and Stelmach, G. E. (2001). Parkinsonian patients reduce their stroke size with increased pro-cessing demands. Brain and cognition, 47(3), 504–512.

Zhi, N., Jaeger, B.K., Gouldstone, A., Sipahi, R., Frank, S., (2017). Toward monitoring Parkinson's though analysis of static handwriting samples: A quantitative analytical framework. *IEEE Journal of biomedical and health informatics*, vol. 21, No. 2, 2017, pp. 488–495.

https://doi.org/10.1142/9789811226830_0010

# Chapter 10

# Stroke Gesture Synthesis in Human-Computer Interaction

Luis A. Leiva[1,*], Daniel Martín-Albo[2,†], Radu-Daniel Vatavu[3,‡] and Réjean Plamondon[4,§]

[1] *Department of Communications and Networking, Aalto University*
[2] *Independent researcher, WIRIS Math SL, Barcelona, Spain*
[3] *MintViz Lab, MANSiD Research Center, University "Stefan cel Mare" of Suceava, Romania*
[4] *Laboratoire Scribens, Département de génie électrique, Polytechnique Montréal*
[*] *luis.leiva@aalto.fi*, [†] *danielmalbos@gmail.com*, [‡] *radu.vatavu@usm.ro*, [§] *rejean.plamondon@polymtl.ca*

Gesture recognizers usually require a large number of examples to achieve good accuracy. To achieve this goal, a series of time-consuming and expensive experiments must be followed, e.g. preparing the lab, recruiting participants, data collection and labeling, and often reporting to review boards. Fortunately, the Kinematic Theory allows to easily bootstrap gesture data generation. The synthesized data, in turn, may enable further applications of interest. In this chapter, we review the foundations of synthetic *stroke gestures* generation; i.e. the synthesis of data sequences comprising 2D points and associated timestamps, derived e.g. from electronic pens and touchscreens. We show that synthesized gestures not only perform equally similar to gestures generated by human users, but also they "look and feel" the same. We also discuss how the synthesized gestures can be used to estimate *production time*, which is a fundamental measure of performance in Human-Computer Interaction. Ultimately, this work benefits researchers and designers who wish to create gesture-driven prototypes or use the synthesized data to build more sophisticated applications.

## 1. Introduction

Gestures are increasingly becoming a predominant input modality in today's graphical user interfaces (UIs). Gesture interaction is possibly one of the most researched areas in Human-Computer Interaction (HCI), with a long history that started as early as 1960, with the Sketchpad project [Sutherland (1963)] and the RAND tablet [Davis and Ellis (1964)]. Gestures can be mid-air (more prominent in gaming applications) or stroke based (more prominent in mobile applications). We are particularly interested in the latter type, since stroke gestures are becoming more and more relevant to mainstream products such as touchscreen-capable devices like smartphones and tablets; see Fig. 1.

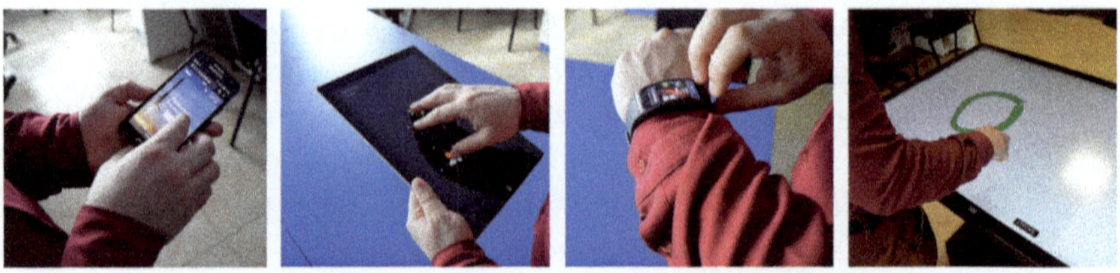

Fig. 1. Stroke gestures input is common on many devices with a wide variety of form factors, from smartphones and tablets to tiny touchscreens featured by some wearables, such as smartwatches, and to large interactive surfaces.

Stroke gestures represent the movement trajectory of one or more contact points on a sensitive surface. Stroke gestures are sometimes also called "pen gestures", "hand drawn marks", "hand drawn gestures", "hand markings", or "markings" [Zhai *et al.* (2012)]. Appert and Zhai (2009) demonstrated the cognitive advantage of stroke gestures in the area of command shortcuts, concluding that stroke gestures tend to give richer perceptual cues to the user, to form an association between the shape of the gesture and the meaning of the command. Stroke gestures also may improve the usability of UIs, by replacing standard shortcuts by more accessible triggers.

Stroke gestures have existed in the market for decades. Early examples of commercial products that successfully incorporated gestures are, e.g., PDAs like the Palm Pilot or the Apple Newton, and the Windows Tablet. These devices featured the Graffiti and Unistroke shorthand writing systems, which used a single stroke Roman letter-like gesture vocabulary. Stroke gestures are increasingly being incorporated to facilitate random access to smartphone contents, such as invoking a command hidden in an advanced settings menu or quickly searching for a friend's email in the contacts list. For example, drawing a letter "S" on a mobile phone screen can be used to search in the address book [DroidByDesign (2014)] or speed-dial some contacts [Li (2010a)]. Similarly, in a video game [POW Studios (2008)] players can draw circles to create shields, arcs to launch fireballs, and hearts to drink potions. More recently, the massive online game 'Harry Potter: Wizards Unite' requires players to draw stroke gestures to create spells and defeat enemies [WB Games and Niantic (2019)]. In addition, many modern tablets incorporate a stylus,[a] which allows for more precise and enriched gestures. Therefore, it is expected that stroke gestures will make a notable impact in consumers' lives.

In general, any application that is driven by gestures must rely on some recognition-based techniques. These techniques often require expert knowledge in pattern recognition or machine learning, something that is typically beyond the reach of many developers and UI designers. Furthermore, these techniques require a large pool of labeled training data, which is usually both time-consuming and expensive to acquire. Thus, it is important to investigate how to empower developers to (1) quickly collect and label gesture samples and (2) create accurate recognizers; both for improving UI usage and user experience.

This book chapter provides a compilation of our previous results on the application of the Kinematic Theory to stroke gestures synthesis [Leiva *et al.* (2016); Martín-Albo and Leiva (2016)], and summarizes for readers empirical results regarding the articulation characteristics of synthetic stroke gestures [Leiva *et al.* (2017a)], the perception of human observers comparing synthetic and authentic gesture articulations [Leiva (2017)], as well as applications to predicting users' gesture input performance [Leiva *et al.* (2018a,b)] and extensions towards modeling the articulation characteristics of gestures produced by various categories of users [Leiva *et al.* (2017b); Ungurean *et al.* (2018b,a)]. By compiling all these previous results in one informative and instructional book chapter, we hope to deliver readers a clear understanding of the application potential of synthetic gestures to support innovations and advances in stroke gestures recognition and analysis.

[a]For example: Apple iPad Pro, Microsoft Surface Pro 6, Samsung Galaxy Tab S4, Huawei MediaPad M5 Pro, Wacom Cintiq 22HD, and Lenovo Yoga Book, among others.

## 2. Related Research

In this section we review core areas that resemble the most to our work: approaches to gesture recognition and gesture bootstrapping.

### 2.1. *Gesture recognition*

Gesture recognition has its own roots in sketching and handwriting recognition [Connell and Jain (2000); Deepu *et al.* (2004); Marzinkewitsch (1991); Rubine (1991)]. Classification methods include, among others: linear discriminant analysis [Rubine (1991)], template matching [Connell and Jain (2000)], decision trees [Belaid and Haton (1984)], neural networks [Marzinkewitsch (1991)], hidden Markov models [Koschinski *et al.* (1995)], parsing grammars [Costagliola *et al.* (2004)], support vector machines [Bahlmann *et al.* (2001)], principal component analysis [Deepu *et al.* (2004)], or *ad-hoc* recognizers [Leiva *et al.* (2013, 2014)].

In HCI, most gesture recognizers for prototyping UIs are based on the template matching (or instance-based) approach [Leiva *et al.* (2014)]: a query gesture is geometrically compared against a number of stored templates, using 1 nearest-neighbor for classification and either Euclidean distance or a Mean Square Error (MSE) score as dissimilarity measures. Template matchers are a very viable and a relatively simple solution for recognizing gestures, and can be adapted to personalized user gestures.

Popular examples of these template-based recognizers among the HCI literature are part of the so-called "$ family": $1 [Wobbrock *et al.* (2007)], $N [Anthony and Wobbrock (2010)], and their improved versions Protractor [Li (2010b)] and $N-Protractor [Anthony and Wobbrock (2012)], respectively. Vatavu *et al.* (2012) introduced $P, an articulation-invariant gesture recognizer that represents stroke gestures as clouds of 2D points, discarding thus stroke count, order, and direction, with the most recent instantiation being $Q, a quick and accurate recognizer for point clouds [Vatavu *et al.* (2018)].

For personalized, gesture-based interaction, it is hard to foresee what gestures an end-user would specify and what the distribution of these gestures will look like [Li (2010b)]. Nonetheless, they can be both time and space consuming on the computational side, given the size of the gesture vocabulary and the number of stored templates to define each gesture.

### 2.2. *Gesture boostrapping*

Training a high-quality recognizer requires providing examples that illustrate sufficient variation to enable robust inference on unseen, future data. Example-based approaches like GRANDMA [Rubine (1991)], Agate [Landay and Myers (1993)], or Gesture Studio [Lü and Li (2013)] allow developers to create and test gestures by recording examples. There are a number of similar systems tailoring end-users, like EventHurdle [Kim and Nam (2013)], A CAPpella [Dey *et al.* (2004)], or GestIT [Spano *et al.* (2013)]. They support designers' explorative prototyping through programming by demonstration environments. Another strand of research is aimed at simplifying the process of designing gesture sets. For example, Gesture Script [Lü *et al.* (2014)], Gesture Marks [Ouyang and Li (2012)], Gestalt [Patel *et al.* (2010)], or CrowdLearner [Amini and Li (2013)]. Finally, we can find a number of competing systems aimed at creating synthetic 3D gestures as a means to improve gesture recognizers, including e.g. MAGIC [Ashbrook and Starner (2010); Kohlsdorf and Starner (2013)] and Gesture Follower [Caramiaux *et al.* (2014)]. MAGIC performs local perturbations to the resampled points of a gesture, whereas Gesture Follower introduce some variations to a gesture template using Viviani's curve formulation.

Overall, training data is the key factor to build a competitive gesture recognizer, for which most of the previously reviewed approaches have contributed to generating their own, without having to recruit participants and perform time-consuming user evaluations. They also have contributed to decreasing the number of iterations needed to build a fast and stable gesture recognition interface. However, there is no evidence that any of the previous works can produce human-like samples. Further, artificially generated samples usually achieve suboptimal results since they do not illustrate sufficient variation required for high-quality training [Almaksour *et al.* (2011); Plamondon *et al.* (2014); Reznakova *et al.* (2015)] and therefore the achieved error rates are typically higher when compared to training exclusively with human-generated data.

Besides the above-mentioned error rates, another important factor worth mentioning toward the adoption of one gesture recognizer over another is performance, typically represented by processor time and memory usage. It is here where the above-mentioned template matchers usually excel, and the main reason why we chose them to conduct our recognition experiments in the next sections. In addition, template matchers are very easy to convey, implement, and deploy on any platform for non-specialists whose objective is quickly enhancing interactivity and not dealing with the complexity of the underlying recognition algorithms.

## 3. Synthesizing Gestures

Many models have been proposed to study human movement production; e.g., models relying on neural networks [Bullock and Grossberg (1988)], equilibrium point models [Feldman (1966)], behavioral models [Thomassen *et al.* (1983)], coupled oscillator models [Hollerbach (1981)], kinematic models [Meyer *et al.* (1990)], or models exploiting minimization principles [Flash and Hogan (1985)]. Other models exploit the properties of various functions to reproduce human movements; e.g., exponentials [Plamondon and Lamarche (1986)], second order systems [van der Gon and Thuring (1965)], beta functions [Alimi (2003)], splines [Morasso *et al.* (1983)], Viviani's curves [Viviani and Flash (1995)], and trigonometrical functions [Maarse (1987)]. Among these approaches, the Kinematic Theory [Plamondon (1995)] provides a well-established and solid framework for the study of the production of human movements. This framework takes into account different psychophysiological features, such as the neuromuscular response time, and has been shown to outperform many other approaches [Plamondon *et al.* (1993)]. The Sigma-Lognormal ($\Sigma\Lambda$) model [Plamondon and Djioua (2006)] is the latest instantiation of this framework, and very recently has been used to explore gesture recognition.

At a high-level representation, the Kinematic Theory assumes that a complex handwritten trace (such as a stroke gesture) is composed of a series of *primitives*[b] (circular arcs) connecting a sequence of *virtual targets.* This series of primitives conform the "action plan" of the user, which is fed through the neuromuscular network to produce a trajectory that leaves a handwritten trace; see Fig. 2.

Under this framework, each $i$th gesture primitive is modeled according to a lognormal function representing their corresponding velocity profile. Each primitive is defined by a set of *central* parameters $(D, t_0, \theta)$ and *peripheral* parameters $(\mu, \sigma)$ [Plamondon (1995)]. The central parameters describe the internal cognitive processes that will lead the user to articulate a handwritten trajectory (i.e. *Where to move?*), whereas the peripheral parameters describe the motor implementation of such articulation.

[b]In the gesture recognition literature, the term "stroke" denotes the trajectory between two consecutive pointer-down and pointer-up events. In the Kinematic Theory literature, a "stroke" is what we call "primitive" in this chapter.

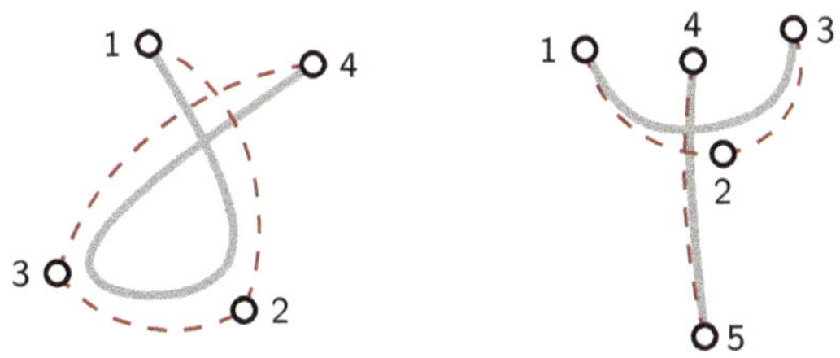


Fig. 2. A gesture stroke (solid lines) is described by the temporal overlap of a series of primitives (dashed arcs) connecting a sequence of virtual targets (numbered circles). Each primitive is described by a lognormal velocity profile.

Concretely, $D$ represents the overall size (amplitude) of the primitive, $t_0$ accounts for the handwriting start time, $\theta$ informs about rotation and direction, and finally $\mu$ and $\sigma$ denote the mean and variance of the lognormal function, respectively, reflecting the neuromuscular response.

$$\begin{aligned} \|\vec{v}_i(t)\| &= D_i \Lambda(t; t_{0_i}, \mu_i, \sigma_i^2) \\ &= \frac{D_i}{\sigma_i \sqrt{2\pi}(t - t_{0_i})} \exp\left(\frac{-[\ln(t - t_{0_i}) - \mu_i]^2}{2\sigma_i^2}\right) \end{aligned} \tag{1}$$

$$\vec{v}(t) = \sum_{i=1}^{N} \vec{v}_i(t) = \sum_{i=1}^{N} \begin{bmatrix} \cos \phi_i(t) \\ \sin \phi_i(t) \end{bmatrix} D_i \Lambda(t; t_{0_i}, \mu_i, \sigma_i^2) \tag{2}$$

$$\phi_i(t) = \theta_{s_i} + \frac{\theta_{e_i} - \theta_{s_i}}{2} \left[1 + \operatorname{erf}\left(\frac{\ln(t - t_{0_i}) - \mu_i}{\sigma_i \sqrt{2}}\right)\right] \tag{3}$$

Then, a $\Sigma\Lambda$ extractor computes the parameter values that best explain the observed velocity profiles [Martín-Albo *et al.* (2015)]. Once the gesture primitives are modeled, perturbations can be added to the model parameters in order to produce different gesture variations [Leiva *et al.* (2017a)]:

$$p_i^* = p_i + n_{p_i} \tag{4}$$

where $p_i = \{\mu_i, \sigma_i, D_i, \theta_i\}$ denote the $\Sigma\Lambda$ parameters, with $n_{p_i} = \mathcal{U}(-n_i, n_i)$ being the noise applied to each primitive, according to a uniform distribution (i.e., a rectangular distribution with constant probability) centered around the expected human variability ranges [Galbally *et al.* (2012b); Leiva *et al.* (2017a)]: $n_\mu = n_\sigma = 0.1$, $n_D = 0.15$, $n_\theta = 0.06$. Figure 3 shows a series of actual examples of the synthetic gestures produced with the Kinematic Theory.

Previous works have demonstrated the connection between the distortion of the $\Sigma\Lambda$ model parameters and the intra-variability found in human

**Human sample** **Synthetic samples**

Fig. 3. Examples of gestures synthetized with the Kinematic Theory, using a single human example as input.

handwriting [Djioua and Plamondon (2009)]. Combining both types of variations reflects real-life situations like performing the same movement under different psychophysiological conditions. For example, perturbations in $\mu$ and $\sigma$ mimic peripheral noise, e.g., a user who articulates the same gesture slightly different each time; perturbations in $D$ and $\theta$ refer to central fluctuations that occur in the position of the virtual targets of the action plan from one articulation to the next [Leiva *et al.* (2016, 2017a)]. Finally, we should mention that we decided to not add perturbations to the $t_0$ parameter, since it is very sensitive to fluctuations [Djioua and Plamondon (2009); Leiva *et al.* (2016)]. Nevertheless, perturbations in $t_0$ have been suggested to reflect changes in the sequence of command instantiation e.g. due to a decrease in attention or cognitive neuromotor fatigue [Djioua and Plamondon (2009)]. Therefore, further analysis of the $t_0$ parameter is left as an opportunity for future work.

## 4. Gesture Performance Analysis

The first experiment compared the performance of synthetic gestures with that of human samples under user-independent tests in terms of articulation speed, input device, and gesture variability. The user-independent setting is the more generalizable scenario, however the interested reader may consult user-dependent tests and a follow-up evaluation in our previous work [Leiva *et al.* (2016)].

We synthesized two popular datasets in HCI: GDS [Wobbrock *et al.* (2007)] and MMG [Anthony and Wobbrock (2012)]. Both datasets include examples of simple and complex gestures produced with different devices and under different execution speeds. Therefore these datasets are a relevant testbed for conducting replicable research on stroke gestures.

On the one hand, the GDS dataset comprises 5,280 *unistroke* gestures (16 classes). Ten users provided 10 samples per class at 3 articulation speeds (slow, medium, fast) using an iPAQ Pocket PC (stylus as input device). On the other hand, the MMG dataset comprises 9,600 *multistroke* gestures (16 classes). Twenty users provided 10 samples per class at 3 articulation speeds (same as in GDS) using either finger (half of the users) or stylus as input device on a Tablet PC. Both GDS and MMG datasets are available at `http://depts.washington.edu/madlab/proj/dollar/`. The synthesized versions of both datasets are available at `https://luis.leiva.name/g3/#datasets`.

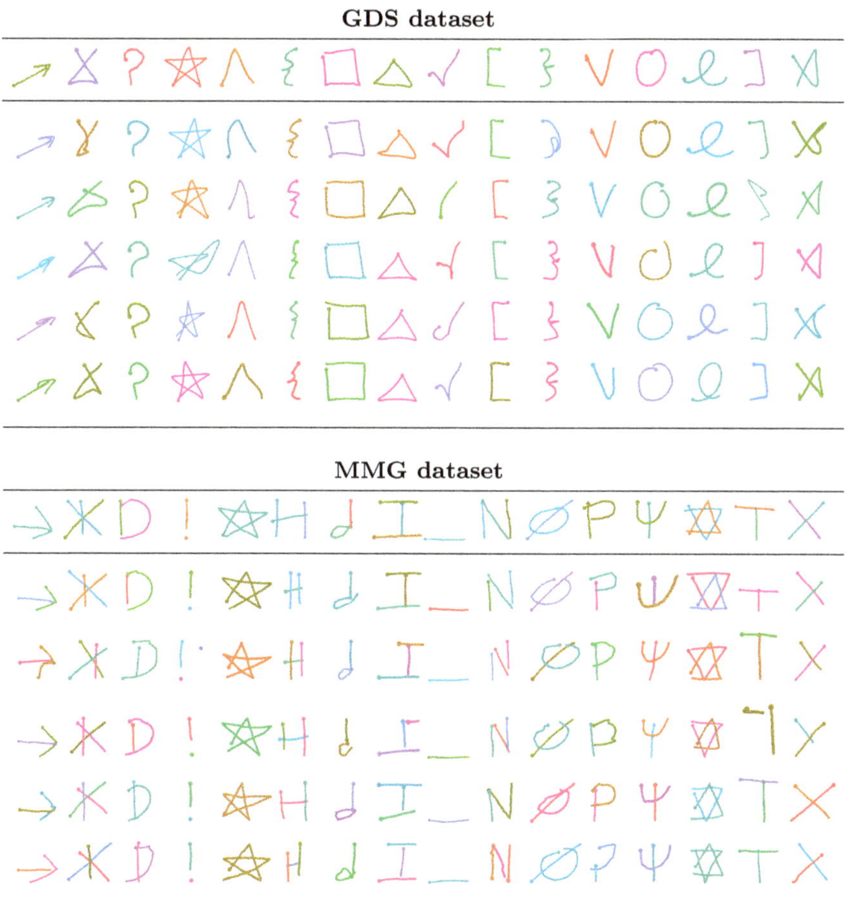


Fig. 4. Five synthesized gestures using one human example (top row) as input.

### 4.1. *Impact of articulation speed*

We sought to analyze whether gesture articulation speed leads to a difference in classification error rates between human and synthetic templates. The GDS dataset (unistroke gestures) was analyzed with the $1 recognizer, whereas the MMG dataset (multistroke gestures) was analyzed with the $P recognizer. Both recognizers were initialized with 10 templates, as suggested in previous work [Anthony and Wobbrock (2010); Li (2010b); Wobbrock *et al.* (2007)]. Table 1 summarizes this experiment.

As can be observed, error rates are very small ($< 1\%$), indicating that the recognizers were successful in classifying correctly each gesture instance. The error rates for the synthesized gestures were slightly higher, therefore we ran two-tailed paired $t$-tests to investigate further these differences. We applied the Bonferroni correction, to counteract Type I errors as a result of multiple comparisons. The statistical tests revealed no statistically significant differences on recognition errors for any of the articulation speeds, suggesting thus that synthetic gestures can be recognized with the same confidence as their human counterparts, regardless the execution speed.

Table 1. Effect of articulation speed on error rates (in %).

| **Type** | | **GDS** | | | **MMG** | |
|---|---|---|---|---|---|---|
| | slow | med. | fast | slow | med. | fast |
| Human | 0.09 | 0.06 | 0.18 | 0.03 | 0.02 | 0.06 |
| Synthetic | 0.20 | 0.27 | 0.90 | 0.17 | 0.36 | 0.56 |

### 4.2. *Impact of input device*

We also sought to analyze whether the input device leads to a difference in classification error rates between human and synthetic templates. This analysis was performed over the MMG dataset, which is the one that provides two data splits (finger and stylus). We used the $P recognizer with 10 templates, as in the previous experiment. Table 2 summarizes this experiment.

As can be observed, error rates are very small ($< 0.5\%$), indicating that the recognizers were successful in classifying correctly each gesture instance. The error rates for the synthesized gestures were slightly higher, therefore we ran two-tailed paired $t$-tests (Bonferroni corrected) to investigate further these differences. The statistical tests revealed no statistically significant differences on recognition errors for any of the articulation speeds,

Table 2. Effect of input device on error rates (in %).

| Human | | Synthetic | |
|---|---|---|---|
| finger | stylus | finger | stylus |
| 0.03 | 0.04 | 0.43 | 0.29 |

suggesting thus that gestures can be successfully synthesized using either a stylus or the human finger as input.

## 4.3. *Impact of gesture variability*

Finally, we sought to analyze whether an increase in the amount of noise $\xi$ introduced to the $\Sigma\Lambda$ model parameters leads to more variable synthetic gestures. We computed the mean squared error[c] between human and synthetic gestures for different number of synthesized samples using $\xi$ from 0.0 (no variability) to 1.0 (maximum variability, in the allotted human ranges [Leiva *et al.* (2016)]). Table 3 summarizes this experiment.

Table 3. Gesture variability in term of mean squared error (standard errors in parentheses).

| N | GDS | | | | MMG | | | |
|---|---|---|---|---|---|---|---|---|
| | $\xi = 0.0$ | | $\xi = 1.0$ | | $\xi = 0.0$ | | $\xi = 1.0$ | |
| 10 | 554.9 | (56.6) | 593.6 | (57.8) | 169.7 | (9.5) | 490.4 | (38.2) |
| 100 | " | (17.8) | 622.1 | (20.5) | " | (3.0) | 493.4 | (10.7) |
| 1000 | " | (5.6) | 621.4 | (6.4) | " | (0.9) | 498.7 | (3.5) |

As expected, it was found that synthetic samples are more variable as $\xi$ increases. As expected, within-samples variability was found to increase as the number of requested synthetic samples increases. In general, we observed that requesting a small number of synthetic samples (10 samples per gesture) provides slightly less variable samples. For a given value of $\xi$, variability was found to increase as the number of requested synthetic samples increases, though we suspect it is because the MSE is underestimated for small batch sizes. Indeed, the standard error (Table 3) gets smaller as the number of samples gets larger, because the mean of a large sample is likely to be closer to the true population mean.

[c]To ease computation, strokes were resampled in such a way that a human sample and its synthesized samples had the same length.

## 5. Gesture Similarity Analysis

To provide further evidence on the value of the Kinematic Theory as a means to generate synthetic stroke gestures, we conducted an online survey that measured the user perception toward gestures' human likeness. We used the same datasets depicted in the previous section, both in their original and synthesized form. The survey is still available online at `https://g3.prhlt.upv.es/guessit/`. Eventually, 236 participants took part in this study.

We defined the guessing accuracy as the user's ability to distinguish between human and synthetic samples; i.e., the proportion of gestures that were successfully classified by the user. Then, two types of errors[d] can be committed [Galbally *et al.* (2012a,b)]: (i) a synthetic gesture is mistaken with a real sample, measured by the False Real Rate (Type I error); and (ii) a real gesture is marked as synthetic, measured by the False Synthetic Rate (Type II error). The results are presented in Table 4.

Table 4. Accuracy and error rates (in %).

| **Accuracy** | **False Real Rate** | **False Synthetic Rate** |
|---|---|---|
| 49.65 | 29.84 | 20.40 |

A paired two-sample $t$-test (two-tailed alternative hypothesis) revealed that there is no difference between classified and misclassified gestures, proving thus the human-like appearance of synthetic samples.[e] In sum, participants could not tell human and synthetic gestures apart. Follow-up analyses [Leiva *et al.* (2017a); Leiva (2017)] provided further evidence that synthesized gestures are actually reflective of how users produce stroke gestures.

We also examined the intra-class variability of human gestures, distance-wise; i.e., how variable is a human gesture sample as compared to the rest of the human samples that belong to the same gesture class. The Pearson's correlation coefficient was found to be $\rho > 0.9$ in all datasets, and we observed that $\rho$ decreased as $\xi$ increased. This indicates that, as expected, samples synthesized with a low variability degree look much more similar

[d]The sum of False Real Rate (Type I error) and False Synthetic Rate (Type II error) yields the overall error rate (100 - Accuracy).
[e]We assume that humans are reasonable judges of realism, but it might not always be the case.

to the human samples from which they were generated. Taken together, these experiments suggest that the visual appearance of synthetic gestures is very similar and close to that of their human counterparts.

## 6. Production Times Analysis

As a practical application of the synthesized gestures with the Kinematic Theory, we report results on a fundamental topic in HCI. The production time of a stroke gesture, i.e., how long it takes users, on average, to produce a 2D handwritten trace on a touch-sensitive surface, is one essential aspect of user performance with gesture input [Castellucci and MacKenzie (2008); Cao and Zhai (2007); Vatavu *et al.* (2011); Rekik *et al.* (2014)]. Such insightful information about users' performance represents a valuable asset for practitioners to inform gesture design directly, e.g., what are the fastest gestures to produce [Appert and Zhai (2009); Castellucci and MacKenzie (2008)] or indirectly, e.g., what are the easiest gestures to execute from a given set [Vatavu *et al.* (2011); Rekik *et al.* (2014)]. Moreover, gesture production time turned out to be an excellent predictor of users' subjective perceptions of the difficulty to articulate stroke gestures [Rekik *et al.* (2014); Vatavu *et al.* (2011)]. In this context, it is important for user interface designers to be able to estimate a priori, as accurately as possible, users' input performance in order to save considerable time and effort demanded by subsequent user evaluations and/or gesture set redesigns.

For this experiment, we compared the synthesized gesture production times against their human counterparts [Leiva *et al.* (2018a,b, 2020)] using the same two datasets described in the previous experiments (GDS and MMG). The performance of $\Sigma\Lambda$ as time predictor was evaluated with the following accuracy measures:

**Rank accuracy** evaluates the extent to which the synthetic gestures deliver the correct *ranking* (Spearman's correlation) of gestures according to their production times. The closer the rank to 1, the better.

**Absolute accuracy** evaluates the extent to which the synthetic gestures deliver the correct *magnitude* (in ms) of the expected production time of a given gesture type. The closer the magnitude to the true production time, the better.

Production times were computed using a user-independent leave-one-out cross-validation procedure, which considers each execution from each gesture produced by each participant as the representative gesture sample to compare against the $\Sigma\Lambda$ model. The results of this experiment are shown in Table 5. In any case, we found no statistically significant differences between true and estimated times, which builds our confidence that synthesized gestures are on par with users' actual time performance with stroke gesture input.

Table 5. Production times analysis. Time in ms. SDs denoted in parentheses.

| Dataset | Device | Speed | True time | Rank acc. | Abs. acc. |
|---|---|---|---|---|---|
| GDS | stylus | slow | 1607 (679) | .979*** | 1587 (678) |
| | | med. | 1055 (438) | .992*** | 1078 (425) |
| | | fast | 618 (259) | .991*** | 670 (273) |
| MMG | stylus | slow | 1011 (683) | .932*** | 1120 (638) |
| | | med. | 668 (511) | .900*** | 672 (408) |
| | | fast | 537 (440) | .788*** | 609 (339) |
| MMG | finger | slow | 1038 (694) | .950*** | 940 (665) |
| | | med. | 688 (504) | .914*** | 805 (528) |
| | | fast | 553 (434) | .775*** | 730 (464) |

Statistical significance levels are $p < .001$ in all cases.

We are thus confident that synthetic samples look like real ones at the geometric level. However, velocity profiles comprise subtle time-dependent relationships that so far have not been studied for stroke gestures. Therefore, how realistic are the velocity profiles synthesized? To answer this question, the plots below display the velocity profiles of a set of gestures chosen at random from each dataset. We can observe that the synthesized velocity profiles are often in line with the human velocity profiles, however the velocity range is typically smaller for the synthesized gestures. This happened especially to the synthesized samples from the GDS dataset, as shown in Fig. 5, and is a consequence of our current implementation of the $\Sigma\Lambda$ model, which transforms the original stream into constant frequency; i.e., the resulting synthesized coordinates are uniformly distributed in time. This resampling process allows to "fix" downsampled strokes that were acquired with under-resourced hardware [Leiva *et al.* (2016)] but perhaps an undesired side effect is that synthetic velocity profiles look smoother than usual. In addition, the human velocity profiles have large peaks that are typically caused by sensor noise from the input device. Various types of

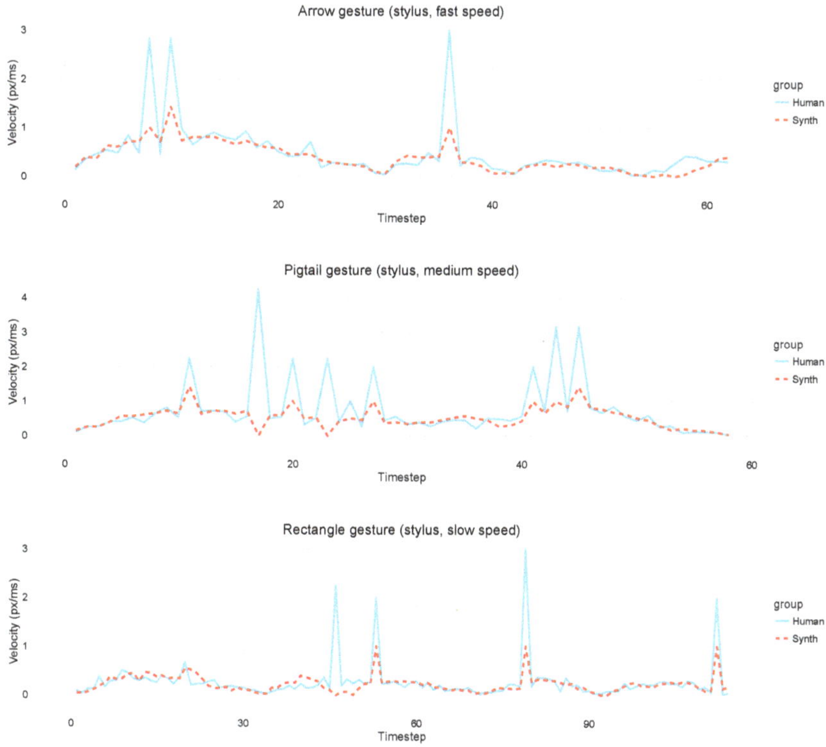


Fig. 5. Comparison of velocity profiles from samples in the GDS dataset.

low-pass filters can be applied to remove such noise, but we decided to analyze the raw signal to get more fundamental insights.

The MMG is a particularly bad dataset to analyze velocity profiles, because many gesture points have duplicated timestamps [Leiva *et al.* (2017a)], which makes it challenging to estimate the actual velocity values. For example, if two consecutive points at timestep $i-1$ and $i$ have the same timestamp, the time derivative at timestep $i$ is zero and therefore the velocity value is infinite. In consequence, careful preprocessing must be performed. Figure 6 shows a couple of velocity profiles corresponding to gestures articulated with stylus and finger, respectively.

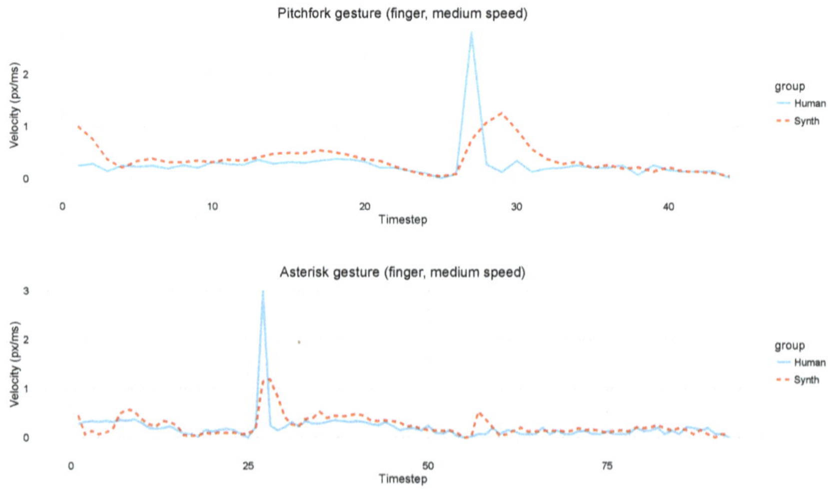


Fig. 6. Comparison of velocity profiles from samples in the MMG dataset.

## 7. Discussion

Users tend to be reluctant to invest time and effort upfront to train or adjust software before using it [Appert and Zhai (2009)]. Further, users are unwilling to provide more than a small set of samples for training [Li (2010b)]. Consequently, synthesizing techniques like ours are of high value, as they help to lower time and costs associated to recruiting users and subsequent data labeling. Furthermore, researchers can focus exclusively on UI design rather than fret over machine learning concepts or toolkits that may not be available for their platform. Eventually, our web application [Martín-Albo and Leiva (2016)] can be used for rapid prototyping, allowing developers to define new gestures on demand. The interested reader can access it at `https://g3.prhlt.upv.es`.

Until now the "human likeness" of synthesized gestures was measured indirectly, intermediated by classification/recognition accuracy performance. Our studies are important because recent research has shown how different users produce different gesture articulations in various conditions. Thus, researchers and practitioners can be confident that synthesized gestures using the Kinematic Theory are actually reflective of how users produce stroke gestures.

One requirement for the ΣΛ model to produce proper results is that the user-provided gesture example should be reconstructed with high quality, as defined by the signal-to-noise ratio (SNR). Previous work suggested that SNR values below 15 dB denote poor execution quality [Almaksour *et al.* (2011); Leiva *et al.* (2016, 2017a)] and, in such cases, the input gesture should be discarded. However, we found this situation to appear extremely rarely in practice: out of the 14, 240 stroke gestures that we evaluated in our experiments, only 22 samples had SNR$<$ 15 dB, which represents merely 0.15% of the data.

On a side note, the concepts concerning internal models of human movements have been well supported by behavioral studies in the field of sensory motor control. Overall, it is assumed that users are "ideal" motion planners who choose movement trajectories to minimize an expected loss [Trommershäuser *et al.* (2003); Quinn and Zhai (2018)]. Currently, we can find two compelling theories to describe those movements: the Minimization Theory [Flash and Hogan (1985)] and the Kinematic Theory [Plamondon (1995)]. Actually, it has been shown that their concepts are linked and describe, with different arguments, a model of velocity profiles [Djioua and Plamondon (2010); Leiva *et al.* (2017a)], the Minimization Theory being as a very good approximation of the lognormal description provided by the Kinematic Theory.

The previous instantiation of the Kinematic Theory (the Delta-Lognormal model) assumed that the production of a stroke requires the synergetic activation of two neuromuscular systems, one agonist and the other antagonist to the direction of the movement. These synchronous commands propagate in parallel across the two neuromuscular systems, each of which is described by a lognormal impulse response and has its own timing properties. On the contrary, the ΣΛ model does not assume that the two neuromuscular systems are working in precisely opposite directions. The output velocity is thus described by a vectorial summation of the contribution of each neuromuscular system involved in the production of a stroke. This model is actually very general, and is not limited to a single stroke description [Plamondon and Djioua (2006); O'Reilly and Plamondon (2009)]. This corroborates the prediction of the Kinematic Theory, where it is theorized that the ideal impulse response of the human neuromotor system follows a lognormal response that results from the limiting behavior of a large number of interdependent neuromuscular networks, as stated in the first chapter of this book and referred to as the lognormality principle.

The fundamental advantage of our approach over others is that the $\Sigma\Lambda$ model only needs *one* user example to start synthesizing more gestures. Although using only one gesture example could be seen as a limitation (i.e., the results are bound to the sample gesture provided as seed), our experiments revealed that synthetic gestures are on par with their human counterparts. This performance is explained by the fact that our synthesizer [Leiva *et al.* (2016); Martín-Albo and Leiva (2016)] uses generic, user-independent value ranges for the $\Sigma\Lambda$ parameters, which were empirically derived from and validated for many user categories by prior work [Galbally *et al.* (2012b); Leiva *et al.* (2016, 2017b); Martín-Albo *et al.* (2014)]. The interested reader can refer to these prior works to know other range values and how they may impact recognition performance. Although we should note that different values may be needed for different user categories, such as gestures articulated by visually impaired users [Leiva *et al.* (2017b)] or users with motor impairments [Ungurean *et al.* (2018b,a)].

We also have shown that the $\Sigma\Lambda$ model delivers accurate predictions of users' stroke gesture production times with no effort required from designers. This is a practical application to enable effective gesture sets design. We should stress the fact that it is important to provide designers with both measures of central tendency (i.e., the expected production time of a gesture) and, equally important, measures of variation as well; i.e., how much are users expected to deviate their production times from the mean? Given that users are known to vary their gesture articulations [Anthony *et al.* (2013); Vatavu *et al.* (2013)], it also causes variation in their production times. With synthesized gestures, we are able to deliver the extra information given by location and dispersion-based measures that can tell the practitioner the range in which the mean time is likely to lie and also how much to expect individual times to deviate from the mean.

Finally, in light of the analysis of the synthetic velocity profiles, we believe there is still room for improvement in how synthetic gestures are produced by the $\Sigma\Lambda$ model. While the articulation of synthetic gestures, as reflected by the shape of their velocity profiles, is very much in line with their corresponding human velocity profiles, the $\Sigma\Lambda$ model is often unable to capture *all* the compensatory micro-movements observed in human gesturing, very much like it happens in human handwriting. These subtle micro-movements are explained by the isochrony principle [Viviani and McCollum (1983)], which states that the velocity of a movement is proportionally linked to its linear extension (or trajectory) so as to permit the execution time to be maintained approximately constant [Freund (1986)].

Stroke gestures captured on commodity touchscreens, such as the ones we have analyzed in this work, have usually low temporal resolution and *asynchronous* timestamps. Therefore, their velocity profiles are much more challenging to reconstruct than, say, handwritten signatures performed on a high-resolution tablet with a high and fixed sampling rate. More research is thus needed to understand the articulation of stroke gestures on touch-capable devices.

## 8. Conclusion

Our experiments provide evidence against the implied alternate hypothesis of a difference between human and synthesized stroke gestures. Researchers and practitioners can be finally confident that the Kinematic Theory generates stroke gestures that not only perform equally similar to their human counterparts but also they look and feel the same. And while there is still room for improving how synthetic strokes gestures are articulated, it is reliable to generate synthetic datasets this way, since the overall performance and behavior of gesture samples will be consistently similar to that of actual users. This can be useful to researchers and practitioners in many ways.

As a practical application of the synthesized gestures, we have discussed how they can be used to estimate production times, which is one of the fundamental performance measures in HCI. We have left out other practical applications that also might be of interest to practitioners, such as synthesizing gestures across populations [Leiva *et al.* (2017b); Ungurean *et al.* (2018b)], evaluating the effect of hardware resolution on handwriting analysis [Martín-Albo *et al.* (2016b)], or detecting the "hidden" user intent in mouse cursor movements [Martín-Albo *et al.* (2016a)]. However, space precludes a complete treatment of all possible applications of synthetic gestures in HCI. Instead, the interested reader is redirected to the works referenced above.

In summary, the Kinematic Theory provides the HCI community with a reliable way to synthesize stroke gesture sets without having to expressly collect them from a large pool human subjects. However, we do not to encourage the substitution of human gestures by synthetic ones, but rather to provide an automated way to lower the need of recruiting a large number of users and subsequent data labeling.

## References

Alimi, A. M. (2003). Beta neuro-fuzzy systems, in W. Duch and D. Rutkowska (eds.), *TASK Quarterly J., Special Issue on Neural Networks*, Vol. 7.

Almaksour, A., Anquetil, E., Plamondon, R. and O'Reilly, C. (2011). Synthetic handwritten gesture generation using Sigma-Lognormal model for evolving handwriting classifiers, in *Proceedings of Biennial Conf. of the Intl. Graphonomics Society (IGS)*.

Amini, S. and Li, Y. (2013). CrowdLearner: Rapidly creating mobile recognizers using crowdsourcing, in *Proceedings of Annual ACM Symp. on User Interface Software and Technology (UIST)*.

Anthony, L., Vatavu, R.-D. and Wobbrock, J. O. (2013). Understanding the consistency of users' pen and finger stroke gesture articulation, in *Proceedings of Graphics Interface (GI)*.

Anthony, L. and Wobbrock, J. O. (2010). A lightweight multistroke recognizer for user interface prototypes, in *Proceedings of Graphics Interface (GI)*.

Anthony, L. and Wobbrock, J. O. (2012). $N-protractor: a fast and accurate multistroke recognizer, in *Proceedings of Graphics Interface (GI)*.

Appert, C. and Zhai, S. (2009). Using strokes as command shortcuts: Cognitive benefits and toolkit support, in *Proceedings of SIGCHI Conf. on Human Factors in Computing Systems (CHI)*.

Ashbrook, D. and Starner, T. E. (2010). MAGIC: A motion gesture design tool, in *Proceedings of SIGCHI Conf. on Human Factors in Computing Systems (CHI)*.

Bahlmann, C., Haasdonk, B. and Burkhardt, H. (2001). On-line handwriting recognition with support vector machines: A kernel approach, in *Proceedings of Intl. Workshop on Frontiers in Handwriting Recognition (IWFHR)*.

Belaid, A. and Haton, J.-P. (1984). A syntactic approach for handwritten formula recognition, *IEEE Trans. Pattern Anal. Mach. Intell.* **6**, 1.

Bullock, D. and Grossberg, S. (1988). The VITE model: A neural command circuit for generating arm and articulator trajectories, in *Dynamic Patterns in Complex Systems*.

Cao, X. and Zhai, S. (2007). Modeling human performance of pen stroke gestures, in *Proceedings of SIGCHI Conf. on Human Factors in Computing Systems (CHI)*.

Caramiaux, B., Montecchio, N., Tanaka, A. and Bevilacqua, F. (2014). Adaptive gesture recognition with variation estimation for interactive systems, *ACM Trans. Interact. Intell. Syst.* **4**, 4.

Castellucci, S. J. and MacKenzie, I. S. (2008). Graffiti vs. Unistrokes: An empirical comparison, in *Proceedings of SIGCHI Conf. on Human Factors in Computing Systems (CHI)*.

Connell, S. D. and Jain, A. K. (2000). Template-based on-line character recognition, *Pattern Recogn.* **34**, 1.

Costagliola, G., Deufemia, V., Polese, G. and Risi, M. (2004). A parsing technique for sketch recognition systems, in *Proceedings of Symp. on Visual Languages and Human-Centric Computing (VLHCC)*.

Davis, M. R. and Ellis, T. O. (1964). The RAND tablet: A man-machine graphical communication device, in *Proceedings of American Federation of Information Processing Societies (AFIPS).*

Deepu, V., Madhvanath, S. and Ramakrishnan, A. G. (2004). Principal component analysis for online handwritten character recognition, in *Proceedings of Intl. Conf. on Pattern Recognition (ICPR).*

Dey, A. K., Hamid, R., Beckmann, C., Li, I. and Hsu, D. (2004). A CAPpella: Programming by demonstration of context-aware applications, in *Proceedings of SIGCHI Conf. on Human Factors in Computing Systems (CHI).*

Djioua, M. and Plamondon, R. (2009). Studying the variability of handwriting patterns using the Kinematic Theory, *Hum. Mov. Sci.* **28**, 5.

Djioua, M. and Plamondon, R. (2010). The limit profile of a rapid movement velocity, *Hum. Mov. Sci.* **29**, 1.

DroidByDesign (2014). Oftseen gestures, URL `https://play.google.com/store/apps/details?id=com.proofbydesign.oftSeenGestures`, accessed on May 2019.

Feldman, A. (1966). Functional tuning of the nervous system with control of movement or maintenance of a steady posture, *Biophysics* **11**, 1.

Flash, T. and Hogan, N. (1985). The coordination of arm movements: An experimentally confirmed mathematical model, *J. Neurosci.* **5**, 7.

Freund, H.-J. (1986). Time control of hand movements, *Prog. Brain Res.* **64**.

Galbally, J., Plamondon, R., Fierrez, J. and Ortega-García, J. (2012a). Synthetic on-line signature generation. Part I: Methodology and algorithms, *Pattern Recogn.* **45**, 7.

Galbally, J., Plamondon, R., Fierrez, J. and Ortega-García, J. (2012b). Synthetic on-line signature generation. Part II: Experimental validation, *Pattern Recogn.* **45**, 7.

Hollerbach, J. M. (1981). An oscillation theory of handwriting, *Biol. Cybern.* **39**, 2.

Kim, J.-W. and Nam, T.-J. (2013). EventHurdle: Supporting designers' exploratory interaction prototyping with gesture-based sensors, in *Proceedings of SIGCHI Conf. on Human Factors in Computing Systems (CHI).*

Kohlsdorf, D. and Starner, T. E. (2013). MAGIC Summoning: Towards automatic suggesting and testing of gestures with low probability of false positives during use, *J. Mach. Learn. Res.* **14**, 1.

Koschinski, M., Winkler, H. J. and Lang, M. (1995). Segmentation and recognition of symbols within handwritten mathematical expressions, in *Proceedings of Intl. Conf. on Acoustics, Speech, and Signal Processing (ICASSP).*

Landay, J. A. and Myers, B. A. (1993). Extending an existing user interface toolkit to support gesture recognition, in *Proceedings of INTERCHI'93 Conf. on Human Factors in Computing Systems.*

Leiva, L. A. (2017). Large-scale user perception of synthetic stroke gestures, in *Proceedings of the ACM Conf. on Designing Interactive Systems (DIS).*

Leiva, L. A., Alabau, V., Romero, V., Toselli, A. H. and Vidal, E. (2014). Context-aware gestures for mixed-initiative text editing UIs, *Interact. Comput.* **27**, 1.

Leiva, L. A., Alabau, V. and Vidal, E. (2013). Error-proof, high-performance, and context-aware gestures for interactive text edition, in *Proceedings of Extended Abstracts on Human Factors in Computing Systems (CHI EA).*

Leiva, L. A., Martín-Albo, D. and Plamondon, R. (2016). Gestures à Go Go: Authoring synthetic human-like stroke gestures using the kinematic theory of rapid movements, *ACM Trans. Intell. Syst. Technol.* **7**, 2.

Leiva, L. A., Martín-Albo, D. and Plamondon, R. (2017a). The Kinematic Theory produces human-like stroke gestures, *Interact. Comput.* **29**, 4.

Leiva, L. A., Martín-Albo, D., Plamondon, R. and Vatavu, R.-D. (2018a). KeyTime: Super-accurate prediction of stroke gesture production times, in *Proceedings of SIGCHI Conf. on Human Factors in Computing Systems (CHI).*

Leiva, L. A., Martín-Albo, D. and Vatavu, R.-D. (2017b). Synthesizing stroke gestures across user populations: A case for users with visual impairments, in *Proceedings of SIGCHI Conf. on Human Factors in Computing Systems (CHI).*

Leiva, L. A., Martín-Albo, D. and Vatavu, R.-D. (2018b). GATO: Predicting human performance with multistroke and multitouch gesture input, in *Proceedings of ACM Conf. on Human-computer interaction with mobile devices and services (MobileHCI).*

Leiva, L..A, Vatavu, R.D., Martín-Albo, D. and Plamondon, R. (2020). "Omnis Prædictio: Estimating the Full Spectrum of Human Performance with Stroke Gestures" *Int. J. Hum. Comput. Stud.* 142, DOI:10.1016/j.ijhcs.2020.102466.

Li, Y. (2010a). Gesture Search: A tool for fast mobile data access, in *Proceedings of Annual ACM Symp. on User Interface Software and Technology (UIST).*

Li, Y. (2010b). Protractor: a fast and accurate gesture recognizer, in *Proceedings of SIGCHI Conf. on Human Factors in Computing Systems (CHI).*

Lü, H., Fogarty, J. A. and Li, Y. (2014). Gesture Script: Recognizing gestures and their structure using rendering scripts and interactively trained parts, in *Proceedings of SIGCHI Conf. on Human Factors in Computing Systems (CHI).*

Lü, H. and Li, Y. (2013). Gesture Studio: Authoring multi-touch interactions through demonstration and declaration, in *Proceedings of SIGCHI Conf. on Human Factors in Computing Systems (CHI).*

Maarse, F. J. (1987). *The Study of Handwriting Movement: Peripheral Models and Signal Processing Techniques*, Swets & Zeitlinger.

Martín-Albo, D. and Leiva, L. A. (2016). G3: bootstrapping stroke gestures design with synthetic samples and built-in recognizers, in *Proceedings of ACM Conf. on Human-computer interaction with mobile devices and services (MobileHCI).*

Martín-Albo, D., Leiva, L. A., Huang, J. and Plamondon, R. (2016a). Strokes of insight: User intent detection and kinematic compression of mouse cursor trails, *Inform. Process. Manag.* **56**, 6.

Martín-Albo, D., Leiva, L. A. and Plamondon, R. (2016b). On the design of personal digital bodyguards: Impact of hardware resolution on handwriting analysis, in *Proceedings of Intl. Conf. on Frontiers in Handwriting Recognition (ICFHR).*

Martín-Albo, D., Plamondon, R. and Vidal, E. (2014). Training of on-line handwriting text recognizers with synthetic text generated using the Kinematic Theory of rapid human movements, in *Proceedings of Intl. Conf. on Frontiers in Handwriting Recognition (ICFHR)*.

Martín-Albo, D., Plamondon, R. and Vidal, E. (2015). Improving sigma-lognormal parameter extraction, in *Proceedings of Intl. Conf. on Document Analysis and Recognition (ICDAR)*.

Marzinkewitsch, R. (1991). Operating computer algebra systems by hand-printed input, in *Proceedings of Intl. Symp. on Algorithms and Computation (ISAAC)*.

Meyer, D. E., Keith-Smith, J. E., Kornblum, S., Abrams, R. A. and Wright, C. E. (1990). Speed-accuracy tradeoffs in aimed movements: Toward a theory of rapid voluntary action, *Atten. and Perform.* **13**, 23.

Morasso, P., Ivaldi, F. A. M. and Ruggiero, C. (1983). How a discontinuous mechanism can produce continuous patterns in trajectory formation and handwriting, *Acta Psychol.* **54**, 1.

O'Reilly, C. and Plamondon, R. (2009). Development of a Sigma-Lognormal representation for on-line signatures, *Pattern Recogn.* **42**, 12.

Ouyang, T. and Li, Y. (2012). Bootstrapping personal gesture shortcuts with the wisdom of the crowd and handwriting recognition, in *Proceedings of SIGCHI Conf. on Human Factors in Computing Systems (CHI)*.

Patel, K., Bancroft, N., Drucker, S. M., Fogarty, J., Ko, A. J. and Landay, J. (2010). Gestalt: Integrated support for implementation and analysis in machine learning, in *Proceedings of Annual ACM Symp. on User Interface Software and Technology (UIST)*.

Plamondon, R. (1995). A kinematic theory of rapid human movements. Part I: Movement representation and control, *Biol. Cybern.* **72**, 4.

Plamondon, R., Alimi, A. M., Yergeau, P. and Leclerc, F. (1993). Modelling velocity profiles of rapid movements: A comparative study, *Biol. Cybern.* **69**, 1.

Plamondon, R. and Djioua, M. (2006). A multi-level representation paradigm for handwriting stroke generation, *Hum. Mov. Sci.* **25**, 4–5.

Plamondon, R. and Lamarche, F. (1986). Modelization of handwriting: A system approach, in H. S. R. Kao, G. P. van Galen and R. Hoosain (eds.), *Graphonomics: Contemporary Research in Handwriting*.

Plamondon, R., O'Reilly, C., Galbally, J., Almaksour, A. and Anquetil, E. (2014). Recent developments in the study of rapid human movements with the Kinematic Theory: Applications to handwriting and signature synthesis, *Pattern Recogn. Lett.* **35**.

POW Studios (2008). Mr. spiff's revenge, URL `https://youtu.be/nqeIuNTbT-8`, accessed on May 2019.

Quinn, P. and Zhai, S. (2018). Modeling gesture-typing movements, *Hum.-Comput. Interact.* **33**, 2.

Rekik, Y., Vatavu, R.-D. and Grisoni, L. (2014). Understanding users' perceived difficulty of multi-touch gesture articulation, in *Proceedings of Intl. Conf. on Multimodal Interaction (ICMI)*.

Reznakova, M., Tencer, L., Plamondon, R. and Cheriet, M. (2015). The generation of synthetic handwritten data for improving on-line learning, in *Proc. IGS.*

Rubine, D. (1991). Specifying gestures by example, in *Proceedings of annual Conf. on Computer graphics and interactive techniques (SIGGRAPH).*

Spano, L. D., Cisternino, A., Paternò, F. and Fenu, G. (2013). GestIT: A declarative and compositional framework for multiplatform gesture definition, in *Proceedings of ACM SIGCHI Symp. on Engineering Interactive Computing Systems (EICS).*

Sutherland, I. E. (1963). Sketchpad: A man-machine graphical communication system, Tech. Rep. 296, Lincoln Laboratory, MIT.

Thomassen, A. J., Keuss, P. J. and van Galen, G. P. (1983). Motor aspects of handwriting, *Acta Psychol.* **54**, 1–3.

Trommershäuser, J., Maloney, L. T. and Landy, M. S. (2003). Statistical decision theory and trade-offs in the control of motor response, *Spat. Vis.* **16**, 3–4.

Ungurean, O.-C., Vatavu, R.-D., Leiva, L. A. and Martín-Albo, D. (2018a). Predicting stroke gesture input performance for users with motor impairments, in *Proceedings of ACM Conf. on Human-computer interaction with mobile devices and services (MobileHCI).*

Ungurean, O.-C., Vatavu, R.-D., Leiva, L. A. and Plamondon, R. (2018b). Gesture input for users with motor impairments on touchscreens: Empirical results based on the Kinematic Theory, in *Proceedings of Extended Abstracts on Human Factors in Computing Systems (CHI EA).*

van der Gon, J. J. D. and Thuring, J. P. (1965). The guiding of human movements, *Kybernetik* **14**, 1.

Vatavu, R.-D., Anthony, L. and Wobbrock, J. O. (2012). Gestures as point clouds: A $P recognizer for user interface prototypes, in *Proceedings of Intl. Conf. on Multimodal Interaction (ICMI).*

Vatavu, R.-D., Anthony, L. and Wobbrock, J. O. (2013). Relative accuracy measures for stroke gestures, in *Proceedings of Intl. Conf. on Multimodal Interaction (ICMI).*

Vatavu, R.-D., Anthony, L. and Wobbrock, J. O. (2018). $Q: A super-quick, articulation-invariant stroke-gesture recognizer for low-resource devices, in *Proceedings of ACM Conf. on Human-computer interaction with mobile devices and services (MobileHCI).*

Vatavu, R.-D., Vogel, D., Casiez, G. and Grisoni, L. (2011). Estimating the perceived difficulty of pen gestures, in *Proceedings of IFIP Intl. Conf. on Human-computer Interaction (INTERACT).*

Viviani, P. and Flash, T. (1995). Minimum-jerk, two-thirds power law, and isochrony: converging approaches to movement planning, *J. Exp. Psychol.* **21**, 1.

Viviani, P. and McCollum, G. (1983). The relation between linear extent and velocity in drawing movements, *Neuroscience* **10**.

WB Games and Niantic (2019). Harry potter: Wizards unite, URL `https://wizardsunitehub.info/spells/`, accessed on Jun 2019.

Wobbrock, J. O., Wilson, A. D. and Li, Y. (2007). Gestures without libraries, toolkits or training: A $1 recognizer for user interface prototypes, in *Proceedings of Annual ACM Symp. on User Interface Software and Technology (UIST)*.

Zhai, S., Kristensson, P. O., Appert, C., Anderson, T. H. and Cao, X. (2012). Foundational issues in touch-surface stroke gesture design — an integrative review, in *Foundations and Trends in Human-Computer Interaction*, Vol. 5.

https://doi.org/10.1142/9789811226830_0011

# Chapter 11

# Kinematics Reconstruction of Static Calligraphic Traces from Curvilinear Shape Features*

Daniel Berio[1,†], Frederic Fol Leymarie[1,‡] and Réjean Plamondon[2,§]
[1] *Computing Dept., Goldsmiths, University of London, UK*
[2] *Laboratoire Scribens, Département de génie électrique, Polytechnique Montréal, Canada*
[†] *drand48@gmail.com*, [‡] *ffl@gold.ac.uk*, [§] *rejean.plamondon@poymtl.ca*

Our goal is to be able to reproduce computationally calligraphic traces, such as found in the art practices of graffiti and various forms of more traditional calligraphy, while mimicking their production process. To this end, we propose a method that allows to reconstruct kinematics solely from the geometric samples of handwritten traces in the form of parameters of the *Sigma-Lognormal model*. We ignore the kinematics possibly embedded in the data in order to treat online data and vector patterns with the same procedure.

At the heart of our method, we develop a robust procedure to identify curvilinear shape features based on an analysis of local symmetry axes. These features determine the segmentation of a trace into circular arcs and guide an iterative reconstruction of the input kinematics and geometry in the form of Sigma-Lognormal parameters. We demonstrate how this parametrisation can be used to generate plausible kinematics for a static input trace, and how parameter variations can be exploited to generate traces that resemble the ones seen in real instances of human made calligraphy and graffiti.

## 1. Introduction

Many handwriting analysis methods rely on a prior segmentation of the written trace into constituent *primitives* or *strokes*. Some methods exploit the kinematics of the movement and segment the trajectory in correspondence with minima or other features of velocity [O'Reilly and Plamondon, 2008; Plamondon *et al.*, 2014]. Using the known inverse relation between speed and curvature [Viviani and Schneider, 1991], other methods rely on the identification of curvature extrema along the pen-trace [De Stefano *et al.*, 2005; Ferrer *et al.*,

*An initial shorter version of this chapter appeared as: "Kinematic Reconstruction of Calligraphic Traces from Shape Features", by D. Berio, F.F. Leymarie and R. Plamondon, *Proceedings of ICPRAI*, Concordia University, Montréal, Canada, May 2018, pp. 762–767.

2018], where movement slows down or has to stop momentarily to allow for a discrete change in orientation. This fits with a modelization of handwriting in the form of ballistic "stroke" primitives, in which curvature extrema will typically correspond with velocity minima and are indicative of the initiation of a new stroke.

Most of the existing works are aimed either at a precise analysis of the kinematics of a digitised input or at the segmentation of a handwriting trace into components for biometric or pattern recognition purposes. On the other hand, our specific aim is perceptually and artistically driven, and we seek to *infer a physiologically plausible motion* from an input trace, the kinematics of which may be unavailable, such as when using vector graphics inputs, or may be degraded or unreliable due to the poor quality of a digitisation device, such as when using low cost tablets or trackpads. The motivation for this approach is grounded on the hypothesis that the visual perception of marks made by a drawing hand triggers activity in the motor areas of the brain [Freedberg and Gallese, 2007; Longcamp *et al.*, 2003], and further induces an approximate mental recovery of the (likely) movements and gestures underlying the artistic production [Freyd, 1983; Pignocchi, 2010]. We argue that this is particularly true for certain art forms such as expressed in calligraphy [Fong, 2003] and graffiti art [Berio *et al.*, 2017a; Mediavilla, 1996], in which the mastery of a skilful movement in large part determines the aesthetic quality of the resulting artefact.

Table 1. Main symbols

| Symbol | Meaning |
|---|---|
| $\boldsymbol{z}(s)$ | Input trace (parametrised) |
| $\boldsymbol{z}_i(s)$ | Trace segment |
| $\boldsymbol{z}_i^{\kappa max}$ | Curvature maximum locus (on trace) |
| $\boldsymbol{z}_i^{lhs}(s)$ | Left segment (for a given $\boldsymbol{z}_i^{\kappa max}$) |
| $\boldsymbol{z}_i^{rhs}(s)$ | Right segment (for a given $\boldsymbol{z}_i^{\kappa max}$) |
| $CC_i$ | Contact circle |
| $\overset{\frown}{CC}_i$ | Contact circular arc |
| $N_{CSF}$ | Number of Curvilinear Shape Features |
| $N_S$ | Number of strokes (drawn gestures) |
| $\Sigma\Lambda$ | Sigma Lognormal (SL) function |
| $\boldsymbol{slt}(t)$ | SL generated trace |
| $\boldsymbol{p}_i$ | Virtual or Aiming target per stroke |

In this chapter we take advantage of the duality between curvature and symmetry axes [Leyton, 1987] in order to extract more robustly curvilinear shape features, such as those based upon extrema (of some curvature measure or approximation) along a handwriting or drawing trace. The method is also directly adaptable to open contours (with ends), to contours with breaks in curvature (non differentiable at some finite set of loci), and can further be used to identify loops (such as where a trace overlaps itself). Each feature is also explicitly paired with corresponding contact gauges (in the present case: circles) and a pair of curvilinear support regions: contour traces on each side of an identified extremum, where curvature is approximately monotonic. Given such a robust and rich feature description of an handwritten trace, we show how to exploit this spatial and structural geometric representation to infer the kinematics of a likely generative movement — as would be performed by a skilled human expert or artist, as predicted by the lognormality principle. To do so, we rely on the *Kinematic Theory of Rapid Human Movements* [Plamondon, 1995; Plamondon *et al.*, 2014], a family of models of reaching and handwriting motions, in which a movement is described as the result of the parallel and hierarchical interaction of a large number of coupled neuromuscular components. The resulting method allows the reconstruction of physiologically plausible velocity profiles for the geometric trace of an input movement given as an *ordered sequence of points.* While state of the art methods exist for the parameter reconstruction of Kinematic Theory based models from digitised traces of handwriting [O'Reilly and Plamondon, 2008; Plamondon *et al.*, 2014; Fischer *et al.*, 2014; Ferrer *et al.*, 2018], we design our method with the goal of targeting applications in graphonomics, computer aided design (CAD) and computer graphics. As a result, we purposely ignore the kinematics of the input in order to seamlessly handle online handwriting with arbitrary sampling quality as well as vector art in which only the sequential ordering of points may be available. We also choose this approach with the future aim of combining our method with one that recovers temporal information from bitmap images such as the one presented by Plamondon and Privitera [1999]. We also argue that – from a computer graphics perspective – the ability to represent curves and traces through parameters that directly reflect the kinematics of a physiologically plausible movement opens up many possibilities, such as (i) more expressive rendering techniques, (ii) the generation of artificial variations that reflect the ones that would be made by a human, (iii) artistically driven fairing or beautification [Thiel *et al.*, 2011; McCrae and Singh, 2009; Zitnick, 2013], as well as (iv) potential applications in procedural content generation for movies and games.

In the following sections, we summarise first our approach to identify curvilinear shape features or CSF (Sec. 2). We then describe the trajectory segmentation method (Sec. 3) based on CSF and Euler spiral fitting. In Sec. 4 we summarise the Sigma Lognormal ($\Sigma\Lambda$) model used to describe the motions of a pen-tip. In Sec. 5 we conside the recovery of $\Sigma\Lambda$ parameters *from geometry only*. In Sec. 6 we discuss some applications before concluding.

## 2. Curvilinear Shape Features

### 2.1. *Background*

The study of the curvature along a contour — being the trace of an object's boundary — has been the focus of decades of research in various fields, including that of visual perception and cognitive science, as well as computer vision and pattern recognition. Significant extrema of curvature (i.e. with associated contour segments having a role of support) have been shown empirically to be the most salient loci along piecewise smooth contours [Attneave, 1954; Feldman and Singh, 2005; De Winter and Wagemans, 2008] and to play an important role in the perceptual decomposition of objects into parts [Richards and Hoffman, 1985; Brault and Plamondon, 1993b,a; De Winter and Wagemans, 2006]. Curvature also plays an important role in the study of human movement and handwriting by relating the kinematics of a movement to its trace. It is well known that tangential speed and curvature of human hand movements are inversely proportional [Viviani and Schneider, 1991] and this relation can take the form of a power law [Viviani and Schneider, 1991; Plamondon and Guerfali, 1998]. Furthermore, curvature extrema are indicative of speed minima and therefore useful for the segmentation of complex movements into simpler primitives, for the study of traces in both the structural and kinematic domains.

A robust identification of curvature extrema from the curvature function alone can be difficult as it requires the evaluation of a second order differential quantity which tends to amplify the effects of noise in the input as an outcome of the digitisation process. One popular method to overcome this problem is to first smooth the digitised signal using a filter (e.g. convolving with a Gaussian) or interpolating with some analytic function (e.g. smoothing splines). However, smoothing risks removing perceptually important features of an outline and choosing reasonable parameters remains a difficult task. To overcome this fundamental issue, one possible avenue is to generate an intermediate *scale-space* in which features are identified and tracked at different scales [Witkin, 1983]. Such a scale-space is very often produced by iterative Gaussian filtering in the spatial domain, or via the frequency domain using wavelets [De Stefano *et al.*, 2005]. An alternative to such traditional filtering — which tends to blur away details, especially in the vicinity of corners with sharp changes in orientation — is to use a *structural* notion of scale, e.g. by associating a support metric along a contour with each curvature feature being tracked, such as when performing morphological operations on the curvature function [Leymarie and Levine, 1989]. Nevertheless, such methods operating directly on the curvature function (along a contour) still suffer from poor localisation, and do not capture well singularities which can be perceptually significant, such as curvature discontinuities which are typical of corner features.

An alternative to working directly with the curvature function is to exploit the correspondence existing between symmetry axes and the curvature behavior of a contour [Leyton, 1987]. Originally pioneered by Harry Blum in the 1960's for the study of biological shape [Blum, 1967], the Symmetry Axis Transform (SAT) – also known as Medial Axis or simply "skeleton" for closed contours – is a shape representation that provides a bridge between geometry and topology. The SAT is commonly viewed as the set of centers of "maximally inscribed" disks, or with the "prairie grassfire" or wave-front analogy, in which the symmetry axes are given by the "quench" points at which fire fronts or waves propagating from the object boundary meet and stop expanding [Leymarie and Levine, 1992]. Variants of the SAT have also been widely used in pattern recognition applications, especially to extract topological and structural knowledge from the outline of handwritten or printed characters [Li and Plamondon, 2006]. Contrary to a common misinterpretation, the SAT is not only defined for closed shapes, but is valid also for open contours or even point samplings [Blum, 1973] — in the latter case becoming similar to the Voronoi graph.

### 2.2. *Our solution*

In order to robustly identify useful curvilinear shape features, we exploit the duality [Leyton, 1987] between the two representations of (2D) contour curvature and SAT, which allows us to identify *significant* curvature extrema and discontinuous breaks along a handwriting or drawing trace through the analysis of its local symmetries. Each feature is paired with a corresponding contact circle (near the extrema) and its support region — where curvature is approximately monotonic and in correspondence with a local symmetry axis. We then exploit this intermediate representation to infer the kinematics of a movement that reconstructs a likely trace.

We recall a result presented by Leyton which links the symmetry axes of an object having a smooth bounding contour to its curvature extrema [Leyton, 1987]:

> *Any segment of a smooth planar curve, bounded by two consecutive curvature extrema of the same type, has a unique symmetry axis, and the axis terminates at the curvature extremum of the opposite type.*

Motivated by this concept,[a] given a symmetry axis, it is then possible to identify and locate a curvature extrema near one axis end. In a traditional setting, the SAT is computed at once for a given final contour or, as in our case, a written trace. This however does not allow to identify all perceptually significant curvature extrema, as part of a contour may forbid or *mask* the existence of a symmetry axis that would otherwise end at a curvature break or corner, or end at the center of curvature of the circular arc associated to a curvature extremum.[b] Furthermore, data from contours or handwritten traces typically are not smooth everywhere, and thus we need a solution which is adapted to discrete samplings, possibly with sharp changes in orientation (corresponding to a break in curvature for the equivalent continuous trace), as well as for open contours (with ends).

Inspired by Leyton's idea, we have explored the following discrete scheme, as illustrated step by step, in Fig. 1: (i) we select a starting point either at one end of an open contour or randomly along a closed contour; (ii) we traverse the contour in one direction (until another end or until we come back to our initial position); (iii) we initiate a local symmetry axis computation from the sequential set of encountered trace samples (e.g. using a well established local Voronoi method [Ogniewicz, 1992]) and identify a first local axis end; (iv) we pursue the local axis reconstruction until we detect a new potential axis end — this indicates the end of one curvilinear side of the associated region of support — (v) we repeat the computation of a local axis for the new axis end, ignoring old samples which are behind, in terms of the traversal order, the previously identified curvature extremum (i.e. associated to the previously identified local axis end) — for this new axis end we already know its first curvilinear region of support, and thus we keep refining the local axis computation by visiting more contour samples, until we hit yet another candidate axis end or run out of samples; (vi) the procedure repeats until no more samples are left to consider.

In summary, a local symmetry axis is evaluated as one travels along a planar contour, such as obtained from a written trace, denoted $\boldsymbol{z}(s)$, which is assumed

[a]Leyton's result [Leyton, 1987] is restricted to smooth contours and to the description of so-called "codons": contour segments delimited by a triplet of curvature extrema such that a pair of concavities (convexities) delimit an intermediate convexity (concavity).

[b]One way to prove this masking effect, is by considering the behavior of cusps of evolutes in relation to a symmetry axis. Belyaev and Yoshizawa [2001] proved that an evolute cusp corresponds with a symmetry axis branch only when the segment going from the cusp to the associated curvature extrema, does not intersect the remaining skeleton of the shape.

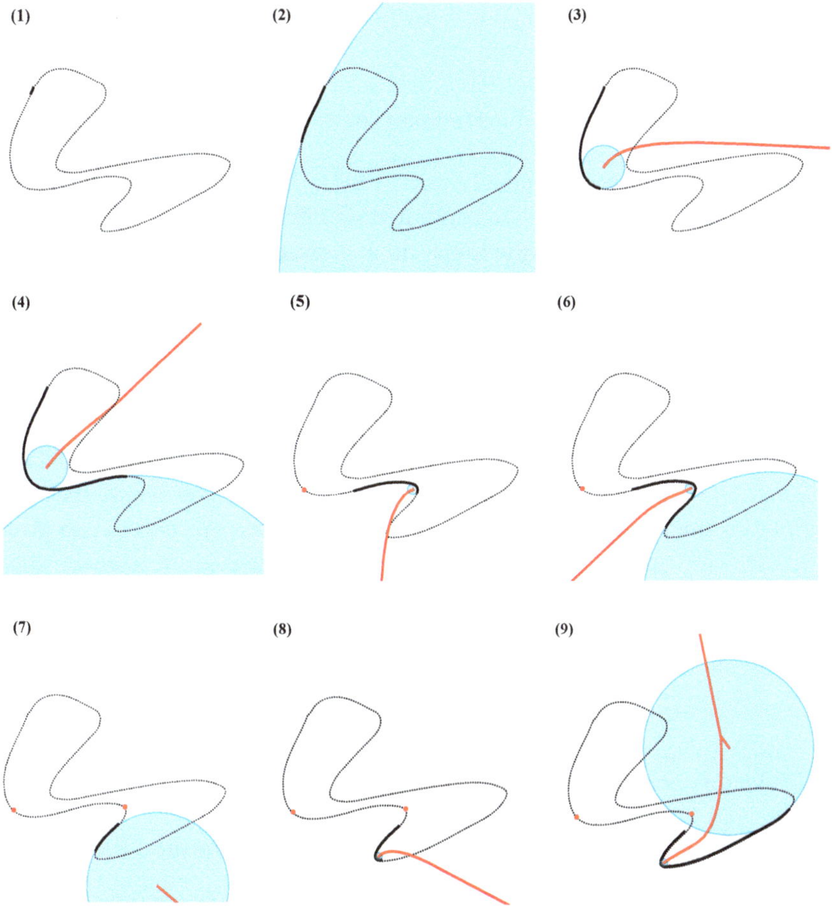


Fig. 1. Significant steps of computation of the localised symmetry axes. In cyan, the disks for the extremities of the candidate axis (in red); in black the current contour segment. The small red circles are the identified . **(1)** The first step of the process. The contour segment is too short so there are no branches. **(2)** A first branch appears (out of the figure) with the corresponding disk. The disk is unstable, i.e. it tends to shrink and moves towards a curvature extrema. **(3)** The disk stabilises in proximity of a curvature extrema. **(4)** A new axis appears; this triggers an event in which **(5)** the extremum corresponding to the disk in (3) is marked and a new start point is set for the contour segment. The new start point corresponds with the first anchor point of the newly appeared disk. The remaining images show steps identical to (3, 4, 5) which are repeated iteratively until all significant symmetry axes have been found. Note that in **(9)** a new symmetry branch appears, which will lead to the identification of the subsequent axis and associated curvature extremum along the outline.

parametrised by arc length $s$.[c] Each *significant* axis is found,[d] such that its existence ends once another significant axis emerges, and the previous written trace already traversed is then "forgotten", so as to not mask other potential extrema of curvature associated to later parts of the trace.

**Definition 1 (Curvilinear Shape Feature (CSF)).** Each CSF is composed of five main elements (Fig. 2): (i) $SA_i$, a computed significant symmetry axis, (ii) $CC_i$, a contact circle centered at the initial tip of a given $SA_i$, (iii) $\overrightarrow{CC}_i$, the associated circular arc overlapping with the contour trace, (iv) $\boldsymbol{z}_i^{\kappa_{max}}$, the associated curvature maximum locus, which we take as the mid-point of $\overrightarrow{CC}_i$,[e] (v) $[\boldsymbol{z}_i^{lhs}(s), \boldsymbol{z}_i^{rhs}(s)]$, a pair of contour segments on each side of $\overrightarrow{CC}_i$. The index $i$ indicates one of $N_{CSF}$ computed CSF: $1 \le i \le N_{CSF}$.

We note that each $SA_i$ is not necesarilly part of the classic Blum medial axis set, nor the Voronoi graph; rather, it is always part of the Full Symmetry Set as defined in singularity theory [Bruce and Giblin, 1992]. Furthermore, a given arc, $\overrightarrow{CC}_i$, may vanish in size when coinciding with a break of curvature or sharp corner tip, becoming identically $\boldsymbol{z}_i^{\kappa_{max}}$. We also emphasise that this definition of a CSF is more general than the older concept of a (contour-based) "codon": a triplet of alternate (in sign) curvature maxima (e.g. concave, convex, concave) [Richards and Hoffman, 1985].

## 3. Trajectory Segmentation into Circular Arcs

The proposed trajectory reconstruction method exploits the prior feature analysis of the input $\boldsymbol{z}(s)$ (Sec. 2), and thus takes as its input the set of computed Curvilinear Shape Features (CSF) which identifies a set of contact circles, $CC_i$, circular arcs, $\overrightarrow{CC}_i$, curvature extremum loci, $\boldsymbol{z}_i^{\kappa_{max}}$, and pairs of contour segments, $[\boldsymbol{z}_i^{lhs}(s), \boldsymbol{z}_i^{rhs}(s)]$, while the corresponding axial branches, $SA_i$, are ignored — their essential role in our method being in identifying the other items.

On the basis of this information, our goal is then to segment the entire trace, $\boldsymbol{z}(s)$, with a series of best fitting circular arcs. The reason being that circular arcs can be directly exploited by the kinematic analysis based on the Sigma

[c]In practice, we sample, for example uniformly, the trace in a set of discrete points at approximately equal distance from each other. Other (adaptive) sampling strategies are possible, e.g. in the presence of corners.

[d]Significance is linked to the particular computation of symmetry axial structure; for the results reported here, we rely on the well established parametric method of Ogniewicz [1992].

[e]Alternatively, $\boldsymbol{z}_i^{\kappa_{max}}$ can be taken as the projection of $SA_i$ intersecting the contour trace, a definition closer to the spirit of the work of Leyton [1987].

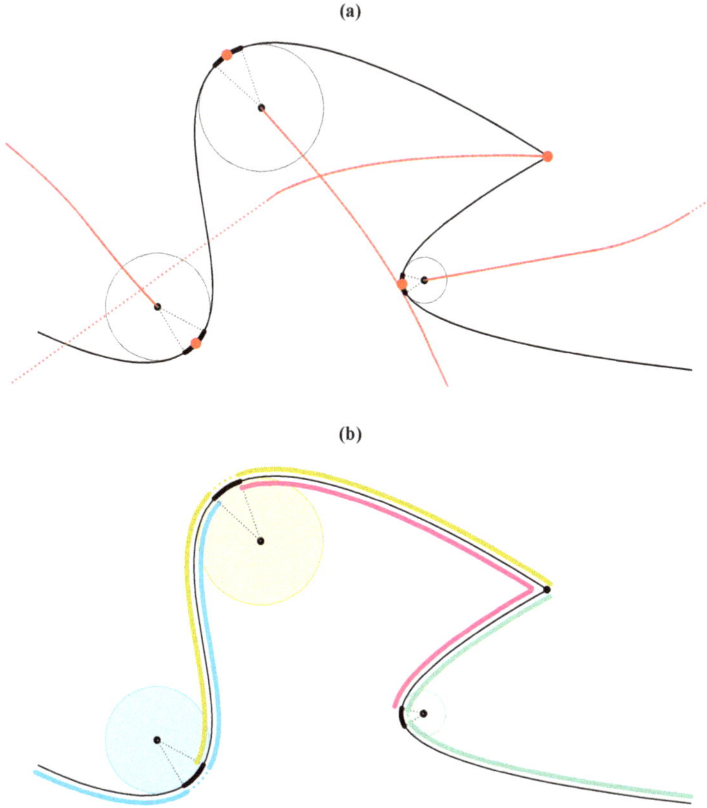


Fig. 2. (a) Four successive Curvilinear Shape Features (CSFs): Each CSF is defined by (i) a symmetry axis, $SA_i$, (ii) a contact circle, $CC_i$, (iii) its corresponding circular arc, $\overrightarrow{CC}_i$, part of the local trace, (iv) a representative curvature extremum locus, $\boldsymbol{z}_i^{\kappa max}$, taken as the mid-point of $\overrightarrow{CC}_i$ (indicated as red dots), (v) a pair of contour segments, $[\boldsymbol{z}_i^{lhs}(s), \boldsymbol{z}_i^{rhs}(s)]$ delimiting the region of influence of a CSF. Note that the third CSF (from the left) corresponds to a typical discontinuity in curvature, a corner where their is a sudden change in orientation, and thus with a contact circle reduced to a point which coincides with the curvature extremum locus. This is also indicated by having the local symmetry axis reaching the input trace ($\boldsymbol{z}(s)$). (b) Each pair of contour segments for each CSF is emphasised: colored and slightly translated away from the original outline.

lognormal modelling (to be described in the following Secs. 4 and 5). To obtain a trajectory segmentation in terms of a series of circular arcs, we start from the set $\overrightarrow{CC}_i$ and then propose in what follows a systematic method to approximate the remaining parts of the trace (given by the set of pairs $[\boldsymbol{z}_i^{lhs}(s), \boldsymbol{z}_i^{rhs}(s)]$) in two steps: (i) first by tightly fitting Euler spiral segments, and then (ii) mapping each Euler spiral segment to up to four best fitting circular arc(s), as a function of the presence of an inflection along the spiral segment and the angular extent of each arc.[f]

### 3.1. *Euler spiral fitting*

Euler spirals (also known as Cornu spirals, or clothoids) [Levien, 2009] are a useful type of curves in which curvature varies linearly with arc length, permitting the description of variably curved segments which may contain an inflection (Fig. 3). An Euler spiral is commonly parameterised by arc length using the cosine ($C(t)$) and sine ($S(t)$) Fresnel integrals:

$$C(t) = \int_0^t \cos(u^2)du\,, \quad S(t) = \int_0^t \sin(u^2)du\,. \tag{1}$$

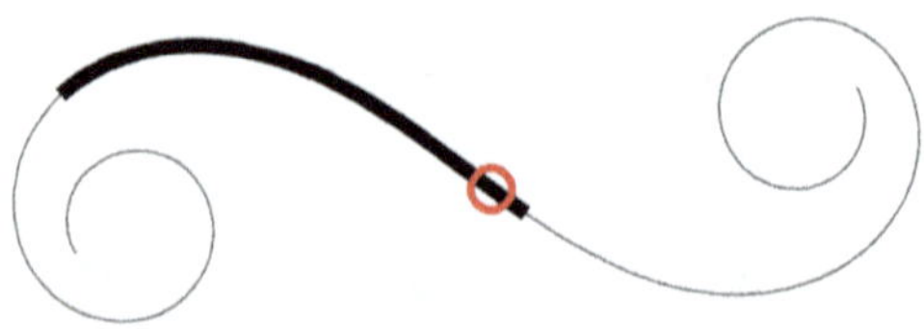

Fig. 3. An Euler spiral, its inflection point (circle) and a Euler spiral segment (black).

Such that the Euler spiral is obtained as:

$$\boldsymbol{q}(t) = (x(t), y(t)) = (C(t), S(t))\,, \tag{2}$$

where $t$ can vary between minus infinity to plus infinity, and where the origin $(0,0)$ corresponds to $t = 0$ which is the inflection point for the spiral; note that $t$ is then identically the (signed) arc length parameter for the spiral curve. In our application, an Euler spiral segment to be fitted to the data is defined between initial ($t = t_1$) and final ($t = t_2$) parameter values. If the values alternate in sign, then we have a segment with an inflection (at $t = 0$). Such a segment can be conveniently computed and sampled in an efficient manner using an approximation method developed by Heald [1985], which results in $n$ samples along the segment from $\boldsymbol{q}(t_1)$ and $\boldsymbol{q}(t_2)$.

In order to fit an Euler spiral segment to one of the segments from our input trace, we first compute approximate tangent directions along the trajectory, for a given segment, $\boldsymbol{z}_i(s)$, *i.e.*, in correspondence with the initial and final points of the segment under examination. This allows to rapidly compute a first estimate

[f] Our use of Euler spirals is also inspired by the work of Leyton who studied contour regions between curvature extrema of opposite sign by using "bi-spiral" segments [Leyton, 1987].

of the spiral segment's initial and final parameter values using a secant method described by Levien [2009].

However, the tangent estimates are likely to be unreliable in the presence of noisy input data, and thus we proceed to refine this initial fit with a least squares optimisation based on the classic Gauss-Newton method. Our method consists then in three additional steps. First, we linearly transform the given trace segment, $\boldsymbol{z}_i(s)$, such that its end points match those of the computed spiral segment in its canonical form. Second, we modify the canonical form of the Euler spiral, by introducing a scaling factor $\alpha$ and a rotation by an angle $\omega$ (in radians) with:

$$\boldsymbol{q}(t) = \begin{bmatrix} \alpha\cos(\omega)C_1(t) - \alpha\sin(\omega)S_1(t) \\ \alpha\sin(\omega)C_1(t) + \alpha\cos(\omega)S_1(t) \end{bmatrix}, \quad \text{where} \tag{3}$$

$$C_1(t) = C(t) - C(t_1) \quad \text{and} \quad S_1(t) = S(t) - S(t_1). \tag{4}$$

Note that the (initial) canonical form is for $\alpha = 1$ and $\omega = 0$. Third, and finally, we proceed with the minimisation:

$$\min_{t_1,t_2,\alpha,\omega} \frac{1}{2}\sum_{j=1}^{n} \| \boldsymbol{q}[j] - \boldsymbol{z}_i[j] \|^2 \quad , \tag{5}$$

where $\boldsymbol{q}[j]$ and $\boldsymbol{z}_i[j]$ both denote $n$ equally spaced samples with a sampling index $[j]$, in the former case for the spiral segment $\boldsymbol{q}(t)$ between $t_1$ and $t_2$, and in the later case along the input trace segment $\boldsymbol{z}_i(s)$.

## 3.2. *Inflections and final circular arc decomposition*

The presence of inflections can easily and robustly be identified by checking the signs of the two spiral parameters $t_1$ and $t_2$. With alternate signs, $t = 0$ gives the inflection position. For each potential inflection, we then check if the ratio $[\min(|t_1|,|t_2|)]/|t_1 - t_2|$ is less than a user defined threshold $\epsilon^{flex}$ (which we empirically set to 0.2 in the accompanying examples), in which case we discard it as a near degenerate case, the inflection being very close to one spiral's end point.

Depending on the presence of an inflection, we fit either one or two circular arcs to each Euler spiral segment. A final simple refinement step is then executed by considering the angular extent of each computed arc. For relatively wide arcs, we split these in two halves. The heuristic used here being that arcs should not be too wide such that the later computation of the kinematics converges more easily.

The internal angle of any circular arc derived from Euler spiral segments is easily estimated by integrating the curvature of the spiral and distinguishing between three cases: *(i)* For the case of two arcs separated by an inflection, the pair of internal angles are given by $t_1 \times |t_1|$ and $t_2 \times |t_2|$ (Fig. 4(a)). *(ii)* In the case of a degenerate inflection, we use the same method to fit a single arc and choose only the parameter with the greatest absolute value and consequently higher curvature. *(iii)* When no inflection is present the internal angle is given by: $|(t_2 \times |t_2| - t_1 \times |t_1|)| \times \text{sgn}(t_1)$. In each of these three cases, and for each arc, we then check if the absolute internal angle is greater than $(3/4)\pi$ — heuristically set, so that no derived circular arc is close to half a circle, similarly to Li *et al.* [1998] — in which case we further subdivide such a wide arc in two halves (Fig. 4(b)).

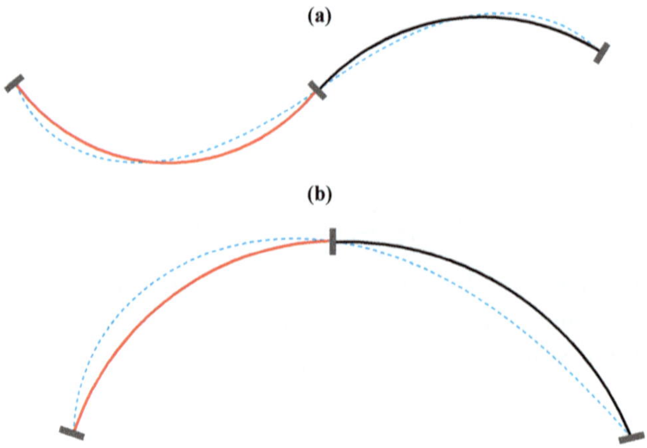


Fig. 4. Decomposing Euler spirals (stippled cyan) into arcs. **(a)** two arcs delimiting an inflection. **(b)** one segment with internal angle > $(3/4)\pi$ divided into two equal arcs.

In summary, we have that the trace is now represented by a sequence of $N_{CSF}$ contact circular arcs, $\overrightarrow{CC}_i$, with intermediate Euler spiral segments each with or without an inflection. Each spiral segment is then mapped to either a pair of arcs (with a separating inflection) or a single arc. Each such circular arc derived from a spiral may then be further split in two if wide enough. A little inspection shows that, in the case of a closed trace (no endpoints), we have a maximum of $5 \times N_{CSF}$ circular arcs fitted, while for an open trace (with a pair of endpoints) the maximum is $N_{CSF} + 4 \times (N_{CSF} + 1)$. Figure 5(a) shows the results of (i) identifying (here, five) curvature maxima, $\boldsymbol{z}_i^{\kappa_{max}}$, followed by (ii) fitting (five) Euler spirals, and (iii) finding (two) inflections and corresponding circular arcs. This approximate reconstruction of the original trajectory in the form of circular arc segments, is related to the method originally proposed by Li *et al.* [1998], but with the following main three differences: (i) We have found experimentally that our method to identify curvature maxima is more robust — in particular as it does not rely on an explicit *a priori* estimation of the curvature signal. (ii) We use Euler spirals to fit intermediate data which gives a simpler and more robust method to identify inflections. (iii) We explicitly obtain contact circle arcs, $\overrightarrow{CC}_i$, which results in a more accurate reconstruction of the original trace (Fig. 5).

This representation of the input trace, $\boldsymbol{z}(s)$, as a series of circular arcs, is now ready to be exploited (in §5) together with the Signal Lognormal model which we first summarise in the next section.

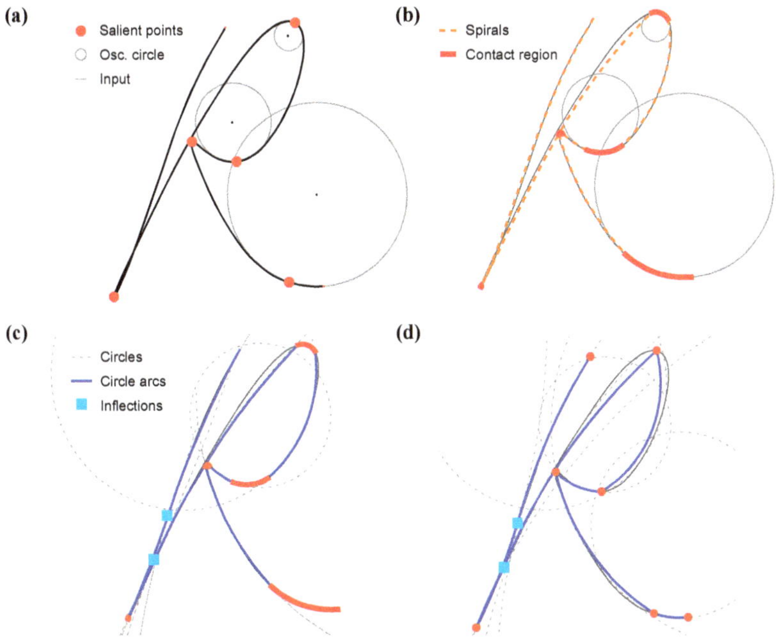


Fig. 5. Sample "R" character from the UJI handwritten character dataset [Llorens *et al.*, 2008]. **(a)** Significant curvature maxima, $\boldsymbol{z}_i^{\kappa max}$. **(b)** Euler spiral fitting; and **(c)** circular arcs decomposition, where the arcs in red indicate the contact circular arcs, $\overrightarrow{CC}_i$. **(d)** Demonstrative example of least-squares fitting of circular arcs to the segments defined between (the same) consecutive curvature maxima, $\boldsymbol{z}_i^{\kappa max}$. Conclusion: Not considering the contact region results in a much less precise reconstruction of the input.

## 4. The Sigma Lognormal ($\Sigma\Lambda$) Model

On the basis of the Kinematic Theory [Plamondon, 1995], we study handwriting and drawing movements via the the Sigma Lognormal ($\Sigma\Lambda$) model [Plamondon *et al.*, 2014], which describes complex trajectories on the plane via the vectorial superimposition of $N_S$ time shifted stroke primitives.

Each stroke is described as an aiming drawing gesture, and the sequence of strokes is modelled by a scaffold-like action plan, starting from an initial position, and defining a curvilinear graph made of vertices as aiming positions with connecting circular arcs. The integration of the $\Sigma\Lambda$ along this scaffold provides for a drawing engine where variations of parameters give different possible (but related) approximating traces, more or less smooth, more or less reaching the various aiming vertices (aka "virtual targets", Fig. 6). We summarise the main algorithmic elements of the $\Sigma\Lambda$ model in the remaining of this section.

The velocity function of each stroke is given by a classic lognormal function:

$$\Lambda_i(t) = \frac{1}{\sigma_i\sqrt{2\pi}(t - t_{0i})} \exp\left(-\frac{(ln(t - t_{0i}) - \mu_i)^2}{2\sigma_i{}^2}\right) \quad , \tag{6}$$

which characterises the impulse response of each stroke to a centrally generated command occurring at time $t_{0i}$. The parameters $\mu_i$ and $\sigma_i$ represent the stroke delay and response time in a logarithmic scale; they also determine the shape and asymmetry of the lognormal profile.

### 4.1. *Circular arc assumption*

An important practical assumption is typically made when initiating the $\Sigma\Lambda$ model: *handwriting* movements are mostly made with rotations of the elbow or wrist. The corollary is then that the curvilinear evolution of a drawing stroke can be approximated by a *circular arc*. This has for consequence to simplify the computation of the *angular evolution* of a stroke by using the time integral of Eq. (6):

$$w_i(t) = \int_0^t \Lambda_i(u)du = \frac{1}{2}\left[1 + \text{erf}\left(\frac{\log(t - t_{0i}) - \mu_i}{\sigma_i\sqrt{2}}\right)\right] \quad , \tag{7}$$

which results in

$$\phi_i(t) = \theta_i - \frac{\delta_i}{2} + \delta_i w(t) \quad , \tag{8}$$

such that $\theta_i$ is the *direction* of the stroke and $\delta_i$ is the *stroke curvature* parameter which determines the internal angle of the assumed *circular* arc.

The planar pen-tip velocity is then calculated with:

$$\dot{x} = \sum_{i=1}^{N} \hat{D}_i \Lambda_i(t) cos(\phi_i(t)) \quad \text{and} \quad \dot{y} = \sum_{i=1}^{N} \hat{D}_i \Lambda_i(t) sin(\phi_i(t)) \quad , \tag{9}$$

where $\hat{D}_i = D_i h(\theta_i)$ is the length $D_i$ of the stroke scaled by:

$$h(\theta_i) = \begin{cases} \frac{\delta_i}{2\sin(\delta_i/2)} & \text{if } |\sin\theta_i| > 0 \quad , \\ 1 & \text{otherwise} \quad , \end{cases} \tag{10}$$

which compensates for the stroke curvature based on the ratio between the perimeter and the chord length of a circular arc. The acceleration components of the lognormal trajectory are then given by [Plamondon and Guerfali, 1998]:

$$\ddot{x} = \sum_{i=1}^{N} \hat{D}_i \dot{\Lambda}_i(t) \cos(\phi_i(t)) - \hat{D}_i \delta_i \Lambda_i^2(t) \sin(\phi_i(t)) \quad , \tag{11}$$

$$\ddot{y} = \sum_{i=1}^{N} \hat{D}_i \dot{\Lambda}_i(t) \sin(\phi_i(t)) + \hat{D}_i \delta_i \Lambda_i^2(t) \cos(\phi_i(t)) \quad , \tag{12}$$

with

$$\dot{\Lambda}_i(t) = \Lambda_i(t) \frac{\mu_i - \sigma_i^2 - \log(t - t_{0i})}{\sigma^2(t - t_{0i})} \quad , \tag{13}$$

which allows us to compute the curvature function at time $t$ with the classic formula from differential geometry:

$$\kappa(t) = (\dot{x}\ddot{y} - \dot{y}\ddot{x})/(\dot{x}^2 + \dot{y}^2)^{3/2} \quad . \tag{14}$$

Given Eq. (7) the planar displacements along the $\Sigma\Lambda$ trajectory can then be computed using the error function and thus not requiring the numerical integration of Eq. (9), which simplifies to:

$$x = \sum_{i=1}^{N} \begin{cases} \frac{\hat{D}_i}{\delta_i} (\sin(\delta_i w(t) + \theta_i - \delta_i/2) - \sin(\theta_i - \delta_i/2)) & \text{if } \delta_i > 0 \quad , \\ \hat{D}_i w_i(t) \cos\theta_i & \text{otherwise} \quad , \end{cases} \tag{15}$$

$$y = \sum_{i=1}^{N} \begin{cases} \frac{\hat{D}_i}{\delta_i} (\cos(\delta_i w(t) + \theta_i - \delta_i/2) - \cos(\theta_i - \delta_i/2)) & \text{if } \delta_i > 0 \quad , \\ \hat{D}_i w_i(t) \sin\theta_i & \text{otherwise} \quad . \end{cases} \tag{16}$$

### 4.2. *Targets*

The previous pair of equations for displacement give us a way to efficiently compute a new position, $\boldsymbol{slt}(t)$, at a given time $t$ from an initial position $\boldsymbol{p}_0$:

$$\boldsymbol{slt}(t) = \boldsymbol{p}_0 + [x, y]^\top . \tag{17}$$

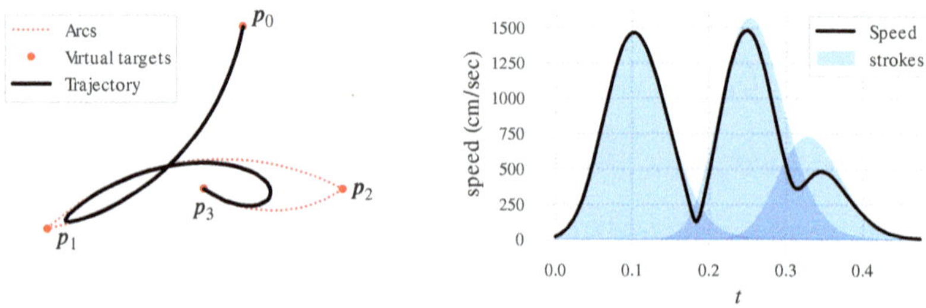


Fig. 6. Left: Lognormal trajectory (in black) with the corresponding action plan (targets as red dots, and scaffold of circular arcs as dotted red curves. Right: The corresponding speed profile with different time overlaps between lognormals.

The *initial* sequence of curvilinear strokes describes an *action plan* consisting of the initial position $\boldsymbol{p}_0$ followed by a sequence of $N_S$ *targets* $\boldsymbol{p}_1, \ldots, \boldsymbol{p}_{N_S}$ each corresponding to an aiming locus per stroke (Fig. 6, Left).

The degree of *time overlap* between lognormal components is used to characterise the degree of smoothness of the trajectory in correspondence with each target, where a greater overlap results in a smoother trajectory (Fig. 6, Right). In order to facilitate interactive applications [Berio *et al.*, 2017b] and to simplify the subsequently described parameter reconstruction method, we compute the directions $\theta_i$ and length $D_i$ for all $N_S$ pairs of consecutive targets $(\boldsymbol{p}_{i-1}, \boldsymbol{p}_i)$. Furthermore, we explicitly define the time overlap of each lognormal through an intermediate parameter $\Delta t_i \in [0, 1]$ where $t_{0i} = t_{0i-1} + \Delta t_i \sinh(3\sigma_i)$ for $i > 1$ and $t_{01} = 0$.

## 5. Iterative Reconstruction of ΣΛ Parameters

Having introduced the essential elements of the ΣΛ model in the previous section, and given the earlier trajectory segmentation derived from identified CSFs (§2) and circular arcs derived from Euler spiral fitting (§3), we have the necessary information to describe how we reconstruct the input trajectory with an approximate associated kinematics given only information about its (static) sampled geometry. Hence, we will be able to seamlessly process on-line handwriting data as well as vector art in which only the sequential ordering and coordinates of trace samples is required. The method is a development and improvement over our prior efforts [Berio and Fol Leymarie, 2015; Berio *et al.*, 2017a]. We re-emphasise that, although a number of methods exist for the accurate reconstruction of ΣΛ parameters from digitised traces [O'Reilly and

Plamondon, 2008; Plamondon *et al.*, 2014; Fischer *et al.*, 2014], these require as input the kinematics of the original trajectory.

### 5.1. *Initialisation: Features, strokes, initial targets*

The initial set of aiming *targets* (aka "virtual targets") consists of three types of *feature points*, or *features* for short: from CSF analysis (i) recovered curvature maxima loci, $\boldsymbol{z}_i^{\kappa_{max}}$, and from Euler spiral analysis (ii) inflections, and (iii) splits (of wide angled circular arcs). We can either directly used these loci or find their nearest neighbors, $\boldsymbol{z}(\hat{s}_i)$, on the original input trace, $\boldsymbol{z}(s)$, which leads to slightly more accurate reconstructions. We follow the later approach in results reported thereafter.

The initial set of $N_S$ *strokes* is defined from only those circular arcs derived from the Euler spiral segment fitting; i.e. we do not generate strokes for each contact circular arc, $\overrightarrow{CC}_i$, associated to each $\boldsymbol{z}_i^{\kappa_{max}}$. We have found experimentally that if we include strokes for such arcs and compute trajectories via the $\Sigma\Lambda$ model we easily overshoot the original trace, $\boldsymbol{z}(s)$, in the vicinity of each $\boldsymbol{z}_i^{\kappa_{max}}$. Also, removing these contact circular arcs, $\overrightarrow{CC}_i$, from the $\Sigma\Lambda$ modeling gives us more flexibility in aiming at curvature maxima loci when generating smooth variants (using Eq. (17)), and also it reduces the overall possible number of strokes considered.

However, the set $\overrightarrow{CC}_i$ can be used for defining stopping criteria for our iterative scheme, as defined below. Intuitively, as a generated trace, $\boldsymbol{slt}(t)$ aims at reaching towards a curvature maximum locus, $\boldsymbol{z}_i^{\kappa_{max}}$, we can stop locally pulling in that direction once the generated equivalent curvature maximum locus enters a zone such as the angular sector defined by a $\overrightarrow{CC}_i$ or even as it enters the interior of the contact circle itself, $CC_i$.

The used circular arcs — all derived from Euler spiral segments — give us a set of internal angles $\hat{\theta}_{i\,i=1}^{\,N_S}$, centers $\boldsymbol{c}(\hat{\theta}_i)$ and radii $r(\hat{\theta}_i)$. The circular arcs are delimited by $N_S$ feature points $\{\hat{s}_i\}_{i=1}^{N_S}$, plus additional starting and end loci if modeling an open trace (with endpoints, $\boldsymbol{p}_0$ and $\boldsymbol{p}_{N_{S+1}}$). If modeling a closed trace, we randomly pick one feature point as both the start and end positions.

An initial estimation of the trajectory parameters is given by a target sequence $\boldsymbol{p}_i = \boldsymbol{z}(\hat{s}_i)$, stroke curvature parameters $\delta_i = \theta(\hat{s}_i)$ and time overlap parameters $\Delta t_i = 0.5$. For the sake of simplicity, we consider the remaining parameters $\sigma_i$ and $\mu_i$ as typical properties of the neuromuscular system of a writer and keep these set to a user configurable value. The initial trajectory estimate is likely to differ from the original, $\boldsymbol{z}(s)$, and to be much smoother due to the initial lognormal stroke overlap (Fig. 7(a)).

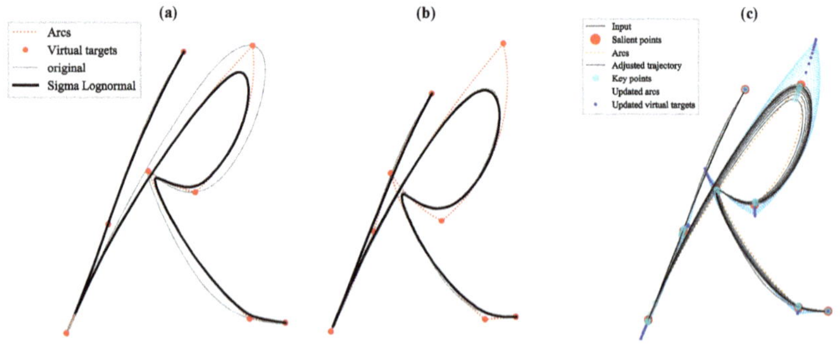


Fig. 7. ΣΛ parameter reconstruction using features from CFSs and Euler spiral derived arcs. **(a)** First guess (in black) of the stroke parameters and action plan from features. **(b)** Reconstruction of the input after iterative refinement steps. **(c)** Iterative refinement steps. The initial action plan has targets corresponding with the features along the input (large red circles). At every iteration, the targets are shifted (small blue circles) in order to reduce the distance between keys along the generated trajectory (cyan circles) and the features of the original trace.

### 5.2. *Iterative scheme: Keys, max speeds, moving targets*

To improve the reconstruction, we adopt an iterative refinement scheme (Fig. 7) in which we adjust the curvature and time overlap parameters together with the target positions in order to minimise the difference between the reconstructed, $\boldsymbol{slt}(t)$, and original, $\boldsymbol{z}(s)$, trajectories. At each iteration, we rely on the estimation of a series of *key points*, or *keys* for short, which approximate the initial feature point loci. We compute $N_S$ keys $\{\tau_i\}_{i=1}^{N_S}$ along the trajectory (Fig. 8) where $\tau_1,\ldots,\tau_{N_S}$ indicates the time occurrence at which the influence of one stroke exceeds the previous one and the generated curvature is locally maximal, while $\tau_0$ and $\tau_{N_{S+1}}$ are the starting and ending time of the trajectory. To compute the keys, we intersect the scaled profiles of consecutive stroke pairs, using an iterative or Newton scheme, by solving:

$$\hat{D}_i\Lambda_i(t) - \hat{D}_{i+1}\Lambda_{i+1}(t) = 0 \quad .$$

In addition, we compute $N_S$ *maximum speed points*, or *max speeds* for short, $\{\gamma_i\}_{i=1}^{N_S}$, which indicate the time occurrence of the maximum speed for each stroke (Fig. 8); this is explicitly obtained by the mode of the corresponding lognormal: $t_{0i} + \exp(\mu_i - \sigma_i^2)$.

The iterative refinement scheme was designed based on three observations:

**Observation 1.** The time parameter $\Delta t_i$ is proportional to the curvature $\kappa(\tau_i)$ at the time of the corresponding key point. Thus, a higher value of $\Delta t_i$ will decrease the amount of overlap of successive lognormals. This will result in a

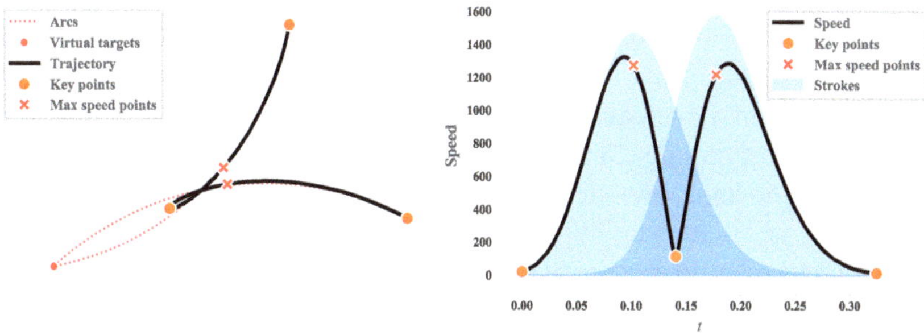


Fig. 8. Keys (orange circles) and max speeds (red crosses) overlaid on the trace (a) and speed profile (b) of a trajectory with two strokes.

lower speed and higher curvature $\kappa(\tau_i)$ at the time occurrence of the key. Since we have a good approximation of the curvature $\kappa(\hat{s}_i)$ in the original trajectory, the relation between the two can be exploited in order to adjust $\Delta t_i$ proportionally at each iteration. We observe that changes in $\Delta t_i$ are not linearly related to changes in the curvature $\kappa(\tau_i)$ at the corresponding key. In order to compensate for this, we assume a $1/3$ power relation [Viviani and Schneider, 1991] which has often been observed in human movement and particularly holds for elliptical portions of the trajectory [Plamondon and Guerfali, 1998], which is often the case near keys. The reasoning is that given the relations

$$\Delta t \propto \kappa \quad \text{and} \quad \Delta t \propto 1/v \quad ,$$

where $v$ denotes speed, we have the proportions relating desired and generated curvature and velocity:

$$\rho_\kappa = \hat{\kappa}/\kappa \quad \text{and} \quad \rho_v = \hat{v}/v \quad .$$

As a result, given the *power law* [Viviani and Schneider, 1991] $v = \kappa^{-1/3}$ and because velocity and $\Delta t$ are inversely proportional, we finally get the relation:

$$\rho_{\kappa v} = v/\hat{v} = (\kappa/\hat{\kappa})^{-1/3} = (\hat{\kappa}/\kappa)^{1/3} \quad .$$

**Observation 2.** Moving targets play a role similar to control points in spline analysis. Shifting a target $\boldsymbol{p}_i$ in a given direction will cause the generated key $\boldsymbol{slt}(\tau_i)$ to move in a similar direction. As a result, shifting the target $\boldsymbol{p}_i$ along the vector $\boldsymbol{z}(\hat{s}_i) - \boldsymbol{slt}(\tau_i)$ will decrease the distance between successive generated keys and original features (Fig. 7(c)).

**Observation 3.** The distance $D_i$ between successive targets $\boldsymbol{p}_i$ and $\boldsymbol{p}_{i-1}$ will influence the curvature of the resulting stroke. Augmenting this distance will increase the radius of curvature of the circular arc defined by the parameter $\delta_i$ and will result in a decrease of curvature for the stroke. While the trajectory tends to depart from the circular arc near the keys at $t = \tau_i$ due to the smoothing effect of the lognormal time overlap, it tends to pass closer to the circular arc at $t = \gamma_i$ where the amplitude of the lognormal is maximal. As a result, we use this locus to evaluate the deviation from the desired arc $\hat{\theta}_i$ and correct the parameter $\delta_i$ accordingly.

As a result of these observations, we define each iteration of our method to consist of the following ordered steps:

$$\Delta t_i \leftarrow \Delta t_i + \lambda_\Delta(\varsigma(\hat{\Delta t_i}, \Delta t_{min}, \Delta t_{max}) - \Delta t_i) \quad , \tag{18}$$

$$\delta_i \leftarrow \delta_i + \lambda_\delta(\hat{\delta}_i - \delta_i) \quad \text{and} \tag{19}$$

$$\boldsymbol{p}_i \leftarrow \boldsymbol{p}_i + \lambda_p(\boldsymbol{z}(\hat{s}_i) - \boldsymbol{slt}(\tau_i)) \quad . \tag{20}$$

Here $\lambda_\Delta$, $\lambda_\delta$ and $\lambda_p$ are damping parameters that we experimentally tune to 0.1 and 0.1 and 0.5 to avoid excessive adjustments at each iteration. The target time offset parameter for each iteration is computed by assuming a 1/3 power relation to curvature and is given by

$$\hat{\Delta t_i} = \varsigma\left((\kappa(\hat{s}_i)/\kappa(\tau_i))^{1/3}, \Delta t_-, \Delta t_+\right) \quad , \tag{21}$$

which is restricted to a user specified range $[\Delta t_-, \Delta t_+]$ by using a sigmoid function $\varsigma$. We observe that this restricted range improves convergence of our method and permits to apply smoothing effects to the trajectory during the reconstruction step (examples in §6).

The desired internal angle of an arc is given by:

$$\hat{\delta}_i = 4\tan^{-1}\left[\frac{h}{a}\tan\left(\frac{\delta_i}{4}\right)\right] \quad \text{with} \tag{22}$$

$$a = \| \boldsymbol{p}_i - \boldsymbol{p}_{i-1} \| \quad \text{and} \tag{23}$$

$$h = \left(r(\hat{\theta}_i) - \| \boldsymbol{p}(\gamma_i) - \boldsymbol{c}(\hat{\theta}_i) \|\right)\operatorname{sgn}(\hat{\theta}_i) \quad , \tag{24}$$

where the term $h$ determines the amount to shift the curvature parameter $\delta_i$ by comparing the radius of the circular arc $\hat{\theta}_i$, initially fitted to the input, to the distance between its center and the lognormal max speed point $\boldsymbol{p}(\gamma_i)$.

### 5.3. *Stopping criteria, SNR*

A few different stopping criteria for the iterative scheme are possible, depending on the user needs. The simplest — and most practical for experimenting with the approach — is to let the user define an overall maximum iteration. Other more sophisticated criteria we have experimented with include: (i) let keys reach each associated $CC_i$ or $\overrightarrow{CC}_i$; (ii) minimise the overall distance between the generated, $\boldsymbol{slt}(t)$, and the input, $\boldsymbol{z}(s)$, traces, by either selecting a threshold value or letting the algorithm reach a local minimum; (iii) optimise the quality of the reconstruction by maximising an error criterion such as the SNR (defined next). We have found in practice the latter SNR-based criterion gives a good compromise between reconstruction quality and computational complexity.

Because we do not take into consideration the kinematics of the input, we evaluate the quality of the reconstruction using the Signal to Noise Ratio (SNR) computed between the reconstructed and input trajectory. While this could be done by uniformly sampling the two trajectories at a constant distance step, this will result in a propagation of errors along the reconstructed trajectory, which leads to unreliable SNR measurements. To overcome this problem, we exploit our initial estimation of features $\boldsymbol{z}(\hat{s}_i)$ in the input and the segmentation given by the keypoints $\tau_i$ of the reconstructed trajectory, $\boldsymbol{slt}(t)$, and uniformly sample $m$ segments for the original and generated trajectory, where the $j$th point for the $i$th segment are respectively denoted as $\boldsymbol{z}_{i,j}$ and $\boldsymbol{p}_{i,j}$ and the mean of an input segment is denoted by $\bar{\boldsymbol{z}}_i$. The trajectory SNR is then:

$$SNR = 10\log_{10}\frac{\sum_i\sum_j\left(\boldsymbol{z}_{i,j}-\bar{\boldsymbol{z}}_i\right)\cdot\left(\boldsymbol{z}_{i,j}-\bar{\boldsymbol{z}}_i\right)}{\sum_i\sum_j\left(\boldsymbol{z}_{i,j}-\boldsymbol{p}_{i,j}\right)\cdot\left(\boldsymbol{z}_{i,j}-\boldsymbol{p}_{i,j}\right)}\quad, \tag{25}$$

which trivially generalises to the case of multiple disconnected trajectory segments, such as when the writer lifts-up their pen or brush.

We tested the iterative refinement on different inputs ranging from vector traces with no a priori kinematic information (Fig. 9), to online data — including the Graffiti Analysis database [F.A.T. Lab, 2009] (Fig. 10 and Fig. 11) and the UJI handwritten character dataset [Llorens *et al.*, 2008] (Fig. 7) — and it consistently produces visually accurate reconstructions of the input. We observe that, while fluctuations may appear during iterations, the refinement scheme consistently and rapidly converges towards a reduction of the error between the input and the generated trajectories and an increase in SNR (Eq. (25)).

The iterative scheme can be applied in a *batch* manner, in which all the $\Sigma\Lambda$ parameters for all strokes are updated at each iteration, or similarly to the iDeLog framework [Ferrer *et al.*, 2018] by traversing the trajectory in an *incremental* manner and adjusting pairs of strokes ordered in time. In our experiments both approaches present similar convergence properties and produce reconstructions with similar SNR.

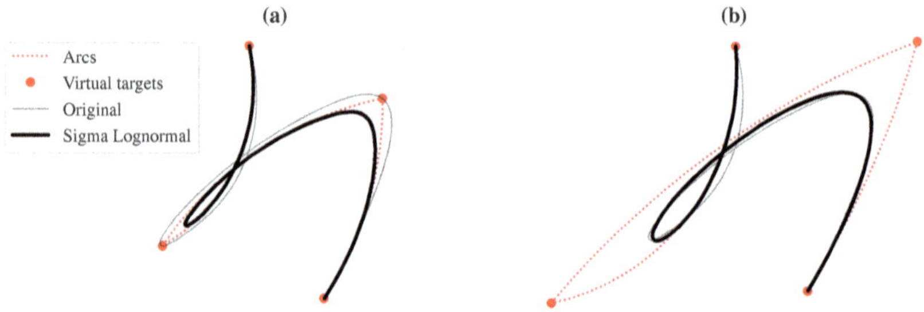


Fig. 9. Reconstruction of vector input initially built with piecewise Bézier curves. Our method reconstructs the (originally only guaranteed to be $C^0$ continuous) input with smooth kinematics given by the $\Sigma\Lambda$ model. **(a)** First guess of the parameters from features. **(b)** Reconstruction of the input after iterative refinement steps.

## 6. Discussion and Applications

The reconstructed $\Sigma\Lambda$ parameters provide a concise and easily manipulable representation of a geometric trace. This can be exploited in a number of applications that are relevant to our desired use cases in CAD [Berio *et al.*, 2017b] and procedural content generation [Berio *et al.*, 2017a]. We list a few such applications in the remaining of this section.

### 6.1. *Artificial variability*

The $\Sigma\Lambda$ model directly reflects the characteristics of a smooth human movement at the planning (targets and scaffold) and neuromotor level (the remaining parameters). We therefore expect and observe that parameter perturbations result in variations of a trace that are similar to the one that would be seen in multiple instances of handwriting or drawing made by one or more subjects (Fig. 12). This property has been exploited to create artificial data for handwriting recognizers [Fischer *et al.*, 2014], signature verifiers [Galbally *et al.*, 2012; Diaz-Cabrera *et al.*, 2018], gesture graphical input [Leiva *et al.*, 2017, 2016], and can be used for artistic oriented applications as well [Berio and Fol Leymarie, 2015; Berio *et al.*, 2017a]. For such a task, it is convenient to apply perturbations at the level of the intermediate parameters $\Delta t_i$ and $\delta i$ and (explicitly defined) target positions $\boldsymbol{p}_i$, which avoids the excessive distortions that we observe by directly perturbing relative angles and displacements as defined in the original formulation of the model [Plamondon *et al.*, 2014].

In our experiments, when perturbing targets, we have found that applying

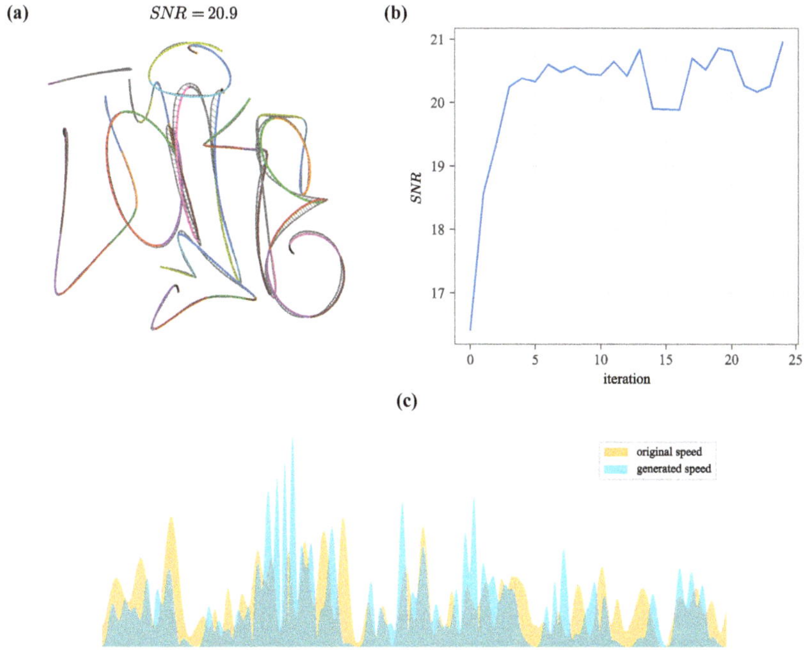


Fig. 10. Reconstruction of a graffiti signature "JANKE" from the Graffiti Analysis database [F.A.T. Lab, 2009]. **(a)** The reconstructed trajectory, subdivided into segments for comparison (color coded), overlaid on the original trace (light grey). The short grey segments mark the errors and correspondences between uniformly distanced samples for each trajectory segment. **(b)** Plot of the $SNR_t$ for each iteration of the iterative optimization scheme. **(c)** The speed profiles of the original (light grey) and reconstructed (dark grey) trajectories, scaled for comparison.

the perturbation with a variance inversely proportional to the temporal overlap parameters $\Delta t_i$ (Fig. 12, top left) improves the legibility of the variations. This corresponds to imposing a higher precision requirement at trajectory locations with higher curvature, locations that are known to be the most informative of a trace/contour [Feldman and Singh, 2005]. From a motor control perspective, this effect is consistent with the "minimum intervention principle" [Todorov, 2004], suggesting that human movement variability is higher where it does not interfere with the performance required for a task.

## 6.2. *Smoothing and stylisation*

Another application of the $\Sigma\Lambda$ parametrisation is to perform kinematic smoothing of a given trajectory. The result is a method that bears similarities to computer graphic approaches for curve fairing or neatening [Thiel *et al.*,

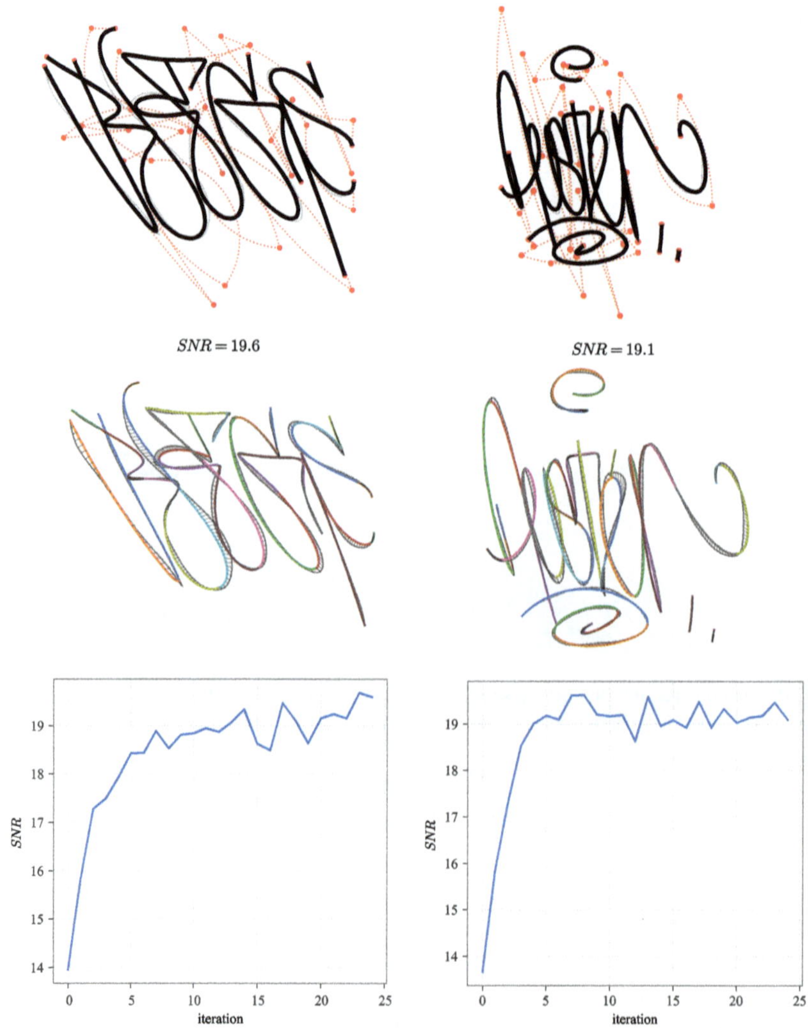


Fig. 11. Additional examples of graffiti tag [F.A.T. Lab, 2009] reconstruction together with the corresponding SNR plots.

2011; McCrae and Singh, 2009] as well as curve stylisation approaches [Lang and Alexa, 2015; Lu *et al.*, 2012]. In particular, the Euler spiral decomposition step of our reconstruction method is similar to a few previously proposed methods [Baran *et al.*, 2010; McCrae and Singh, 2009], which exploit the decomposition of an input curve into Euler spiral segments to remove discontinuities and

Fig. 12. Parametric variations of a reconstructed graffiti instance from the Graffiti Analysis database. Top left, the original reconstruction (black trace) overlaid with 30 variations. Note that variability is higher in proximity of smooth segments of the trajectory. The remaining traces are randomly perturbed samples, with the corresponding (perturbed) action-plan in red.

guarantee second order ($C^2$) continuity in the output. In our case, we instead rely on the properties of the $\Sigma\Lambda$ model, which ensures the resulting reconstruction is smooth and infinitely differentiable ($C^\infty$).

Another method to smooth the generated trajectory is to simply scale the $\Delta t_i$ parameters (Fig. 13(a)). However, this quickly and coarsely simplifies the original trace — similarly to the smoothing effect of a convolution with a large

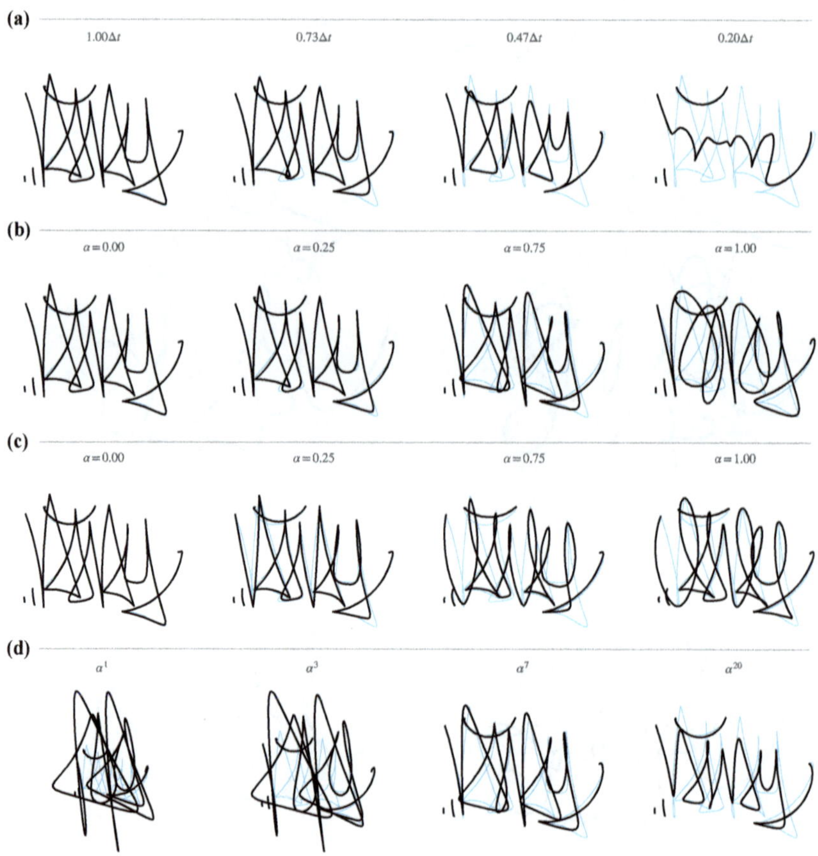


Fig. 13. Comparison of smoothing and stylisation methods. Row **(a)**, smoothing by global scaling of the time offset parameters $\Delta t_i$. Row **(b)**, smoothing by using the parameter $\alpha$ to interpolate between the $\Sigma\Lambda$ parameters of two reconstructions with different values for $\Delta t_+$. Row **(c)**, stylisation effects achieved by interpolating between the $\Sigma\Lambda$ parameters of two reconstructions, where the second is performed with user specified values of $\delta_i$. Row **(d)**, effect of different powers of $\alpha = 0.75$ used to interpolate the virtual target positions between the two reconstructions used in row **(b)**; The examples in row **(b)** and **(c)** use a power of 7.

mask — and can even result in a loss of structure and legibility. One method to mitigate this undesirable effect is to run a second instance of the iterative refinement procedure by lowering the values of both $\Delta t_-$, $\Delta t_+$. As a result, we can achieve a smoothing effect while still preserving the structural similarity of the input, as provided by the original features. Variable degrees of smoothing can then be achieved by interpolating the $\Sigma\Lambda$ parameters between the original reconstruction and the smoothed one with a parameter $\alpha \in [0, 1]$ (Fig. 13(b)).

More flexible stylisation effects can also be achieved with a similar approach, for example by constraining all stroke curvature parameters $\delta_i$ to a user-specified value (Fig. 13(c)) and then running the iterative refinement with $\lambda_\delta = 0$, hence not further affecting the parameters. Although we use linear interpolation for the parameters $\Delta t_i$, $\delta i$, we observe that these are not linearly related to the target positions. While this relation deserves further analysis in future studies, we achieve satisfactory results by specifying a power of $\alpha$ for interpolating targets (Fig. 13(d)) and observe experimentally that a power of 7 works particularly well for our use case (Figs. 13(b), (c)).

### 6.3. *Expressive rendering and interaction*

The smooth kinematics produced by the $\Sigma\Lambda$ model can be exploited to generate expressive brush renderings of the trajectory. This has been explored in previous work to generate realistic renditions of signatures [Ferrer *et al.*, 2015] using a model of ink deposition model [Franke and Rose, 2004]. Here we demonstrate a brush model that builds upon the assumption that the amount of paint deposited is inversely proportional to the speed of the drawing tool. Furthermore, we can sweep a texture along the generated trajectory with width also inversely proportional to speed — refer to Berio *et al.* [2018] for mathematical and implementation details. While this is a simplistic model, it generates patterns that are highly evocative of some instances of calligraphy as well as graffiti made with markers or spray paint (Fig. 14), and accentuates the perceived dynamism of the trace.

The trajectory generated by the reconstruction, as well as the brush rendering parameters can be edited in real time with an intuitive user interface [Berio *et al.*, 2018]. This allows a user to fine tune the rendering results or to apply subsequent modifications to the trajectory by editing the target positions and the stroke parameters through a CAD-like interface. Furthermore, the resulting kinematics reproduce natural human-like movements that can be exploited to create stroke animations of the input as well as to generate smooth motion paths for virtual characters or even humanoid robots [Berio *et al.*, 2016].

Fig. 14. Examples of, on the left, a real graffiti textured over a wall and, on the right, one rendered with kinematics based brushes.

## 7. Conclusion

In this chapter, we developed and explained a systematic method to reconstruct $\Sigma\Lambda$ parameters from solely the static geometric trace (left by handwriting or drawing gestures), which relies on an initial segmentation at perceptually salient points. The method is capable of producing an accurate geometric reconstruction of the input, while inferring plausible kinematics underlying its generation on the basis of just an ordered sequence of points as an input.

We achieved our goal of a plausible reconstruction of the kinematics by designing a method exploiting a notion of Curvilinear Shape Features (or CSF) to incrementally adjust the temporal and spatial parameters of the $\Sigma\Lambda$ model. The method consistently produces accurate ($> 15Db$ SNR) reconstructions of the input, while providing flexibility for the use of additional constraints that can be exploited in order to generate interactive stylisations and variations.

One possible line of extension of our work is to exploit the availability of higher order derivatives of the $\Sigma\Lambda$ model to develop solutions using constrained optimisation methods with stronger convergence guarantees. Another interesting future line of work is to develop more sophisticated methods of fixing the $\Sigma\Lambda$ parameters $\mu_i$ and $\sigma_i$, which are currently experimentally set and kept constant. This shall require to explore in depth how the inferred kinematics relate to human data and are perceived by human observers.

## References

Attneave, F. (1954). Some informational aspects of visual perception. *Psychological Review* **61**, 3, p. 183.

Baran, I., Lehtinen, J. and Popović, J. (2010). Sketching clothoid splines using shortest paths, *Computer Graphics Forum* **29**, 2, pp. 655–664.

Belyaev, A. and Yoshizawa, S. (2001). On evolute cusps and skeleton bifurcations, in *Proceedings of the International Conference on Shape Modeling and Applications (SMI)*, pp. 134–140, https://doi.org/10.1109/SMA.2001.923384.

Berio, D., Akten, M., Fol Leymarie, F., Grierson, M. and Plamondon, R. (2017a). Calligraphic stylisation learning with a physiologically plausible model of movement and recurrent neural networks, in *ACM Proceedings of the 4th International Conference on Movement Computing (MOCO)*, London, UK, https://doi.org/10.1145/3077981.3078049.

Berio, D., Calinon, S. and Fol Leymarie, F. (2016). Learning dynamic graffiti strokes with a compliant robot, in *Proceedings of the IEEE/RSJ International Conference on Intelligent Robots and Systems (IROS)*, Daejeon, South Korea, pp. 3981–86.

Berio, D. and Fol Leymarie, F. (2015). Computational Models for the Analysis and Synthesis of Graffiti Tag Strokes, in P. Rosin (ed.), *Computational Aesthetics*, Eurographics Association, pp. 35–47.

Berio, D., Fol Leymarie, F. and Plamondon, R. (2017b). Computer aided design of handwriting trajectories with the kinematic theory of rapid human movements, in *Proceedings of the 18th International Graphonomics Society Conference (IGS)*, Gaeta, Italy, https://graphonomics.net/igs2017/.

Berio, D., Leymarie, F. F. and Plamondon, R. (2018). Expressive Curve Editing with the Sigma Lognormal Model, in O. Diamanti and A. Vaxman (eds.), *Proceedings of Eurographics — Short Papers*, Delft, The Netherlands, pp. 33–36.

Blum, H. (1967). A transformation for extracting new descriptors of shape, in W. Wathen-Dunn (ed.), *Models for the perception of speech and visual form*, pp. 362–380, MIT Press, proceedings of a symposium held in Boston, MA, November 1964.

Blum, H. (1973). Biological shape and visual science (part I), *Journal of Theoretical Biology* **38**, 2, pp. 205–287.

Brault, J.-J. and Plamondon, R. (1993a). A complexity measure of handwritten curves: Modeling of dynamic signature forgery, *IEEE Transactions on Systems, Man, and Cybernetics* **23**, 2, pp. 400–413.

Brault, J.-J. and Plamondon, R. (1993b). Segmenting handwritten signatures at their perceptually important points, *IEEE Transactions on Pattern Analysis and Machine Intelligence* **15**, 9, pp. 953–957.

Bruce, J. W. and Giblin, P. J. (1992). *Curves and Singularities: A Geometrical Introduction to Singularity Theory*, 2nd edn., Cambridge University Press, doi:10.1017/CBO9781139172615.

De Stefano, C., D'Elia, C., Garruto, M., Marcelli, A. and Di Freca, A. S. (2005). A wavelet based curve decomposition method for on-line handwriting, *Advances in Graphonomics: Proc. of IGS*, pp. 162–167 URL `https://graphonomics.net/publications.php`.

De Winter, J. and Wagemans, J. (2006). Segmentation of object outlines into parts: A large-scale integrative study, *Cognition* **99**, 3, pp. 275 – 325, URL `http://www.sciencedirect.com/science/article/pii/S0010027705000648`.

De Winter, J. and Wagemans, J. (2008). Perceptual saliency of points along the contour of everyday objects: A large-scale study, *Perception & Psychophysics* **70**, 1, pp. 50–64.

Diaz-Cabrera, M., Fischer, A., Ferrer, M. A. and Plamondon, R. (2018). Dynamic signature verification system based on one real signature, *IEEE Transactions on Cybernetics* **48**, 1, pp. 228–239, doi:10.1109/TCYB.2016.2630419.

F.A.T. Lab (2009). 000000book: An open database for graffiti markup language (gml) files, Web, URL `https://000000book.com`.

Feldman, J. and Singh, M. (2005). Information along contours and object boundaries, *Psychological Review* **112**, 1, pp. 243–252.

Ferrer, M. A., Diaz, M., Carmona, C. A. and Plamondon, R. (2018). iDeLog: Iterative dual spatial and kinematic extraction of sigma-lognormal parameters, *IEEE Trans. on Pattern Analysis and Machine Intelligence (PAMI)*. 42(1):114–125. Https://doi.org/10.1109/TPAMI.2018.2879312.

Ferrer, M. A., Diaz-Cabrera, M. and Morales, A. (2015). Static signature synthesis: A neuromotor inspired approach for biometrics, *IEEE Transactions on Pattern Analysis and Machine Intelligence* **37**, 3, pp. 667–680.

Fischer, A., Plamondon, R., O'Reilly, C. and Savaria, Y. (2014). Neuromuscular representation and synthetic generation of handwritten whiteboard notes, in *Proceedings of the 14th International Conference on Frontiers in Handwriting Recognition (ICFHR)*, pp. 222–7, https://doi.org/10.1109/ICFHR.2014.45.

Fong, W. C. (2003). Why Chinese painting is history, *The Art Bulletin* **85**, 2, pp. 258–280.

Franke, K. and Rose, S. (2004). Ink-deposition model: The relation of writing and ink deposition processes, in *IEEE Proceedings of the Ninth International Workshop on Frontiers in Handwriting Recognition (IWFHR)*, Tokyo, Japan, pp. 173–178.

Freedberg, D. and Gallese, V. (2007). Motion, emotion and empathy in esthetic experience, *Trends in cognitive sciences* **11**, 5, pp. 197–203.

Freyd, J. J. (1983). Representing the dynamics of a static form, *Memory & cognition* **11**, 4, pp. 342–346.

Galbally, J., Plamondon, R., Fierreza, J. and Ortega-Garcia, J. (2012). Synthetic on-line signature generation. Part I: Methodology and algorithms, *Pattern Recognition* **45**, 7, pp. 2610–21, doi:https://doi.org/10.1016/j.patcog.2011.12.011.

Heald, M. (1985). Rational approximations for the Fresnel integrals, *Mathematics of Computation* **44**, 170, pp. 459–461.

Lang, K. and Alexa, M. (2015). The Markov pen: Online synthesis of free-hand drawing styles, in *Proceedings of the Workshop on Non-Photorealistic Animation and Rendering (NPAR)*, Istanbul, Turkey, pp. 203–215.

Leiva, L. A., Martín-Albo, D. and Plamondon, R. (2016). Gestures à Go Go: Authoring synthetic human-like stroke gestures using the kinematic theory of rapid movements, *ACM Transactions on Intelligent Systems and Technology (TIST) — Special Issue on Causal Discovery and Inference Archive* **7**, 2, pp. 15:1–15:29, doi:10.1145/2799648.

Leiva, L. A., Martín-Albo, D. and Plamondon, R. (2017). The kinematic theory produces human-like stroke gestures, *Interacting with Computers* **29**, 4, pp. 552–565, doi: 10.1093/iwc/iww039.

Levien, R. (2009). *From Spiral to Spline: Optimal Techniques in Interactive Curve Design*, Ph.D. thesis, EECS Department, University of California, Berkeley.

Leymarie, F. and Levine, M. D. (1989). Curvature morphology, in *Proceedings of Vision Interface '89*, CIPPRS, London, Ontario, Canada, pp. 102–109.

Leymarie, F. and Levine, M. D. (1992). Simulating the grassfire transform using an active contour model, *IEEE Transacctions on Pattern Analysis and Machine Intelligence* **14**, 1, pp. 56–75, doi:10.1109/34.107013.

Leyton, M. (1987). Symmetry-curvature duality, *Computer Vision, Graphics, and Image Processing* **38**, 3, pp. 327–341, https://doi.org/10.1016/0734-189X(87)90117-4.

Li, X., Parizeau, M. and Plamondon, R. (1998). Segmentation and reconstruction of on-line handwritten scripts, *Pattern Recognition* **31**, 6, pp. 675–684.

Li, X. and Plamondon, R. (2006). Medial axis extraction of planar shapes, *International Journal of Applied Mathematical Analysis and Applications* **1**, 1, pp. 11–27.

Llorens, D. *et al.* (2008). The UJIpenchars database: A pen-based database of isolated handwritten characters, in *Proceedings of the 6th International Conference on Language Resources and Evaluation (LREC)*, pp. 2647–51, http://www.lrec-conf.org/proceedings/lrec2008/.

Longcamp, M., Anton, J. L., Roth, M. and Velay, J. L. (2003). Visual Presentation of Single Letters Activates a Premotor area Involved in Writing, *NeuroImage* **19**, 4, pp. 1492–1500.

Lu, J., Yu, F., Finkelstein, A. and DiVerdi, S. (2012). HelpingHand: Example-based stroke stylization, *ACM Transactions on Graphics (TOG)* **31**, 4, p. 46.

McCrae, J. and Singh, K. (2009). Sketching piecewise clothoid curves, *Computers and Graphics* **33**, 4, pp. 452–461.

Mediavilla, C. (1996). *Calligraphy: From Calligraphy to Abstract Painting*, Scirpus Publications, Wommelgem, Belgium.

Ogniewicz, R. (1992). *Discrete Voronoi Skeletons*, Ph.D. thesis, ETH Zürich, https://doi.org/10.3929/ethz-a-000692549.

O'Reilly, C. and Plamondon, R. (2008). Automatic extraction of Sigma-Lognormal parameters on signatures, in *Proc. of 11th International Workshop on Frontiers in Handwriting Recognition (ICFHR)*, Concordia University, Montreal, Canada.

Pignocchi, A. (2010). How the Intentions of the Draftsman Shape Perception of a Drawing, *Consciousness and Cognition* **19**, 4, pp. 887–898.

Plamondon, R. (1995). A Kinematic Theory of Rapid Human Movements. Part I. Movement Representation and Generation, *Biological Cybernetics* **72**, 4, pp. 295–307, http://dx.doi.org/10.1007/BF00202785.

Plamondon, R. and Guerfali, W. (1998). The 2/3 power law: When and why? *Acta Psychologica* **100**, 1, pp. 85–96.

Plamondon, R., O'Reilly, C., Galbally, J., Almaksour, A. and Anquetil, É. (2014). Recent developments in the study of rapid human movements with the kinematic theory: Applications to handwriting and signature synthesis, *Pattern Recognition Letters* **35**, 1, pp. 225–235, https://doi.org/10.1016/j.patrec.2012.06.004.

Plamondon, R. and Privitera, C. M. (1999). The segmentation of cursive handwriting: An approach based on off-line recovery of the motor-temporal information, *IEEE Transactions on Image Processing* **8**, 1, pp. 80–91, https://doi.org/10.1109/83.736691.

Richards, W. and Hoffman, D. (1985). Codon constraints on closed 2D shapes, *Computer Vision, Graphics, and Image Processing* **31**, 3, pp. 265–281.

Thiel, Y., Singh, K. and Balakrishnan, R. (2011). Elasticurves: Exploiting stroke dynamics and inertia for the real-time neatening of sketched 2D curves, in *Proceedings of the 24th annual ACM symposium on User Interface Software and Technology (UIST)*, Santa Barbara, CA, USA, pp. 383–392, https://doi.org/10.1145/2047196.2047246.

Todorov, E. (2004). Optimality principles in sensorimotor control, *Nature Neuroscience* **7**, 9, pp. 907–915, https://doi.org/10.1038/nn1309.

Viviani, P. and Schneider, R. (1991). A developmental study of the relationship between geometry and kinematics in drawing movements, *Journal of Experimental Psychology: Human Perception and Performance* **17**, 1, pp. 198–218.

Witkin, A. (1983). Scale-space filtering, in *Proceedings of the Eighth International Joint Conference on Artificial Intelligence (IJCAI)*, Vol. 2, Karlsruhe, West Germany, pp. 1019—22.

Zitnick, C. L. (2013). Handwriting beautification using token means, *ACM Transactions on Graphics (TOG)* **32**, 4, p. 53, https://doi.org/10.1145/2461912.2461985.

https://doi.org/10.1142/9789811226830_0012

# Chapter 12

# An Interactive Tablet-based System to Run Neuromuscular Tests

Nadir Faci[1,*], Simon Pierre Boyogueno Bidias[1,†], Réjean Plamondon[1,‡] and Nicolas Bergeron[2,§]
[1]*Laboratoire Scribens, Département de génie électrique, Polytechnique Montréal*
[2]*Scribens Project Inc*
[*]*nadir.faci@polymtl.ca,* [†]*simon-pierre.boyogueno-bidias@polymtl.ca,* [‡]*rejean.plamondon@polymtl.ca,* [§]*nicolas@scribensproject.com*

This chapter describes an integrated and updated version of the human writing movement acquisition system referred to as Sign@medic. This new system, called the Lognometer, takes advantage of all the features of the Wacom Cintiq technology to replace the hardware parts of the Sign@medic system. To this end, the Lognormeter includes the acquisition system, the audio-visual stimulus and the guide sheets in a unique tablet environment. This feature allows the users to work directly on the screen of the Cintiq 13HD tablet. Moreover, this new system is more versatile than the Sign@medic, thanks to an easier integration of new tests. The methodology to compare the two systems is described and presented. The Lognormeter is a more efficient and affordable tool than Sign@medic. In future research, the new system will be used in various domains of pattern recognition and artificial intelligence relying on studies of human movements, for e-Security, e-Health and e-Learning applications.

## 1. Introduction

Human movements can be very complex, and their studies in different areas of medicine and human science has been intensively used. In neurosciences, handwriting strokes, which constitute a specific class of rapid human movements, are used for example to study neurodegenerative processes involved in diseases such as Parkinson [Teulings *et al.*, 1997; Costa *et al.*, 2017; Eichhorn *et al.*, 1996; Phillips *et al.*, 1991], and Alzheimer's diseases [Schroter *et al.*, 2003; Impedovo and Pirlo, 2018; Lebel *et al.*, 2017; Nadeau *et al.*, 2018a]. Moreover, it is also used to study ADHD in young children [Kaiser *et al.*, 2015; Graham *et al.*, 2016;

Goulardins *et al.*, 2017; Laniel *et al.*, 2018, 2019], as well as normal children development [Duval *et al.*, 2015] and adult aging [Woch *et al.*, 2011]. A set of neuromuscular tests has been designed to infer the probability of presence of brain stroke risk factors [O'Reilly *et al.*, 2014]. The early screening of cerebral lesions appears to be possible by detecting slight deviations from the normal health state, which are not evident by simple visual inspection [Faci *et al.*, 2018, 2019].

Many of these examples emphasize the need for building robust and easily workable equipment for movement analysis. Among the various sets present in the literature and market, the Sign@medic acquisition system developed at Scribens laboratory decades ago has been found to be successful in acquiring handwriting movements more accurately. According to the Kinematic Theory of rapid human movement, these strokes constitute the elementary movements necessary to understand how the motor control deal in performing complex movements. Over the last fifteen years, all the neuromuscular studies performed at Scribens laboratory have been done using the Sign@medic acquisition devices, based on Wacom Intuos2 technology under Windows XP operating system. This was done to make sure that comparative results could be obtained from one experiment to another. Such a system is no longer adequate and is about to be unsupported by the manufacturers.

This paper proposes an integrated and updated version of the Sign@medic acquisition device based on the Wacom Cintiq technology. The new system takes advantage of all the recent innovations offered by Wacom Cintiq 13HD technology to provide a versatile and convenient acquisition system, allowing users to work directly on the screen of this compact tablet. Excellent colour displays and screen resolution, professional ergonomics and pressure-sensitive controls all come together in this affordable model. Compared to the other tablets based on Android and Ios, the Wacom Cintiq 13HD offers a more stable sampling frequency and a better spatial resolution. Furthermore, this new tablet offers more freedom for development.

The rest of the paper is organized as follows: Section 2 reviews the old acquisition device and its main features. Section 3 reports some drawback of the old equipment and proposed a method to relieve these deficiencies. The global system which include the lognormal parameter extraction and some new features of the new system is also described. Section 4 presents numerical results and performance comparisons with the previous system. Section 5 summarizes the main contributions of this paper and opens the door to future research.

## 2. Description of the Previous System

Figure 1 represents the block diagram of the old acquisition system. It is composed of the in-house software Sign@medic, which managed the synchronization between the digitizer and the stimulator. A Wacom Intuos2 is used as a digitizer to record the coordinates of the pen tip movement, while the subject is executing a handwriting movement. This tablet records the data at a sampling rate of 200 Hz with a spatial resolution of 100 lines per millimetre. Each movement performed on the tablet follows a test protocol based on different neuromuscular tests as proposed in [O'Reilly *et al.*, 2014]. Guide sheets are placed on the tablet under a transparent plastic fold to indicate the starting point and the target zone in an experiment. They also help in reducing friction while a subject is performing an experiment task. An electronic stimulator designed in our Laboratory initiates the beginning of each test. This simulator contains a visual and an audio stimulus. The visual stimulus is displayed by a matrix of $8 \times 10$ light-emitting diodes (LED) with a delay of 1 ms [O'Reilly *et al.*, 2014]. The auditory stimulus is a 1 KHz beep with a duration of 500 ms. The stimulator is linked to the computer by a serial to parallel converters as presented in Fig. 1. The system takes as input a handwriting movement and returns the Cartesian coordinates of the pen tip of a writer including its pressure and different time stamp.

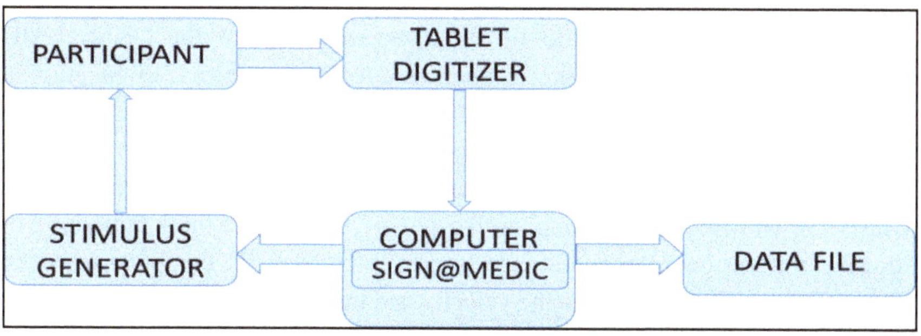


Fig. 1. The block diagram of the old acquisition system.

This system, despite its ability to accurately provide information on a handwritten neuromuscular task, suffers from multiple problems. The compatibility with the new Windows (Windows 10) operating systems is no more possible because of some home-made drivers that are no longer compatible with Windows 10.

The parallel port used by the stimulus generator is outdated, so the stimulus device cannot be connected to modern computers anymore. Moreover, this stimulus generator is cumbersome. Besides these entire problems, the manufacturer no longer supports the Wacom Intuos2 tablet, which is the main piece of that old system. These problems point out the need to update the old system to make it compatible with modern computer environments.

## 3. Description of the New System

### 3.1. *Transfer process*

The goal of this project is to update the old acquisition system such that it will be ergonomic and compatible with Windows 10 operating system. To achieve this goal, we have designed a graphical interface that displays the different functionalities of the visual stimulator on a Wacom Cintiq 13HD tablet. In addition, the various neuromuscular tests present in our database were included as well as provisions for any future new tests. The interface is implemented in C++ using the Microsoft foundation class library and our custom software employing Wacom drivers and libraries does the communication. The tablet interface is divided into three parts. The two parts on each side display the visual stimulator and the middle section displays the guide sheet and the handwriting trajectories. The transparent plastic fold is not needed any longer, but the guide sheets, initially designed to match with the Intuos2 dimensions (250×180 mm), were homothetically resized to fit into the active area of the Cintiq 13HD (220×160 mm). Those guide sheets were included directly as a C++ function in the software. A new button has been added to the interface to switch between right-handed and left-handed guide sheets. Finally, a new element exploiting the electronic ink features offered by the Cintiq 13HD was added to the software. It consists of showing, in real time, on the screen of the tablet the handwriting or drawing patterns being executed by a subject. Moreover, the computer to which the tablet is connected displays the velocity and the pressure profile.

### 3.2. *Description of the lognometer*

As presented in Fig. 2, the new acquisition system is a compact, updated and ergonomic new design elaborated from the old one. Using the Wacom Cintiq 13HD, this system integrates the stimulus generator and the paper guide sheets of the Sign@medic software as a digital data acquisition tool. This make the overall system composed of only a computer containing an in-house software, and the Wacom Cintiq 13HD tablet interface. In addition, the tablet screen acts as an additional monitor for the computer. The visual stimulus is displayed on the tablet,

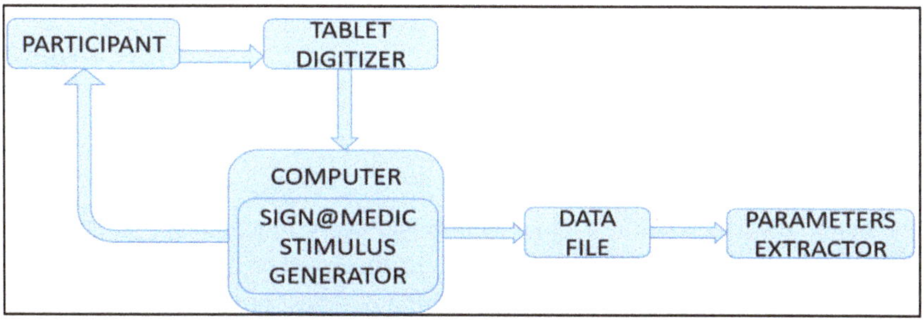


Fig. 2. The block diagram of the global system including the new acquisition system referred to as the Lognometer.

at a response rate of 25 ms. In regard of the audio stimulus, the computer speakers emit an audio cue at 1 KHz during 500 ms, similar to the one previously used in the former stimulus system.

The tablet displays the trajectory made by the participant while being executed. This feature allows the test operator to monitor directly the subject responses. The Wacom Cintiq 13HD records data at 200 Hz with a 200 lines per millimetre spatial resolution. The new acquisition system takes as an input a handwriting movement and delivers as output a data file including the information about this movement.

### 3.3. *Description of the global system*

Figure 2 presents the global system which is composed of the new acquisition software and the lognormal parameters extractors [Djioua and Plamondon, 2009; O'Reilly and Plamondon, 2009, 2012]. Those extractors were developed at the Scribens laboratory to take advantage of the Kinematic Theory of rapid movements [Plamondon *et al.*, 1991; Plamondon R., 1995a, 1995b, 1998; Plamondon *et al.*, 2003]. The global system is able to acquire data related to any simple or complex 1D or 2D movements executed by a participant on the tablet. After that, it allows the extraction of the lognormal parameters by using the different extractors. In conclusion, the global system takes a single or a sequence of stroke movements as input and generates lognormal parameter lists as output, making the whole system acting like a Lognometer.

## 4. Comparative Evaluation

### 4.1. *Advantages and disadvantages of the lognometer as compared to the old Sign@medic*

The major advantage of the Lognometer is that every functionality is integrated in the software. Foremost, the guide sheets are integrated on the software and they can be changed automatically while designing new test protocols. There is also the option to choose between right-handed and left-handed guide sheet. The software also integrates the stimulator generator. The device used before to generate stimulation is now outdated and not needed anymore. Moreover, the stroke trajectory can be seen on the screen of the tablet while being executed. This feature helps the operator to judge if the stroke is well executed. After that, the operator can decide to reject it or not. Finally, the fact that the stimulus is only displayed on the tablet and not on the previous hardware device anymore, gives more focus to the participant. Indeed, the participant only look at the tablet screen and do not need to move his head to see what is happening with the stimulator device anymore. Now, all he needs is to execute the stroke displayed on the tablet. Moreover, the new tablet (5080 lpi) has a better spatial resolution than the old one (2540 lpi).

On the other hand, the new Lognometer presents some disadvantage as compared to the obsolete Sign@medic system. The stimulator device had a 1 ms response time for displaying the visual and audio stimulus. With the new system, the tablet has a 25 ms response rate of to display the visual cue. Knowing that human reaction to visual and auditory stimuli is respectively around 247 ms (18.54) and 228 ms (16.49) for young adults [Jain *et al.*, 2015; Shelton and Kumar, 2012; Ng and Chan, 2012], the difference between the response rate of the new and the old system should not influence the result. This sets a limit to the experimental protocol related to the resolution for observing changes in neuromotor control. In this case, to observe a significant change, the difference should be greater than 25 ms instead of 1 ms. For the sound cue, the auditory stimulus is the same for the two systems.

### 4.2. *Methodology to compare the two systems*

To compare the two systems, we requested a subject to draw fast strokes movement for five days in a row. Each system was connected to different computers. Each day, the subject had to draw 30 valid strokes for the four following tests: visual stimulus tests on the new system, auditory stimulus test on the new system, visual stimulus test on the old system and auditory stimulus test on the old system. The visual stimulus test was always executed before the audio one. Each day, the test order was changed. The first day, the tests were executed on the new system before doing it on the old one. The second day, the tests were executed on the old system before doing it on the new one. The same process was done the other days. Figures 3 and 4 respectively depict the user interface and the new guide sheet used to perform the test while the visual cue is displayed on the tablet screen.

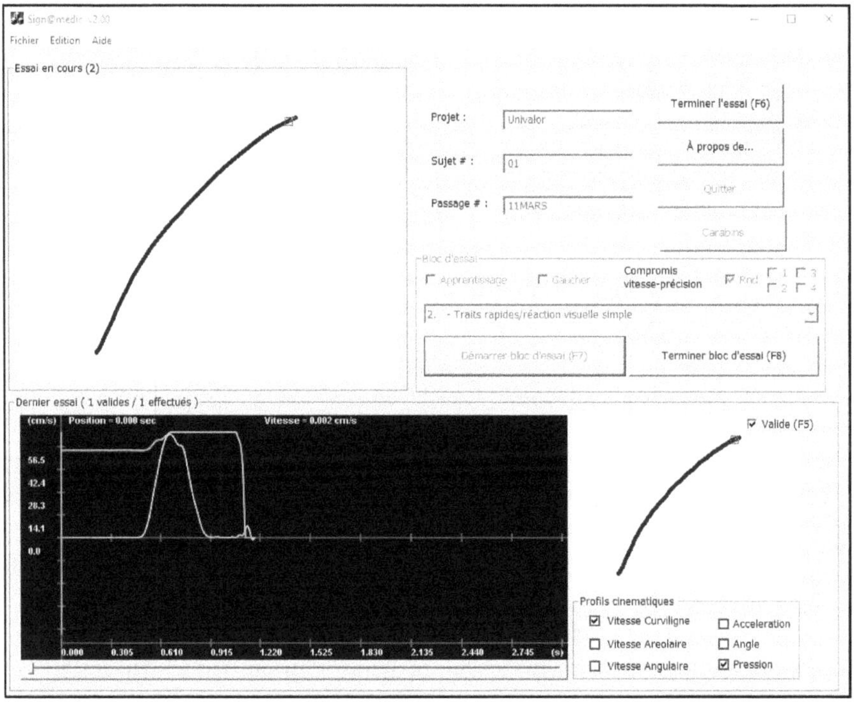


Fig. 3. The user interface displayed on the computer monitors shown to the user.

To test the new system, we wanted to limit the spatial variability of each stroke executed by the subject. For this purpose, a new guide sheet was designed with a narrowed target zone as presented on Fig. 4. This new guide sheet was used to restrict the movement of subjects. Finally, the user interface allows selecting different tests and validating the stroke performed. It also displays several pieces of information, such as the current stroke, the previous stroke, the velocity and the pressure profiles.

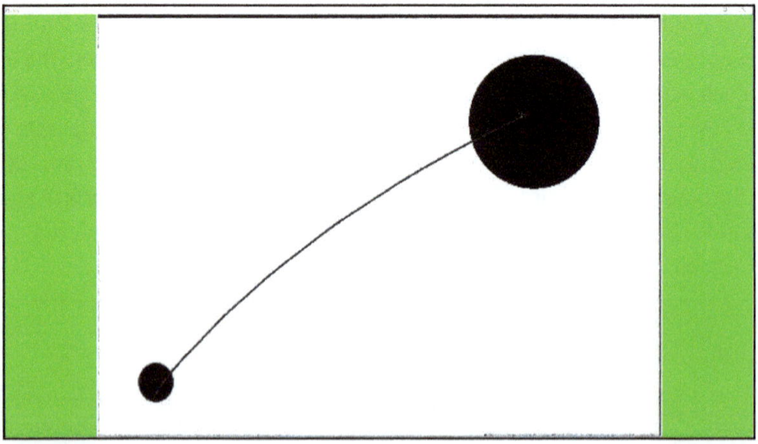

Fig. 4. The interface displayed on the screen of the tablet including the guide sheet on the center with the fast stroke, and the visual cue on each side.

### 4.3. *The different lognormal models*

After recording the data, two different lognormal models were used to extract the parameters from the original data: The Delta-Lognormal and Sigma-Lognormal model. The difference between these two models is that the Delta-Lognormal model will reconstruct the velocity profile using only two lognormals while the Sigma-Lognormal model will use as many lognormals as necessary to reconstruct the signal in order to achieve at least a SNR of 25 dB. The set of parameters extracted from the two models are different. To extract the Delta-Lognormal parameters, we used the XZERO extractor [Djioua and Plamondon, 2009, 2008]. This software extract seven parameters plus the SNR for each stroke. In the case of the Sigma-Lognormal model, the Script Studio software was used to extract the parameters [O'Reilly and Plamondon, 2009; O'Reilly C. *et al.*, 2009; Duval *et al.*, 2014; Bhattacharya *et al.*, 2017]. This software extract eight parameters per lognormal. There are five common parameters:

- $t_0$: Time required (in seconds) for the brain to produce a motor command. $t_0$ should not be confused with the moment a sound cue is emitted ($t = 0$) nor with the response time (RT) measured by moment of movement onset. $t_0$ is the moment when a population of neurons sends a motor command. In other words, $t_0$ occurs after the start cue, given that the central nervous system must perceive the sound cue before sending the command, but before RT, given that a delay is necessary for the motor command to be transmitted and for enough muscle cells to be recruited to begin the movement effectively [O'Reilly *et al.*,

2013]. Longer $t_0$ is indicative of greater delays before a motor command is emitted and, therefore, a longer time to perceive the stimulus and prepare the command. In the case of the Delta-Lognormal model, there is one $t_0$ per stroke, while in the case of Sigma-Lognormal model there is a different $t_0$ for each lognormal used to reconstruct the stroke.

- SNR: The signal-to-noise ratio between the original speed profile and the reconstructed speed profile, computed in decibels (dB). This is a measure of the quality of the sigma-lognormal reconstruction. A higher SNR indicates a better reconstruction.
- $D_{(n)}$: The amplitude of the movement associated with each motor command, in millimeters. Greater D indicates greater amplitude of the movement associated with a command. In the case of the Delta-Lognormal model, there is two Ds ($D_1$ and $D_2$) per stroke, while in the case of Sigma-Lognormal model there is a different D for each lognormal used to reconstruct the stroke. $D_{tot}$ is the total amplitude of the movement equal to the difference between $D_1$ and $D_2$ in the Delta-Lognormal model.
- $\mu_{(n)}$: Logtime delay, the time required to reach the middle of movement distance on a logarithmic scale. It can be associated with the general response speed of the neuromuscular system to a command. In the case of the Delta-Lognormal model, there is two $\mu_s$ ($\mu_1$ and $\mu_2$) per stroke, while in the case of Sigma-Lognormal model there a different $\mu$ for each lognormal used to reconstruct the stroke.
- $\sigma_{(n)}$: Logresponse time, the response time on a logarithmic scale of the neuromuscular system's response to a command. Sigma ($\sigma$) is used to compute an estimate of movement duration, it also reflects the asymmetry of the lognormal. In the case of the Delta-Lognormal model, there is two $\sigma_s$ ($\sigma_1$ and $\sigma_2$) per stroke, while in the case of Sigma-Lognormal model there a different $\sigma$ for each lognormal used to reconstruct the stroke.

In the case of the Sigma-Lognormal model, three supplementary parameters are extracted:

- nbLog: The number of lognormal functions required to reconstruct the signal. This parameter represents fluidity of movement produced by the participant. A higher nbLog, indicates a less fluid movement. The nbLog is not limited to two lognormals as in the case of the Delta-Lognormal model. The reconstruction can be done using one or more lognormals. The direction of the Lognormals are not in perfect opposition and a vector addition is used to reconstruct the original velocity. To this end, $\theta_s$ and $\theta_e$ are used.
- $\theta_s$: The starting angular direction of each stroke in radians.
- $\theta_e$: The ending angular direction of each stroke in radians.

In addition, the SNR/nbLog is also calculated for the Sigma-Lognormal model:

- SNR/nbLog: SNR/nbLog is used as a performance criterion. The capacity to reconstruct a movement's speed profile with lognormals can be interpreted as an indicator of the quality of motor control since the lognormal speed profile corresponds to perfect motor control [Plamondon *et al.*, 2013]. Higher SNR/nbLog indicates that the movement is closer to ideal lognormal behaviour predicted by the lognormality principle.

The nbLog, SNR and SNR/nbLog describe the global state of the neuromotor system lognormality [Plamondon R., 1995a, 1995b; Plamondon *et al.*, 2003; O'Reilly *et al.*, 2013]. The $t_0$ and $D_n$ are indexes of the internal high-level functioning of the neuromotor system and represent the neuromotor action plan. The $\mu_n$ and $\sigma_n$ parameters are peripheral parameters related to motor program execution. Finally, $\theta_s$ and $\theta_e$ are intrinsic characteristics of each lognormal and are related to the motor action plan [Plamondon *et al.*, 2003; Plamondon and Djioua, 2006].

The following figures and tables present the results of the reconstruction of a stroke with both Lognormal models, the Delta-Lognormal and Sigma-Lognormal models. Figure 5 represents an original stroke executed by a subject. Such a stroke is used here to illustrate the parameter extraction process. Tables 1 and 2 represent respectively the Delta-Lognormal and Sigma-Lognormal parameters extracted by both software. Figures 6 and 7 represent respectively the reconstruction of the velocity using the Delta-Lognormal and the Sigma-Lognormal model of the stroke shown above.

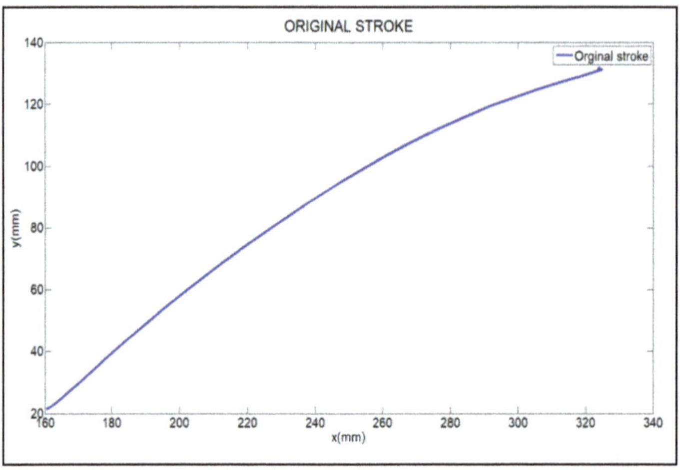


Fig. 5. An original stroke executed by a subject to be reconstructed by the two models.

Table 1. The Delta-Lognormal parameters extracted by the software for the stroke presented on Fig. 5.

| PARAMETERS | | | | | | | |
|---|---|---|---|---|---|---|---|
| $t_0$(s) | $D_1$(mm) | $\mu_1$ | $\sigma_1$ | $D_2$(mm) | $\mu_2$ | $\sigma_2$ | SNR |
| 0,082 | 221,06 | -1,119 | 0,129 | 19,86 | -0,912 | 0,058 | 36,349 |

Table 2. The Sigma-Lognormal parameters extracted by the software for the stroke presented on Fig. 5.

| GLOBAL PARAMETERS | | | | | |
|---|---|---|---|---|---|
| nbLog | | 2 | SNR (dB) | | 30,532 |
| LOGNORMAL PARAMETERS | | | | | |
| $t_0$(s) | D (mm) | $\mu$ | $\sigma$ | $\theta_s$ | $\theta_e$ |
| 0,164 | 218,45 | -1,392 | 0,168 | 0,844 | 0,324 |
| 0,355 | 20,89 | -1,967 | 0,198 | -2,334 | -3,183 |

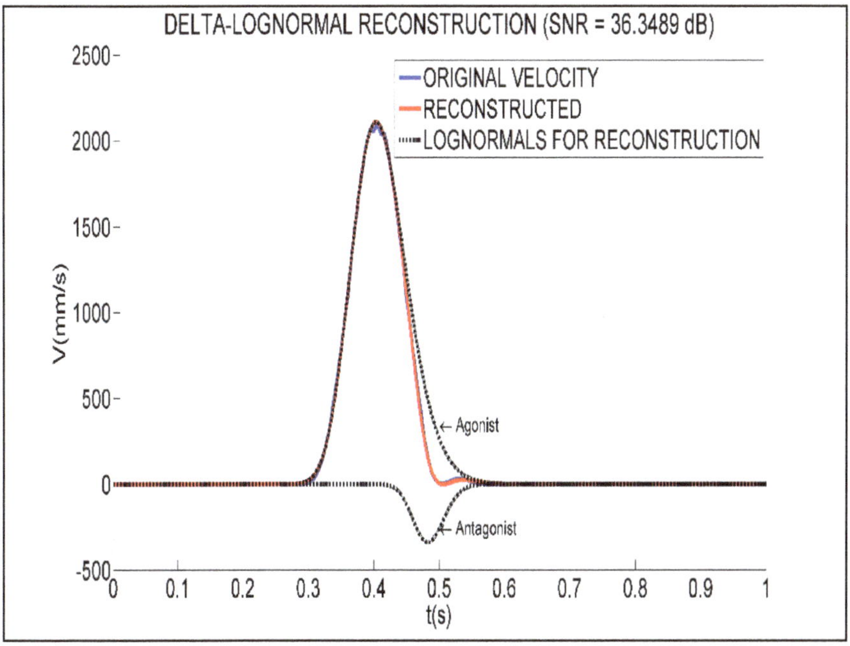


Fig. 6. The Delta-Lognormal reconstruction of the velocity of the stroke shown of Fig. 5.

In this case, the Sigma-Lognormal model reconstructed the stroke using only two lognormals. These two lognormals are not in complete opposition and do not have the same $t_0$.

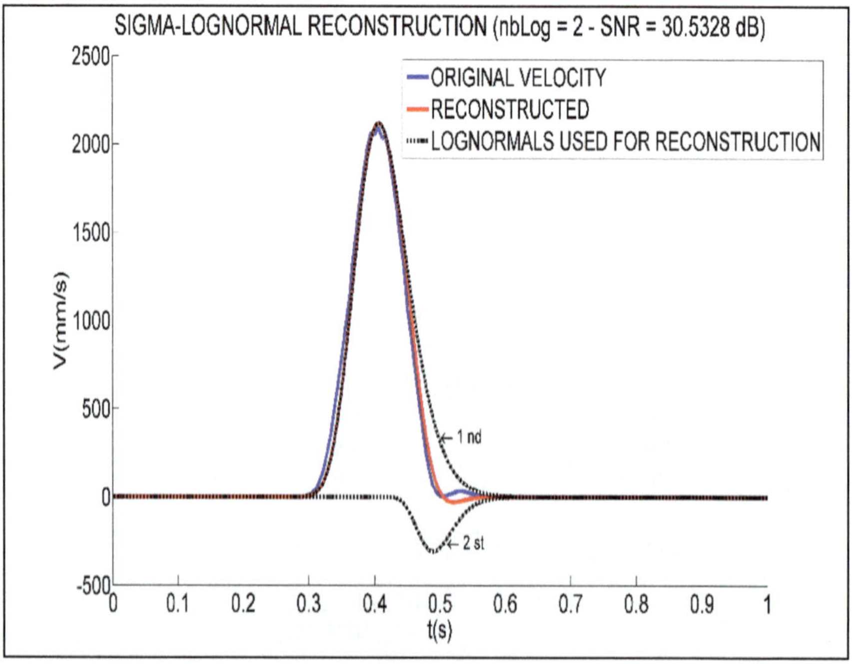


Fig. 7. The Sigma-Lognormal reconstruction of the velocity of the stroke shown of Fig. 5.

### 4.4. *Statistical analysis*

After extracting the parameters, the data for each day were regrouped by tests. Indeed, all the data recorded with the new system using the visual stimulus were regrouped, and so forth. A t-test for independent measures was conducted on the parameters extracted by Script Studio from all the strokes made. The t-test is used to determine if the means of two sets of data are significantly different from each other. Bonferroni correction for multiple comparisons was used to determine if there was a significant difference. This correction is used when there are more than one comparison and to compensate for multiple comparison by decreasing the significant threshold. P-value had to be less than 0.00625 ($\alpha = 0.05/8 = 0.00625$) to have a significant statistical difference between the two groups.

### 4.5. *Preliminary results and discussion*

The following four tables present the results of the statistical analysis comparing the Lognometer to Sign@medic. The first two tables are respectively the results of t-test for independent groups between with the visual and sound cue using the Delta-Lognormal model. The last two tables are respectively the results of t-test for independent groups between with the visual and sound cue using the Sigma-Lognormal model.

Table 3. Result of t-test for independent groups between the new and the old system with the visual cue using the Delta-Lognormal model.

| Parameters | New system M(SD) | Old system M(SD) | t-value | df | p-value |
|---|---|---|---|---|---|
| $t_0$ | 0,144 (0,068) | 0,15 (0,066) | -0,740 | 302 | 0,460 |
| **$D_1$** | **234,45 (30,52)** | **218,62 (31,64)** | **4,44** | **302** | **0,000[a]** |
| **$\mu_1$** | **-1,387 (0,212)** | **-1,447 (0,234)** | **2,348** | **302** | **0,020[b]** |
| **$\sigma_1$** | **0,2 (0,053)** | **0,218 (0,06)** | **-2,886** | **302** | **0,004[a]** |
| **$D_2$** | **39,87 (28,84)** | **28,55 (30,88)** | **3,30** | **302** | **0,000[a]** |
| $\mu_2$ | -1,114 (0,155) | -1,116 (0,184) | 0,134 | 302 | 0,893 |
| $\sigma_2$ | 0,101 (0,033) | 0,095 (0,049) | 1,228 | 302 | 0,221 |
| **SNR** | **32,088 (4,185)** | **30,272 (5,308)** | **3,318** | **302** | **0,001[a]** |
| **$D_{tot}$** | **194,58 (7,33)** | **190,06 (7,77)** | **5,21** | **302** | **0,000[a]** |

Table 4. Result of t-test for independent groups between the new and the old system with the sound cue using the Delta-Lognormal model.

| Parameters | New system M(SD) | Old system M(SD) | t-value | df | p-value |
|---|---|---|---|---|---|
| $t_0$ | 0,1 (0,051) | 0,091 (0,053) | 1,496 | 290 | 0,136 |
| **$D_1$** | **228,16 (26,41)** | **219,47 (26,4)** | **2,811** | **290** | **0,005[a]** |
| $\mu_1$ | -1,415 (0,201) | -1,457 (0,211) | 1,754 | 290 | 0,080 |
| **$\sigma_1$** | **0,196 (0,043)** | **0,215 (0,054)** | **-3,355** | **290** | **0,001[a]** |
| $D_2$ | 32,76 (25,19) | 29 (26,29) | 1,250 | 290 | 0,212 |
| $\mu_2$ | -1,142 (0,165) | -1,149 (0,187) | 0,375 | 290 | 0,708 |
| $\sigma_2$ | 0,097 (0,035) | 0,098 (0,047) | -0,288 | 290 | 0,773 |
| **SNR** | **32,259 (4,311)** | **30,23 (5,055)** | **3,696** | **290** | **0,000[a]** |
| **$D_{tot}$** | **195,4 (8,57)** | **190,47 (9,53)** | **4,648** | **290** | **0,000[a]** |

Table 5. Result of t-test for independent groups between the new and the old system with the visual cue using the Sigma-Lognormal model.

| Parameters | New system M(SD) | Old system M(SD) | t-value | df | p-value |
|---|---|---|---|---|---|
| **SNR** | **30,47 (2,86)** | **28,43 (3,36)** | **5.84** | **320** | **0,0000[a]** |
| **nbLog** | 1,98 (0,37) | 2,03 (0,64) | -0,95 | 320 | 0,3419 |
| **SNR/nbLog** | 16,01 (3,63) | 15,49 (5,43) | 1,00 | 320 | 0,3137 |
| **$t_0$** | **0,169 (0,17)** | **0,211 (0,18)** | **-2,94** | **643** | **0,0036[a]** |
| **$t_0$i** | **0,124 (0,139)** | **0,152 (0,106)** | **-1,97** | **320** | **0,0490[b]** |

Table 5 (*continued*).

| Parameters | New system M(SD) | Old system M(SD) | t-value | df | p-value |
|---|---|---|---|---|---|
| **D** | 120,64(95,89) | 112,03(94,35) | 1,16 | 643 | 0,2452 |
| **$D_{tot}$** | **197,41 (19,0)** | **191,07 (15,5)** | **3,28** | **320** | **0,0011[a]** |
| **μ** | -1,52 (0,65) | -1,59 (0,58) | 1,62 | 643 | 0,1045 |
| **σ** | **0,21 (0,14)** | **0,24 (0,15)** | **-2,47** | **643** | **0,0136[b]** |
| **$\theta_s$** | **0,26 (4,58)** | **-0,89 (4,64)** | **2,47** | **643** | **0,0136[b]** |
| **$\theta_e$** | **-0,08 (4,58)** | **-1,15 (4,64)** | **2,23** | **643** | **0,0256[b]** |

Table 6. Result of t-test for independent groups between the new and the old system with the sound cue using the Sigma-Lognormal model.

| Parameters | New system M(SD) | Old system M(SD) | t-value | df | p-value |
|---|---|---|---|---|---|
| **SNR** | **30,19 (2,60)** | **28,59 (3,23)** | **4,92** | **322** | **0,0000[a]** |
| **nbLog** | **1,88 (0,36)** | **2,17 (0,77)** | **-4,40** | **322** | **0,0000[a]** |
| **SNR /nbLog** | **16,70 (3,73)** | **14,77 (5,57)** | **3,66** | **322** | **0,0003[a]** |
| **$t_0$** | 0,16 (0,14) | 0,18 (0,18) | -1,56 | 655 | 0,1186 |
| **$t_0$i** | 0,09 (0,08) | 0,09 (0,08) | 0,45 | 322 | 0,6519 |
| **D** | **123,63(95,33)** | **106,67(93,02)** | **2,31** | **655** | **0,0214[b]** |
| **$D_{tot}$** | **197,49 (15,4)** | **188,52 (12,8)** | **5,69** | **322** | **0,0000[a]** |
| **μ** | -1,66 (0,6) | -1,67 (0,57) | 0,25 | 655 | 0,7966 |
| **σ** | **0,23 (0,13)** | **0,26 (0,15)** | **-2,18** | **655** | **0,0294[b]** |
| **$\theta_s$** | 0,20 (4,57) | -0,29 (4,62) | 1,38 | 655 | 0,1667 |
| **$\theta_e$** | -0,17(4,58) | -0,59 (5,02) | 1,13 | 655 | 0,2572 |

[a]Significant with Bonferroni corrections for multiple comparisons ($p < 0.00625$).
[b]Significant at $p < 0.05$.

This preliminary test showed that there is a significant difference, with the Bonferroni correction, on the SNR between the two systems in both visual and audio stimuli and for both models. In all cases, the SNR mean is higher for the new system. This could be caused by the higher spatial resolution of the new Wacom Cintiq 13HD as compared to the Wacom Intuos2. The $D_{tot}$ showed a significant difference between the two systems in both visual and audio stimuli and for both models. In both cases, the $D_{tot}$ mean is higher for the new system. Moreover, the Wacom Cintiq 13HD has a surface with less friction than the surface of the Wacom Intuos2. Consequently, drawing on the new tablet is smoother than the old one, so it is more difficult to stop the movement, which leads to higher $D_{tot}$. In the case of the Delta-Lognormal model, there is a significant difference between the two systems for $D_1$, which is caused again by the fact that there is less friction when executing the movement and the subject can reach its real velocity.

For the visual stimulus, there is a significant difference, with the Bonferroni correction, in the $t_0$ parameters for the Sigma-Lognormal model. The $t_0$ mean is lower for the new system as compared to the old one. On the other hand, there is no significant difference for the $t_0$ in the Delta-Lognormal model. A difference exists between the two models for $t_0$, because the Delta-Lognormal model stipulates that the lognormals composing the movement have opposed direction and start at the same time. In contrast, the Sigma-Lognormal model reconstructs the velocity using a vector addition of the different lognormals which their own $t_0$.

## 5. The Battery of Tests Integrated on the New System

Multiple tasks were integrated on the new system to investigate the neuromuscular system [O'Reilly *et al.*, 2014]. Two different stimuli are available: a sound cues and green light. Each row of Fig. 8 represents a different task. In all tasks, the person should execute a fast pen stroke from the black dot (the starting point) to the grey area (the arrival zone). For each task, the red blinking light indicates that the system is ready to use as shown on Fig. 8(a). If the person starts its movement before the cue, an orange cross is displayed as shown on Fig. 8(c). The first task is available with a visual and sound cue (Fig. 8(a)). This task evaluates the ability of the subjects to react as quickly as possible to these stimuli. The second task, Figs. 8(d)–(e), examines the capacity of the person to react quickly and to make a good choice about the direction he draws the stroke. The second row of Fig. 8 shows the visual stimulus showed on each side of the guide sheet indicating the direction to draw the stroke. The task shown on the third row, Figs. 8(f)–(i), is related to the speed-accuracy trade off [Plamondon and Alimi, 1997]. It's used to assess the ability of the participants to coordinate a movement under the speed and spatial accuracy requirements. In this case, each figure represents different sizes of the target areas, and for each target zone, there is four possible starting point. This task is executed with a sound cue. The triangle task, Figs. 8( j)–(l), examines the capacity of a person to produce a sequence of fast pen strokes with a combination of speed and accuracy. There are three different sizes of triangle to evaluate the capacity to drawing sequence of strokes at different sizes. This task is executed with a sound cue. The last task is an oscillation at maximal speed. The oscillation task assesses the rhythmic properties of the subject. All those tests are included in the new system and will be used to investigate the different aspect of the human neuromuscular movement. Moreover, new guide sheet can be added

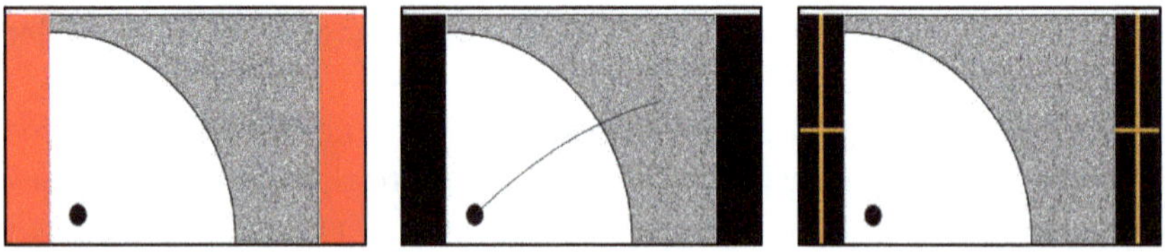

(a) (b) (c)

(a) Guiding sheet for the simple stroke test with the red light, this guide sheet is used with a visual or a sound cue. (b) A simple fast stroke displayed on the guide sheet. (c) The orange cross displayed when the stroke is not valid.

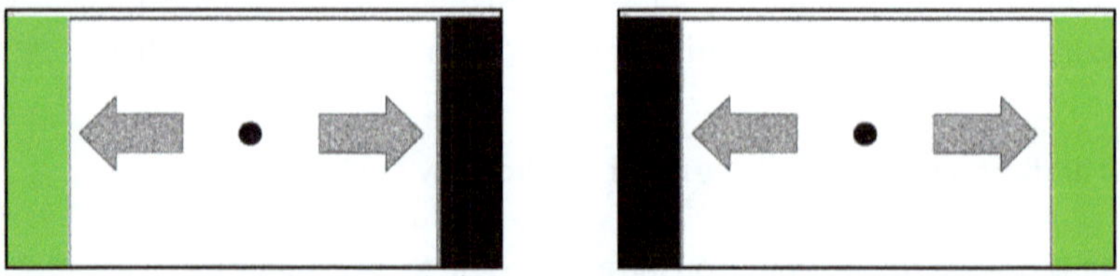

(d) (e)

Guiding sheets for the fast pen stroke on a choice reaction time protocol. The stimulus appears at one side of the guide sheet.

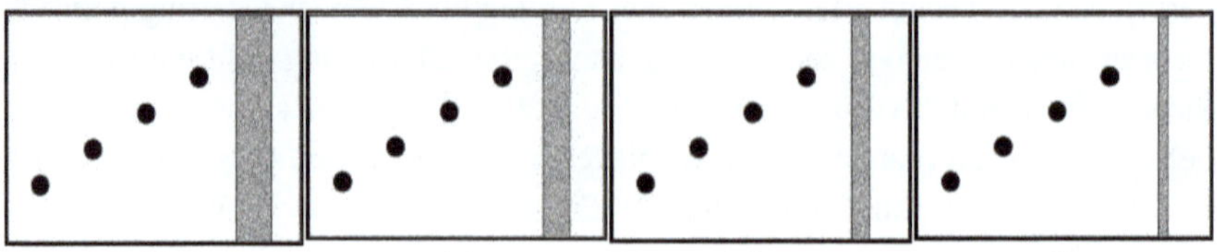

(f) (g) (h) (i)

Guiding sheets for the speed/accuracy trade-off task.

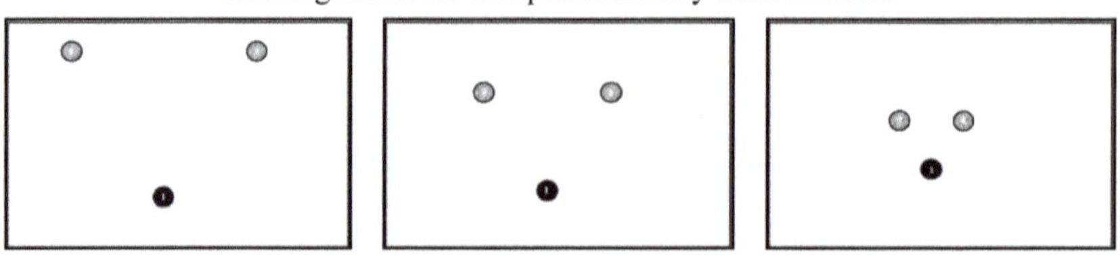

(j) (k) (l)

Guiding sheets for the fast pen stroke test sequence.

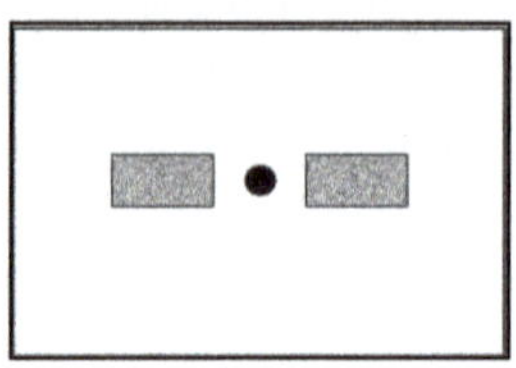

Guiding sheets for the test involving oscillatory movements.

Fig. 8. Different guiding sheet and stimuli integrated in the lognometer.

more easily to the software to replicate existing test, such as the spirals drawing [Saunders-Pullman *et al.*, 2008].

## 6. Conclusion

This project consisted of transferring the old Sign@medic system based on a Wacom intuos2 tablet under Windows XP operating system to the new Wacom Cintiq 13HD Lognometer to operate under Windows 10. Globally, the two systems have equivalent performances with an advantage for the Lognometer considering the fact that the Wacom Cintiq 13HD have better spatial resolution. The lognormeter will be used for future projet. Furthermore, the new system has the advantage of including the external stimulus generator device and the guide sheets in the software. This make the system more compact, ergonomics and user-friendly. Eventually, for more concrete results, a drawing robot will test the two systems to remove the human variability impact. Finally, the proposed system is robust and trusted to be used in different research projects dealing with the human neuromotor system. The system is also available for non-commercial use after signing nondisclosure agreements.

## Acknowledgements

This project was supported by NSERC grant RGPIN-2015-06409 to Réjean Plamondon and MESI and Univalor financial support to Scribens laboratory.

## References

Bhattacharya, U., Plamondon, R., Chowdury, S. D., Goyal, P. and Parui, S. K. (2017). A sigma-lognormal model-based approach to generating large synthetic online handwriting sample databases. *International Journal on Document Analysis and Recognition (IJDAR)*, 1–17.

Costa, F., Marino, S. and Accardo, A. (2017). Kinematic Analysis of Handwriting in Parkinson Disease. *In: IGS2017: 18th Conference of the International Graphonomics Society*, 135–138.

Djioua, M. and Plamondon, R. (2008). *A new algorithm and system for the extraction of delta-lognormal parameters*. Polytechnique Montreal.

Djioua, M. and Plamondon, R. (2009). A new algorithm and system for the characterization of handwriting strokes with delta-lognormal parameters. *IEEE Transactions on Pattern Analysis and Machine Intelligence*, *31*(11), 2060–2072.

Duval, T., Rémi, C., Plamondon, R., Vaillant, J. and O'Reilly, C. (2015). Combining sigma-lognormal modeling and classical features for analyzing graphomotor performances in kindergarten children. *Human Movement Science*, *43*, 183–201.

Duval, T., Rémi, C., Prevost, L., Dorville, A. and Plamondon, R. (2014). Etude de la faisabilité de l'évaluation de l'efficacité des mouvements de tracé du jeune apprenti-scripteur. *IHM'14, 26e conférence francophone sur l'Interaction Homme-Machine*. Lile, France.

Eichhorn, T. E., Gasser, T., Mai, N., Marquardt, C., Arnold, G., Schwarz, J. and Oertel, W. H. (1996). Computational analysis of open loop handwriting movements in Parkinson's disease: A rapid method to detect dopaminergic effects. *Movement Disorders*, *11*(3), 289.

Faci, N., Désiré, N., Beauchamp, M., Gagnon, I. and Plamondon, R. (2018). Lognormality In Children With Mild Traumatic Brain Injury: A Preliminary Pilot Study. *Proceedings of ICPRAI 2018. First International Workshop on the Lognormality Principle and its Applications*, Montréal, Canada pp. 790–794.

Faci, N., Désiré, N., Beauchamps, M., Gagnon, I. and Plamondon, R. (2019). Analysing the Evolution of Children Neuromotor System Lognormality after Mild Traumatic Brain Injury. In Press. In R. Plamondon, A. Marcelli and M. Ferrer, *Lognormality Principle: Applications for e-Security, e-Health and e-Learning*. World scientific Publishing.

Goulardins, J. B., Marques, J. C. and De Oliveira, J. A. (2017). Attention deficit hyperactivity disorder and motor impairment: A critical review. *Perceptual and Motor Skills*, *124*(2), 425–440.

Graham, S., Fishman, E. J., Reid, R. and Hebert, M. (2016). Writing Characteristics of Students with Attention Deficit Hyperactive Disorder: A Meta-Analysis. *Learning Disabilities Research & Practice*, *31*(2), 75–89.

Impedovo, D. and Pirlo, G. (2018). Dynamic handwriting analysis for the assessment of neurodegenerative diseases: A pattern recognition perspective. *IEEE Reviews in Biomedical Engineering*.

Jain, A., Bansal, R. and Kumar, A. (2015). A comparative study of visual and auditory reaction times on the basis of gender and physical activity levels of medical first year students. *International Journal of Applied and Basic Medical Research*, *5*(2), 124.

Kaiser, M. L., Schoemaker, M. M., Albaret, J. M. and Geuze, R. H. (2015). What is the evidence of impaired motor skills and motor control among children with attention deficit hyperactivity disorder (ADHD)? Systematic review of the literature. *Research in Developmental Disabilities*, *25*, 338–357.

Laniel, P., Faci, N., Plamondon, R., Beauchamp, M. H. and Gauthier, B. (2018). Kinematic analysis of fast pen strokes in children with ADHD using the Sigma-lognormal model. *Proceedings of ICPRAI 2018. First International Workshop on the Lognormality Principle and its Applications*, Montréal, Canada, pp.784–789.

Laniel, P., Faci, N., Plamondon, R., Beauchamp, M. H. and Gauthier, B. (2019). Kinematic analysis of fast pen strokes in children with ADHD. *Applied Neuropsychology: Child*, 1–16.

Lebel, K., Nguyen, H., Duval, C., Plamondon, R. and Boissy, P. (2017). Capturing the Cranio-Caudal Signature of a Turn with Inertial Measurement Systems: Methods, Parameters Robustness and Reliability. *Front Bioengineering Biotechnologies*, Vol. 5, p. 51.

Nadeau, A., Lungu, O., Boré, A., Plamondon, R., Duchesne, C., Robillard, M.-È., Doyon, J. (2018a). A 12-Week Cycling Training Regimen Improves Upper Limb Functions in People With Parkinson's Disease. *Frontiers in Human Neuroscience*, *12*.

Ng, A. W. and Chan, A. H. (2012). Finger response times to visual, auditory and tactile modality stimuli. *Proceedings of the International MultiConference of Engineers and Computer Scientists*, (pp. 1449–1454). Hong Kong.

O'Reilly, C. and Plamondon, R. (2009). Development of a Sigma–Lognormal representation for on-line signatures. *Pattern Recognition*, *42*(12), 3324–3337.

O'Reilly, C. and Plamondon, R. (2012). A globally optimal estimator for the Delta-Lognormal modeling of fast reaching movements. *IEEE Transactions on Systems, Man, and Cybernetics, Part B (Cybernetics)*, *42*(5), 1428–1442.

O'Reilly, C., Plamondon, R., Landou, M. K. and Stemmer, B. (2013). Using kinematic analysis of movement to predict the time occurrence of an evoked potential associated with a motor command. *European Journal of Neuroscience*, *32*(2), 173–180.

O'Reilly, C., Plamondon, R. and Lebrun, L.-H. (2014). Linking brain stroke risk factors to human movement features for the development of preventive tools. *Frontiers in Aging Neuroscience*, *6*(3), 150. doi:10.3389/fnagi.2014.00150.

O'Reilly, C., Plamondon, R., Lebrun, L.-H., Clément, B. and Mathieu, P. (2009). Sigma-lognormal analysis of a complex movements neuromuscular test. *14th Conference of the International Graphonomics Society*. Dijon, France.

Phillips, J. G., Stelmach, G. E. and Teasdale, N. (1991). What can indices of handwriting quality tell us about Parkinsonian handwriting? *Human Movement Science*, *10*(2–3), 301–314.

Plamondon, R. (1995a). A kinematic theory of rapid human movements: Part I: Movement representation and generation. *Biological Cybernetics*, *72*, 295–307.

Plamondon, R. (1995b). A kinematic theory of rapid human movements: Part II. Movement time and control. *Biological Cybernetics*, *72*, 309–320.

Plamondon, R. (1998). A kinematic theory of rapid human movements: Part III. Kinetic outcomes. *Biological Cybernetics*, *78*(2), 133–145.

Plamondon, R. and Alimi, A. (1997). Speed/Accuracy tradeoffs in target-directed. *Behavioral and Brain Sciences*, *20*(2), 279–349.

Plamondon, R. and Djioua, M. (2006). A multi-level representation paradigm for handwriting stroke generation. *Human Movement Science*, *25*(4), 586–607.

Plamondon, R., Feng, C. and Woch, A. (2003). A kinematic theory of rapid human movement. Part IV: A formal mathematical proof and new insights. *Biological Cybernetics*, *89*(2), 126–138.

Plamondon, R., O'Reilly, C., Rémi, C. and Duval, T. (2013). The lognormal handwriter: learning, performing, and declining. *Frontiers in Psychology*, *4*.

Plamondon, R., Yu, L.-d., Stelmach, G. E. and Clément, B. (1991). On the automatic extraction of biomechanical information from handwriting signals. *IEEE Transactions on Systems, Man, and Cybernetics*, *21*(1), 90–101.

Saunders-Pullman, R., Derby, C., Stanley, K., Floyd, A., Bressman, S., Lipton, R., Pullman, S. (2008). Validity of Spiral Analysis in Early Parkinson's Disease. *Movement Disorders*, *23*(4), 531–537.

Schröter, A., Mergl, R., Bürger, K., Hampel, H., Möller, H. J. and Hegerl, U. (2003). Kinematic analysis of handwriting movements in patients with Alzheimer's disease, mild cognitive impairment, depression and healthy subjects. *Dementia and Geriatric Cognitive Disorders*, *15*(3), 132–142.

Shelton, J. and Kumar, G. P. (2012). Comparison between auditory and visual simple reaction times. *Euroscience & Medicine*, *1*(1), 30.

Teulings, H.-L., Contreras-Vidal, J. L., Stelmach, G. E. and Adler, C. (1997). Parkinsonism reduces coordination of fingers, wrist, and arm in fine motor control. *Experimental Neurology*, *146*(1), 159–170.

Woch, A., Plamondon, R. and O'Reilly, C. (2011). Kinematic Characteristics of Bidirectional Delta-lognormal Primitives in Young and Older subjects. *Human Movement Science*, *30*(1), 1–17.

https://doi.org/10.1142/9789811226830_0013

# Chapter 13

# Extracting the Motor Program of Handwriting from its Lognormal Representation

Antonio Parziale*, Roberto Parisi† and Angelo Marcelli‡
*DIEM – Natural Computation Lab, University of Salerno*
*Via Giovanni Paolo II, 132, 84084 Fisciano (SA)*
**anparziale@unisa.it, †robertoparisi89@gmail.com, ‡amarcelli@unisa.it*

Lognormality has proven to be an effective way for handwriting modeling. It assumes that handwriting is a time superimposition of a sequence of commands issued by the motor system, each command producing a stroke, i.e. a movement with a lognormal velocity profile. Motor control theories, moreover, suggest that handwriting movements result from both central and peripheral control, thus assuming that some movements of the sequence may not be encoded into the stored motor program but rather generated peripherally to keep the action as close as possible to the intended one. In the light of those observations, we present an algorithm for segmenting handwriting movements into strokes, each of which corresponds to a command stored into the motor program, while disregarding those that may depend on peripheral control. Experiments on handwriting samples show that the proposed algorithm detects the same number of strokes across multiple executions of a handwriting task by the same subject, and this set of strokes provides also a good reconstruction of the action.

## 1. Introduction

The concept of motor program originated in the work of Lashley [1951]. Observing that all skilled movements seem to involve the problem of serial ordering, and that the constituents of such sequences have no intrinsic order of association, he concluded that serial order could not be based on associative connections between elements. Thus, he conjectured the existence of a repertoire of "units of action" that can be preselected and activated by trigger commands specifying the order in which they have to be executed. Since then, the concept of motor program has widely spread across many domains, such as experimental psychology, movement science, and neurophysiology, but it has also been the

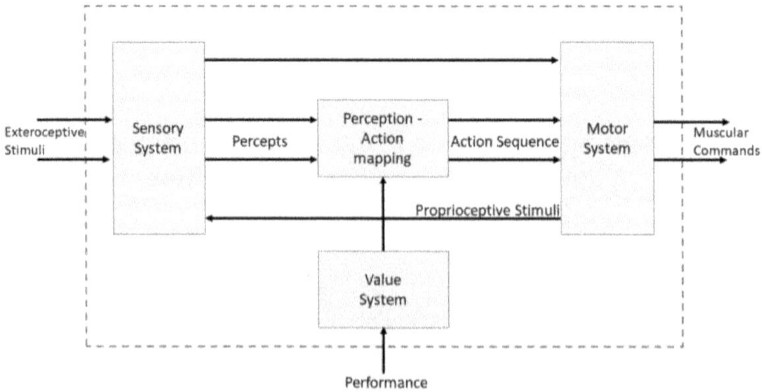


Fig. 1. Model of complex movement learning and execution.

subject of a fierce debate on how the units of actions are represented and their execution controlled. The different hypotheses as well as the debate they have ignited among the researchers are presented and discussed in the excellent review of literature in the fields of experimental psychology, human movement science, and neuroscience by Summers and Anson [2009]. Figure 1 summarizes the most agreed upon model of complex movements learning and execution. The sensory system represents the different areas of the cortex involved in building percepts from sensory stimuli (both exteroceptive and proprioceptive), where percepts may be roughly though as clusters of stimuli to whom the subject associates the same semantics. The motor system represents the brain areas and the spinal cord involved in executing the action by means of muscle commands, so that an action may be represented as a synchronized sequence of units of actions, i.e. a motor program. The association between percepts and actions refers to the brain areas involved in defining the “context of action”, i.e. the task, the goal, the motivation, previous experience and skills, and in planning complex behaviours. During learning, the perception-action cycle is exploited iteratively in a trial-and-error process driven by the performance that, through the value system and depending on the “context of action”, provides rewards for actions. The most rewarding actions form the repertoire of motor skills a subject acquires through repeated practice. When the learning is complete, the initial stimuli trigger the activation of the action, which is automatically mapped to a sequence of muscular commands autonomously, without attention. Eventually, the direct pathway between the sensory and the motor systems implements the innate reflex, while the proprioception pathway provides the feedback signal for adjusting the actual movement to the intended one.

In such a framework, the kinematic theory of rapid human movement seems to offer a suggestion on which ones are the "units of action" we use to encode intended movements [Plamondon, 1995; Plamondon, 1995b; Plamondon and Djioua, 2006]. According to such a theory, these "units of action", called strokes, are movements produced by a synergy of an agonist and an antagonist system, each of which is made up of a feedforward chain of interconnected modules representing the neural and muscular networks recruited for executing the movement. Each system is characterized by a log-normal impulse response and is activated by impulsive commands. Therefore, the velocity profile of a simple human movements shows a lognormal pattern. In the case of complex human movements, such as handwriting, the theory suggests that they can be thought as sequences of strokes [Plamondon and Guerfali, 1998]. To support this view, several algorithms have been proposed to perform the stroke segmentation, i.e. to identify within a handwriting movement its strokes and estimate the values of the parameter describing each of them, as prescribed by the theory [O'Reilly and Plamondon, 2009; Liu *et al.*, 2015]. Along the years, many experiments in many different domains have shown that the proposed algorithms exhibit very good performance in extracting the strokes embedded into the actions and estimating their parameters in such a way that the reconstructed velocity profile resembles very closely the actual one [Plamondon *et al.*, 2013; Djioua and Plamondon, 2009]. According to those findings, one may conclude that the motor program assumes the form of a sequence of strokes, whose parameters are learned by means of a trial-and-error process implemented by the model of Fig. 1, leading to the observed action. However, experiments have also shown that, in contrast to the motor program hypothesis, in case of multiple executions of the same intended action the model may provide sequences with a different number of strokes.

Motor learning and execution studies aimed at modelling the human neuromuscular system, however, have shown that the representation of complex movements such as handwriting is stored in two different forms. One form, called effector-independent, is represented by the sequence of target points, each of which represents the position to reach by each unit of actions. The other form, called effector-dependent, is represented by the commands to the specific neuromuscular system that is recruited for executing the desired movement [Wing, 2000; Raibert, 1977]. More recently, our work on modelling the interactions between the main brain areas involved in handwriting learning suggests that the implementation of an action through the interaction between the central nervous system and the musculoskeletal system can be interpreted as the realization of a motor program stored in the brain [Marcelli and Senatore, 2012]. Further investigations into movement execution have shown that the actual movements result from the interaction between the central nervous system, the spinal cord, the muscles and the proprioceptive receptors [Parziale, 2016]. In a nutshell, to initiate the movement, the central nervous system sends commands to recruit the muscles and to set the forces they have to exert on the bones they are connected to, while, during execution, the spinal cord modulates such command depending on the information received by the proprioceptive receptors in order to keep the execution as close as possible to the intended one.

According to those findings, and assuming the lognormal representation of handwriting movements as described before, the motor program assumes the form of a sequence of strokes embedded in any execution of the action it is meant to encode. The parameters of the strokes and their ordering are learned by a trial-and-error process implemented by the model of Fig. 1.

Following these observations, we propose an algorithm for extracting from handwriting the motor program learned by the subject to produce it, whose preliminary version was presented in [Parisi *et al.*, 2018]. In the next section we describe the algorithm building blocks and provide the rationale and the details of their implementation. Section 3 reports the results of a set of experiments to evaluate the performance of the extracted motor programs in reconstructing both the dynamic and kinematic properties of the action and the robustness of the motor program with respect to the algorithm parameters. In Sec. 4 we discuss the results in the conceptual framework depicted in Fig. 1 and evaluate to which extent the extracted sequence of strokes may actually represent the subject motor program. Eventually, in the conclusion we summarize the experimental findings and highlight the direction they suggest for future investigation.

## 2. Motor Program Extractor

The fine motor control required by handwriting is achieved by learning how to produce the desired movement under different conditions, and this learning is driven by the minimization of the metabolic energy. As mentioned in the Introduction, during handwriting the spinal cord can generate movements that are needed in order to keep the ongoing execution as close as possible to the learned one. Therefore, they are expected to result in small changes on the dynamics and kinematics of the ongoing movement, and the actual strategy adopted for producing the desired changes is the one that requires the smallest amount of metabolic energy to be executed.

Among the lognormal models that have been proposed in the literature, the Sigma-Lognormal Model [O'Reilly and Plamondon, 2009] is particularly suitable for describing what can be considered the motor program, i.e. a time superimposition of strokes aimed at reaching a sequence of target points. It describes each elementary movement in terms of command generation time, magnitude and direction of motion generation, from which we can derive the sequencing of the strokes, the dynamics and kinematics of the movements and the spatial position of the target points.

Three algorithms based upon the Sigma-Lognormal Model have been presented in literature for the automatic extraction of strokes from complex human movements. The algorithm that was introduced first is Robust XZERO [O'Reilly and Plamondon, 2009], hereinafter RX0, and its general structure is shown in Fig. 2. In [Liu *et al.*, 2015] it is presented a stroke parameter extractor based upon the nonnegative matrix factorization that allows to overcome the problem of overlapping strokes. Very recently, the iDeLog algorithm described in [Ferrer *et al.*, 2018] exploits an iterative procedure that moves the virtual target points of the action plan to improve the matching between the observed and reconstructed trajectory and velocity.

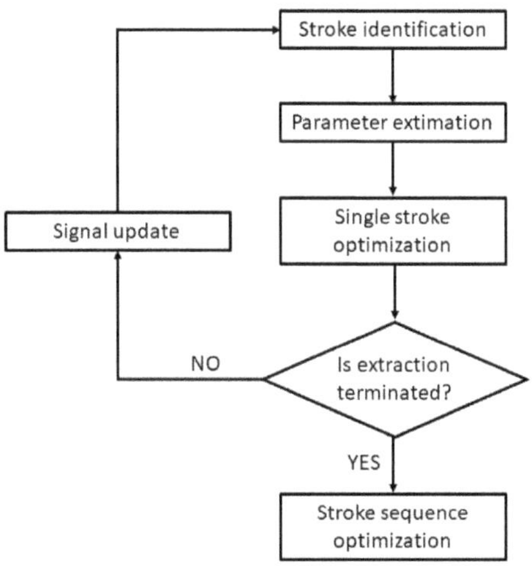


Fig. 2. General procedure adopted by Robust XZERO and Motor Program Extractor for the stroke extraction.

Differently from those algorithms, which aim at reconstructing the actual movement as closely as possible, the algorithm for extracting the motor program, hereinafter MPE, aims at filtering out movements generated by spinal networks. MPE adopts the same structure of RX0 depicted in Fig. 2, for extracting the strokes, but, at each step, MPE introduces heuristic criteria in order to remove from the handwriting representation the strokes corresponding to motor commands generated by the spinal networks. MPE extracts the motor program according to the following iterative procedure:

(1) The j-th stroke is identified by searching for the first peak of the observed velocity magnitude profile $|\overline{v(t)}|$ that satisfies some requirements on its amplitude and time duration, as described in Sec. 2.2;
(2) The six parameters describing the lognormal that fits the j-th stroke are computed, as described in the Secs. 2.3 and 2.4 respectively. We denote with $|\overline{v(t)^j}|$ the velocity magnitude profile of the j-th stroke;
(3) Once the j-th stroke is identified, its velocity is reconstructed and removed from the observed velocity profile, which is updated: $|\overline{v(t)}| = |\overline{v(t)}| - |\overline{v(t)^j}|$;
(4) The procedure returns to point 1, where the new velocity profile is evaluated, until no more strokes can be identified;
(5) Eventually, the parameters of all the strokes identified are optimized in a three-step procedure, as described in the Sec. 2.5.

In the following, we proceed to describe in detail each step of the procedure outlined above.

## 2.1. *Stroke identification*

In the light of the Sigma-Lognormal Model, the velocity profile of a complex movement is the vectorial summation of lognormals, each of which models the velocity profile of a neuromuscular system action.

We hypothesize that both brain and spinal cord contribute to the final movement by generating elementary movements with a lognormal velocity profile. In particular, spinal cord contributes with *fast corrective movements* by integrating commands from brain areas and proprioceptive signals.

According to our hypothesis, MPE discriminates between central and peripheral movements by taking into account that movements generated by the spinal cord should correspond to either velocity peaks whose amplitudes are much smaller than the amplitudes of the velocity peaks corresponding to the movements generated by the commands of the motor program, or whose durations are shorter than the durations of the movements generated by the commands of the motor program.

Strokes are identified by analysing the observed velocity magnitude profile $|\overline{v(t)}|$ from the beginning to the end of the movement, in the order of their occurrence. MPE classifies as strokes, i.e. as movements encoded in the motor program, the ones corresponding to velocity peaks with amplitudes and durations greater than two thresholds, denoted by $V_{th}$ and $T_{th}$, respectively.

The threshold $V_{th}$ is expressed as percentage of the maximum peak amplitude $V_p$ of the velocity profile, while the threshold $T_{th}$ is set to a fixed value. This choice takes into account that the reaction time of a reflexive movement is more or less equal between healthy subjects, while the effectiveness of a reflexive movement may depend on the subject's physical structure and muscle tone.

### 2.2. *Parameter estimation*

At this stage, the lognormal parameters that describe the stroke identified at the previous stage are computed.

MPE implements a simplified version of the procedure used by RX0, which attempts to reduce the complexity of the search by bounding the values of the lognormal parameters depending on the profile of the peak. In particular, MPE adopts a brute force approach to estimate the optimal lognormal parameters $[t_0, D, \mu, \sigma]$: for each $\sigma$ value within the interval [0, 2] it searches for the best fitting lognormal profile with respect to the sample velocity profile. The best fitting parameters are those corresponding to the biggest SNR value between the original velocity signal and the actual reconstruction within the time interval spanning from the velocity minimum that immediately precedes the stroke's velocity peak to the velocity minimum that immediately follows it.

The parameters $[t_0, D, \mu, \sigma]$ of each lognormal and the angles $[\theta_s, \theta_e]$ are estimated by using the equations reported in [O'Reilly and Plamondon, 2009]. For the convenience of the reader, the equations are reported below:

$$D = \sqrt{2\pi} v(t_3) e^{\mu} \sigma e^{\frac{a_3^2}{2\sigma^2} - a_3} \tag{1}$$

$$t_0 = t_3 - e^{\mu} e^{-a_3} \tag{2}$$

$$\mu = \ln\left\{\frac{t_4 - t_2}{e^{-a_4} - e^{-a_2}}\right\} \tag{3}$$

$$\theta_s = \phi(t_3) - \Delta\phi\big(l(t_3) - l(t_1)\big) \tag{4}$$

$$\theta_e = \phi(t_3) + \Delta\phi\big(l(t_5) - l(t_3)\big) \tag{5}$$

where:

$$l(t_i) = \begin{cases} 0 & i = 1 \\ \frac{D}{2}\left[1 + \operatorname{erf}\left(\frac{-a_i}{\sigma\sqrt{2}}\right)\right] & i = 2, 3, 4 \\ D & i = 5 \end{cases} \tag{6}$$

$$a_1 = 3\sigma \tag{7}$$

$$a_2 = 1.5\sigma^2 + \sigma\sqrt{0.25\sigma^2 + 1} \tag{8}$$

$$a_3 = \sigma^2 \tag{9}$$

$$a_4 = 1.5\sigma^2 - \sigma\sqrt{0.25\sigma^2 + 1} \tag{10}$$

$$\Delta\phi = \frac{\phi(t_4) - \phi(t_2)}{l(t_4) - l(t_2)} \tag{11}$$

In the equations above, $t_1, t_2, t_3, t_4, t_5$ are the times of occurence of the left minimum, first inflexion point, maximum, second inflexion point, right minimum of the velocity peak, respectively. Eventually, $\phi$ and $l$ are the angular position and the trajectory lenght as functions of the time, respectively.

### 2.3. *Single stroke optimization*

Single stroke optimization is performed after each lognormal parameter estimation to improve reconstruction quality of the j-th stroke. A non-linear least-square optimization method based upon the Levenberg–Marquardt algorithm is applied first to $[t_0, D, \mu, \sigma]$, for improving the velocity reconstruction, and then to $[\theta_s, \theta_e]$, for improving x-y signal reconstruction. In contrast, RX0 non-linearly optimizes the parameters $[\mu, \sigma, \theta_s, \theta_e]$ and then it computes the parameters $[t_0, D]$ in order to make the estimated lognormal maximum point equal to the actual maximum point. Thus, RX0 is mainly focused on improving the reconstruction of the velocity profile. Our choice follows from the motor program assumption: during the learning process a subject becomes able to produce the single movement toward the desired target first, and then learns how to configure the musculoskeletal system for linking two consecutive target points. On the contrary, the local optimization performed by RX0 assumes that a subject learns at the same time how to execute the basic movement and how to connect it with the following one, while the trajectory to reach the target is improved next.

The main issue with the extraction of the motor program from a learned movement is the effect of time superimposition of two consecutive strokes. The longer the time superimposition between the strokes the larger the overlap between their lognormals, and therefore the parameter estimation of each lognormal becomes less accurate. To overcome this problem, the following procedure is adopted for each stroke:

(1) The j-th stroke is identified, its parameters are estimated and optimized;

(2) The velocity profile of the j-th strokes is reconstructed and removed from the observed velocity profile, $|\overline{v(t)}| = |\overline{v(t)}| - |\overline{v(t)^J}|$;
(3) The (j+1)-th stroke is identified and its parameters are estimated;
(4) The velocity profile of the (j+1)-th strokes is reconstructed and removed from the observed velocity profile while the velocity profile of the j-th strokes is added again, $|\overline{v(t)}| = |\overline{v(t)}| + |\overline{v(t)^J}| - |\overline{v(t)^{J+1}}|$;
(5) The j-th stroke is identified, its parameters are estimated and optimized again.

After each lognormal has been fitted within the sample, the portion of the velocity profile fitted by that lognormal is removed from the velocity signal, but the corresponding portion of the trajectory is not, instead of what happens in the case of RX0, and the procedure is iterated until no more lognormals can be fitted. Differently from the velocity, the trajectory is not updated at each iteration, in order to improve the reconstruction in the regions of the trace where two consecutive strokes are linked.

### 2.4. *Stroke sequence optimization*

During the execution of a complex movement the execution of a stroke affects the execution of most of, if not all, the following ones. In order to mitigate the effect of motor variability in the extraction of the motor program, a final optimization of the sequence of strokes is performed. Three non-linear least-square optimizations are performed:

(1) The parameters $[t_0, D, \mu, \sigma]$ of all the strokes are optimized with the aim of minimizing the difference between the acquired and the reconstructed absolute velocity profile;
(2) The parameters $[\theta_s, \theta_e]$ of all the strokes are optimized with the aim of minimizing the error between the acquired trajectory $[x(t)_{ac}, y(t)_{ac}]$ and the reconstructed one $[x(t)_{rec}, y(t)_{rec}]$, by using the cost function $f = (x(t)_{ac} - x(t)_{rec})^2 + (y(t)_{ac} - y(t)_{rec})^2$;
(3) All the parameters are optimized with the aim of minimizing the errors in both the trace and velocity reconstruction by using the cost function $f = (\dot{x}_{ac}(\mathrm{t}) - \dot{x}_{rec}(\mathrm{t}))^2 + (\dot{y}_{ac}(\mathrm{t}) - \dot{y}_{rec}(\mathrm{t}))^2$;

The sequence of the three optimizations follows from the same assumptions made for the local optimization. In particular, we hypothesize that during the learning process subjects familiarize themselves with the execution of a movement toward a desired target, then learn how to connect two consecutive target points and eventually acquire the motor program skill related to the shape.

## 3. Performance Evaluation

As mentioned in the Introduction, the proposed algorithm aims at extracting from actions the sequence of strokes, discriminating those corresponding to the execution of the motor program stored in the brain from those arising by commands sent by the spinal cord in response to proprioceptive feedback. As at the end of the learning process the effector-independent representation of the action is the *sequence* of target points to reach, one would expect that each execution of that motor program results from the composition of the same number of strokes. As the ground truth, i.e. the number of strokes actually encoded into the motor program, is not available we consider the stability of the number of strokes extracted from repeated executions as metric for evaluating the performance of the proposed method in providing a plausible motor program.

The similarity between the dynamics and the kinematics of movements obtained by executing the motor program provided by the MPE algorithm and the original one is adopted for evaluating the plausibility of the estimated motor program parameters as being the commands for generating the actual movement.

The experiment involved 22 subjects, 14 males and 8 females, whose age ranges in the interval 18–30 years, with a mean value of 23.45 and a standard deviation of 3.07. All of participants were students enrolled at our university who volunteered to take part in the experiment and expressed their formal consent.

Fig. 3. Handwriting pattern used in the experimentation.

During the experiment, each subject was requested to reproduce 10 times the handwriting pattern in Fig. 3. Such a pattern, or very similar ones, has been used in many experiments on handwriting generation modelling [Teulings *et al.*, 2002; Tucha *et al.*, 2005; Ünlü *et al.*, 2006; Frontini *et al.*, 2017; Senatore and Marcelli, 2019], because it is reasonable to believe that its motor program has been learned by the subjects involved in the experiments, independently of their individual characteristics, being at the same time fairly complex to allow a quantitative evaluation of the estimated motor program parameters. An l-shaped loop is generally encoded by a sequence of two strokes, one moving upward and one downward. Therefore, we expected to extract a motor program made up by 8 strokes from each execution of the 4 l-shaped loops represented in Fig. 3.

The handwriting samples were collected by using an ink-and-paper WACOM Intuos 2 digitizing tablet with a 100 Hz sampling rate to record the handwriting movements. We adopted an ink-and-paper digitizing tablet as we aimed at extracting the strokes corresponding to the motor program of the subject, that most likely were learned under the same conditions, to avoid as much as possible the influence of the spinal cord during the movement execution due to unexpected proprioceptive feedback that may arise from different writing condition, as it would have been the case by using a stylus-and-screen digitizing tablet.

We evaluated the performance of MPE in terms of the number of strokes extracted by the algorithm, the Signal-to-Noise Ratio (SNR) and the Root Mean Square Error (RMSE) between the reconstructed specimen and the original one, defined in Eqs. (12) and (13), respectively.

$$SNR = 10\log\left(\frac{\int_{t=0}^{T} v_{ac}(t)^2 dt}{\int_{t=0}^{T}\left(v_{ac}(t)-v_{re}(t)\right)^2}\right) \tag{12}$$

$$RMSE = \sqrt{(x_{ac}-x_{re})^2+(y_{ac}-y_{re})^2} \tag{13}$$

To assess the robustness of the MPE with respect to its parameters, the performance is evaluated by varying $V_{th}$ between the values [0.01, 0.03, 0,05, 0.10, 0.15, 0.20, 0.25, 0.30, 0.35, 0.40, 0.495, 0.50, 0.60]* $V_p$ and $T_{th}$ between the values [20, 40, 50, 60, 70, 80] ms.

The performance of MPE was compared to the performance of RX0 implemented by the software ScriptStudio [O'Reilly and Plamondon, 2009; O'Reilly and Plamondon, 2009b]. Because ScriptStudio provides the best performance when samples are acquired at 200 Hz and have a null velocity at beginning and at the end of the traces, samples were preprocessed as described in [Diaz *et al.*, 2018].

Table 1 reports the number of strokes extracted by MPE, while Tables 2 and 3 report the RMSE and the SNR between the movement from the extracted motor program and the actual one as $V_{th}$ and $T_{th}$ are varied in the ranges defined above. One may notice that the mean number of strokes and the mean SNR decrease as $V_{th}$ increases. On the contrary, given a value of $V_{th}$, there is not a significant variation in either the measure as $T_{th}$ varies. As we are interested in extracting the motor program behind all the executions, we consider that the best configuration of MPE is the one that generates the smallest standard deviation of the number of strokes across all the experiments. Therefore, it was determined that we should use $V_{th} = 0.495 * V_p$ and $T_{th} = 20$ *ms*.

Table 1. Mean number of strokes and standard error of the mean by varying $V_{th}$ and $T_{th}$.

| | $T_{th}$ | | | | | |
|---|---|---|---|---|---|---|
| $V_{th}$ | 20 | 40 | 50 | 60 | 70 | 80 |
| 0.01 | 8.85 ± 1.40 | 8.79 ± 1.40 | 8.72 ± 1.33 | 8.72 ± 1.33 | 8.64 ± 1.29 | 8.61 ± 1.29 |
| 0.03 | 8.77 ± 1.39 | 8.77 ± 1.39 | 8.70 ± 1.32 | 8.70 ± 1.32 | 8.63 ± 1.28 | 8.60 ± 1.28 |
| 0.05 | 8.70 ± 1.31 | 8.70 ± 1.31 | 8.66 ± 1.30 | 8.66 ± 1.30 | 8.60 ± 1.25 | 8.58 ± 1.25 |
| 0.10 | 8.55 ± 1.22 | 8.55 ± 1.22 | 8.53 ± 1.21 | 8.53 ± 1.21 | 8.51 ± 1.17 | 8.49 ± 1.17 |
| 0.15 | 8.50 ± 1.19 | 8.50 ± 1.19 | 8.49 ± 1.19 | 8.49 ± 1.19 | 8.47 ± 1.16 | 8.46 ± 1.16 |
| 0.20 | 8.43 ± 1.16 | 8.43 ± 1.16 | 8.43 ± 1.16 | 8.43 ± 1.16 | 8.42 ± 1.14 | 8.42 ± 1.14 |
| 0.25 | 8.38 ± 1.13 | 8.38 ± 1.13 | 8.38 ± 1.13 | 8.38 ± 1.13 | 8.37 ± 1.13 | 8.37 ± 1.13 |
| 0.30 | 8.33 ± 1.06 | 8.33 ± 1.06 | 8.33 ± 1.06 | 8.33 ± 1.06 | 8.33 ± 1.06 | 8.33 ± 1.06 |
| 0.35 | 8.28 ± 1.04 | 8.28 ± 1.04 | 8.28 ± 1.04 | 8.28 ± 1.04 | 8.28 ± 1.04 | 8.28 ± 1.04 |
| 0.40 | 8.22 ± 0.97 | 8.22 ± 0.97 | 8.22 ± 0.97 | 8.22 ± 0.97 | 8.22 ± 0.97 | 8.22 ± 0.97 |
| 0.495 | 8.01 ± 0.89 | 8.01 ± 0.89 | 8.01 ± 0.89 | 8.01 ± 0.89 | 8.01 ± 0.89 | 8.01 ± 0.89 |
| 0.50 | 7.98 ± 0.92 | 7.98 ± 0.92 | 7.98 ± 0.92 | 7.98 ± 0.92 | 7.98 ± 0.92 | 7.98 ± 0.92 |
| 0.60 | 7.52 ± 1.24 | 7.52 ± 1.24 | 7.52 ± 1.24 | 7.52 ± 1.24 | 7.51 ± 1.22 | 7.51 ± 1.23 |

Table 2. Mean RMSE and standard error of the mean by varying $V_{th}$ and $T_{th}$.

| | $T_{th}$ | | | | | |
|---|---|---|---|---|---|---|
| $V_{th}$ | 20 | 40 | 50 | 60 | 70 | 80 |
| 0.01 | 0.08 ± 0.12 | 0.08 ± 0.12 | 0.08 ± 0.12 | 0.08 ± 0.12 | 0.09 ± 0.13 | 0.09 ± 0.13 |
| 0.03 | 0.08 ± 0.12 | 0.08 ± 0.12 | 0.08 ± 0.12 | 0.08 ± 0.12 | 0.09 ± 0.13 | 0.09 ± 0.13 |
| 0.05 | 0.08 ± 0.11 | 0.08 ± 0.11 | 0.08 ± 0.11 | 0.08 ± 0.11 | 0.09 ± 0.12 | 0.09 ± 0.12 |
| 0.10 | 0.08 ± 0.11 | 0.08 ± 0.11 | 0.08 ± 0.11 | 0.08 ± 0.11 | 0.09 ± 0.12 | 0.09 ± 0.12 |
| 0.15 | 0.08 ± 0.11 | 0.08 ± 0.11 | 0.08 ± 0.11 | 0.08 ± 0.11 | 0.09 ± 0.12 | 0.09 ± 0.12 |
| 0.20 | 0.09 ± 0.13 | 0.09 ± 0.13 | 0.09 ± 0.13 | 0.09 ± 0.13 | 0.09 ± 0.13 | 0.09 ± 0.13 |
| 0.25 | 0.09 ± 0.13 | 0.09 ± 0.13 | 0.09 ± 0.13 | 0.09 ± 0.13 | 0.09 ± 0.13 | 0.09 ± 0.13 |
| 0.30 | 0.10 ± 0.17 | 0.10 ± 0.17 | 0.10 ± 0.17 | 0.10 ± 0.17 | 0.10 ± 0.17 | 0.10 ± 0.17 |
| 0.35 | 0.09 ± 0.12 | 0.09 ± 0.12 | 0.09 ± 0.12 | 0.09 ± 0.12 | 0.09 ± 0.12 | 0.09 ± 0.12 |
| 0.40 | 0.09 ± 0.12 | 0.09 ± 0.12 | 0.09 ± 0.12 | 0.09 ± 0.12 | 0.09 ± 0.12 | 0.09 ± 0.12 |
| 0.495 | 0.14 ± 0.30 | 0.14 ± 0.30 | 0.14 ± 0.30 | 0.14 ± 0.30 | 0.14 ± 0.30 | 0.14 ± 0.30 |
| 0.50 | 0.15 ± 0.35 | 0.15 ± 0.35 | 0.15 ± 0.35 | 0.15 ± 0.35 | 0.15 ± 0.35 | 0.15 ± 0.35 |
| 0.60 | 0.19 ± 0.36 | 0.19 ± 0.36 | 0.19 ± 0.36 | 0.19 ± 0.36 | 0.19 ± 0.36 | 0.19 ± 0.36 |

Table 3. Mean SNR and standard error of the mean by varying $V_{th}$ and $T_{th}$.

| | $T_{th}$ | | | | | |
|---|---|---|---|---|---|---|
| $V_{th}$ | 20 | 40 | 50 | 60 | 70 | 80 |
| 0.01 | 18.99 ± 4.06 | 18.99 ± 4.06 | 18.92 ± 4.12 | 18.92 ± 4.12 | 18.83 ± 4.20 | 18.82 ± 4.20 |
| 0.03 | 18.95 ± 4.10 | 18.96 ± 4.10 | 18.91 ± 4.11 | 18.91 ± 4.11 | 18.82 ± 4.19 | 18.81 ± 4.19 |
| 0.05 | 18.92 ± 4.14 | 18.92 ± 4.15 | 18.89 ± 4.14 | 18.89 ± 4.14 | 18.81 ± 4.23 | 18.79 ± 4.22 |
| 0.10 | 18.83 ± 4.14 | 18.84 ± 4.14 | 18.82 ± 4.14 | 18.82 ± 4.14 | 18.76 ± 4.23 | 18.75 ± 4.23 |
| 0.15 | 18.78 ± 4.15 | 18.78 ± 4.15 | 18.76 ± 4.14 | 18.76 ± 4.14 | 18.71 ± 4.24 | 18.70 ± 4.24 |
| 0.20 | 18.67 ± 4.36 | 18.67 ± 4.36 | 18.67 ± 4.36 | 18.67 ± 4.36 | 18.67 ± 4.36 | 18.66 ± 4.36 |
| 0.25 | 18.67 ± 4.27 | 18.67 ± 4.27 | 18.67 ± 4.27 | 18.67 ± 4.27 | 18.65 ± 4.27 | 18.65 ± 4.27 |
| 0.30 | 18.59 ± 4.40 | 18.59 ± 4.40 | 18.59 ± 4.40 | 18.59 ± 4.40 | 18.59 ± 4.40 | 18.59 ± 4.40 |
| 0.35 | 18.56 ± 4.34 | 18.56 ± 4.34 | 18.56 ± 4.34 | 18.56 ± 4.34 | 18.56 ± 4.34 | 18.56 ± 4.34 |
| 0.40 | 18.41 ± 4.52 | 18.41 ± 4.52 | 18.41 ± 4.52 | 18.41 ± 4.52 | 18.41 ± 4.52 | 18.41 ± 4.52 |
| 0.495 | 17.80 ± 5.30 | 17.80 ± 5.30 | 17.80 ± 5.30 | 17.80 ± 5.30 | 17.80 ± 5.30 | 17.80 ± 5.30 |
| 0.50 | 17.60 ± 5.53 | 17.60 ± 5.53 | 17.60 ± 5.53 | 17.60 ± 5.53 | 17.60 ± 5.53 | 17.60 ± 5.53 |
| 0.60 | 16.04 ± 6.42 | 16.04 ± 6.42 | 16.04 ± 6.42 | 16.04 ± 6.42 | 16.05 ± 6.42 | 16.05 ± 6.41 |

Figure 4 shows, for every subject, the number of strokes of each sample found by MPE in its best configuration. For the sake of comparison, we also report the numbers provided by the RX0 algorithm. Figures 5 and 6 report the same comparison in terms of RMSE and SNR.

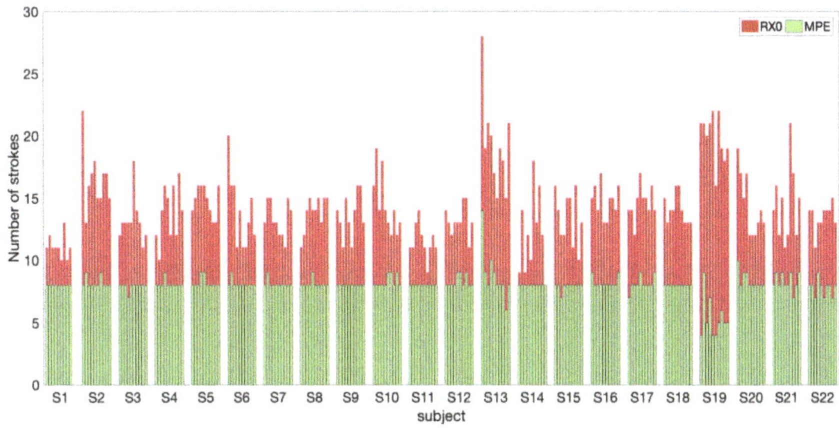


Fig. 4. Number of strokes extracted by RX0 and MPE. Note that user S1 wrote 9 patterns instead of 10. RX0 and MPE bars are overlapped.

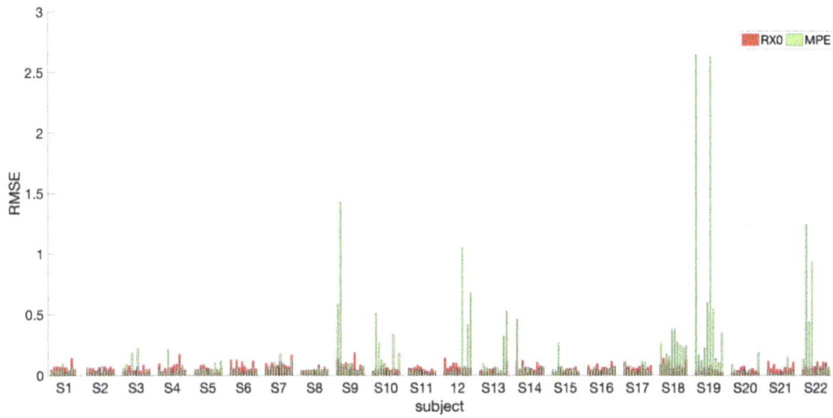


Fig. 5. RMSE values obtained by comparing the reconstructed trajectories and the original ones. Note that user S01 wrote 9 patterns instead of 10. RX0 and MPE bars are overlapped.

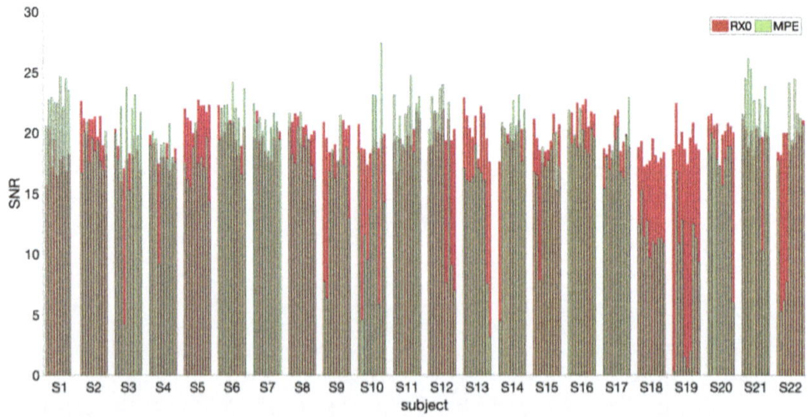


Fig. 6. SNR values obtained by comparing the reconstructed velocity profile and the original ones. Note that user S01 wrote 9 patterns instead of 10. RX0 and MPE bars are overlapped.

## 4. Discussion

Table 4 compares RX0 and MPE in terms of total and mean number of strokes, SNR and RMSE. MPE detects 8.86 ± 1.41 strokes, around 38% less than RX0, when a discrimination between central and peripheral movements is not performed ($V_{th}$ = 0, $T_{th}$ = 0). This result suggests that the way MPE implements the Stroke Identification and the Parameter Estimation stages is less sensitive to the variability among many executions of the same action. When $V_{th}$ = 0.495 * $V_p$, MPE extracts 8.01 ± 0.89 strokes, around 44% less than RX0. Eventually, Table 1 shows that the threshold on the time duration has no effect in distinguishing between central and peripheral movements when $V_{th}$ is greater than 0.20 * $V_p$. The results shown in Fig. 4 reveal that for the large majority of the subjects the motor program is composed by the same number of strokes

Table 4. Number of strokes, mean number of strokes, mean RMSE, mean SNR for RX0 and different configuration of MPE.

| | Strokes Total (Mean ± std) | RMSE | SNR |
|---|---|---|---|
| RX0 | 3127 (14.28 ± 2.84) | 0.07 ± 0.03 | 19.62 ± 1.49 |
| MPE ($V_{th} = 0, T_{th} = 0$) | 1940 (8.86 ± 1.41) | 0.08 ± 0.12 | 18.92 ± 4.23 |
| MPE ($V_{th} = 0, T_{th} = 20$) | | | |
| MPE ($V_{th} = 0, T_{th} = 80$) | 1864 (8.51 ± 1.23) | 0.1 ±0.17 | 18.74 ± 4.32 |
| MPE ($V_{th} = 0.495 * V_p, T_{th} = 0$) | 1754 (8.01 ± 0.89) | 0.14 ± 0.30 | 17.80 ± 5.30 |
| MPE ($V_{th} = 0.495 * V_p, T_{th} = 20$) | | | |
| MPE ($V_{th} = 0.495 * V_p, T_{th} = 80$) | | | |

across most actions. On the contrary, this happens only in few cases with the sequence provided by RX0. The data in Figs. 5 and 6 show that the execution of the motor program provided by MPE leads to actions whose dynamics and kinematics resemble very closely those of the actual actions, although the velocity profiles reconstructed from the strokes provided by the RX0 algorithm is more accurate.

The invariance of the number of strokes across multiple executions of the l-shaped loops by the same subject does not seem to hold for two subjects, namely subject S13 and subject S19. As with regards to subject S13, Fig. 4 shows that the motor programs provided by our algorithm for the first 3 executions contain 14, 9 and 8 strokes, respectively, contrary to what would be expected. To inspect further, we report in Fig. 7 the trajectory (left panel) and the velocity profile (right panel) of the actions corresponding to each execution. Comparing the plots in the panels from top to bottom it can be observed that the trajectory and velocity profile become smoother, while the duration of the movement and the horizontal and the vertical size get shorter. Even more important, in the following executions the number of strokes extracted by the handwriting samples fluctuate around 8. In our opinion, such a behavior can be interpreted as a form of adaptation. In the first few executions, for whatever reason, the subject perceives of doing something different from just writing as he/she has done for more than 15 years. In terms of the general view of voluntary movements learning and execution represented in Fig. 1, the "context of action" was not the same as if the subject was writing the same pattern in the usual context. Because of that the subject has attempted to execute the

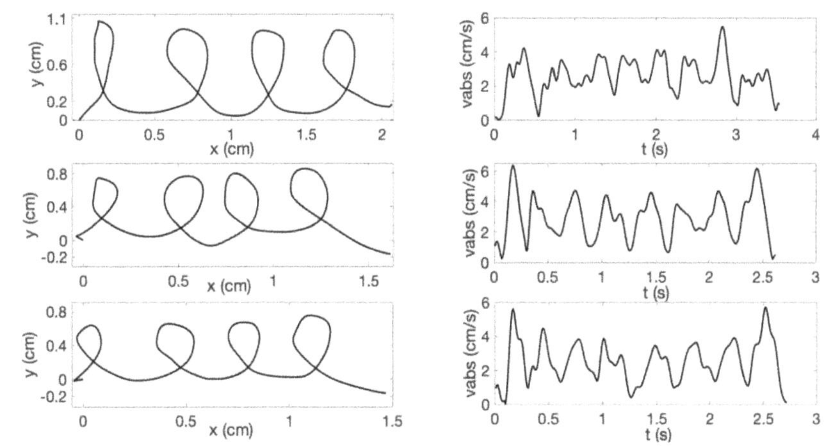


Fig. 7. Trajectories and velocities profiles of the first three samples written by subject S13.

prelearned motor program, but attentively, rather than automatically. However, after few repetitions, the subject restores the usual "context of action", possibly because the feedback he receives, and thus the learned motor program is executed without attention.

As with regards to subject S19, the motor program was made of 4 strokes in most of the executions, which seems to suggest that the "units of actions" encode movements that allow changes of direction, in contrast to the definition of strokes adopted within the kinematic theory of human movements. In Fig. 8, the trajectory and the velocity profiles of the action resulting from the first execution are shown. The velocity profile in the time interval between 0.0 s and 3.0 s shows 4 prominent peaks corresponding to the strokes for drawing the upward parts of the 4 l-shaped loops, while the peaks corresponding to the strokes for moving the pen downward are absent and few bumps are visible. A possible explanation for such a behavior is that this subject is exploiting to a much larger extent than the others a biomechanical advantage. In [Frontini *et al.*, 2017], we have hypothesized that during the execution of a stroke the neuromuscular system alternates an active control stage, during which muscular synergies are properly activated to initiate the movement, and a reactive control stage, during which the biomechanical characteristics of the system are exploited to complete the movement. A systematic investigation of this topic is still going on, but the preliminary results seem to support the conjecture that a u-like shape can be obtained by executing a single command.

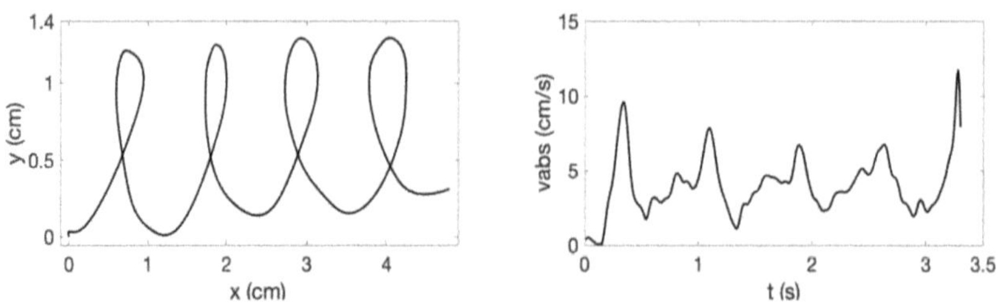


Fig. 8. Trajectory and velocity profile of the first sample written by subject S19.

The results of the experiments, however, also show that, even in the case when the same number of strokes is extracted across many handwriting samples of the same subject, the values of the parameters of the Sigma-Lognormal strokes are not the same across the repetitions, as one would expect from multiple executions of *the* same motor program learned by the subject. This can be explained by considering the neural architecture of the motor system, according to which muscle contraction is regulated by the alpha motor neurons that integrate inputs from both the upper motor neurons of the motor cortex and the interneurons of the spine. As a consequence, the same observed movement can result from many combinations of the signals coming from the two pathways making not possible to separate the two contributions and to compute the actual stroke parameters.

## 5. Conclusions

The research reported in this chapter was motivated by our interest in developing handwriting analysis tools for writer identification [Parziale *et al.*, 2014; Carmona-Duarte *et al.*, 2017; Parziale *et al.*, 2019] and early diagnosis of neurodegenerative diseases [Senatore and Marcelli, 2019; Parziale *et al.*, 2019b]. In particular, as most of neurodegenerative diseases produces difficulties in motor skill learning and execution, it is of paramount importance to be able to understand in which way they affect the motor program, and how the difficulties they produce progress along time. Effective modeling of such difficulties by computational tools that incorporate the features of the neuromuscular system may lead to better understand pathologies and thus provide insights useful for early diagnosis.

In this framework, we have proposed an algorithm for segmenting an action into strokes that attempts to distinguish between movements that result from the execution of the stored motor program and other movements that are generated by the spinal cord in reaction to proprioceptive feedback, as suggested by motor control studies on learning and execution of voluntary movements.

The proposed algorithm builds upon another algorithm proposed in the literature within the framework of the Kinematic Theory of rapid human movements for decomposing the handwriting movements into elementary movements, but, in contrast with such a theory, it assumes that some of the elementary movements originate in the spinal cord and therefore do not correspond to a command embedded into the representation of the motor program stored in the brain.

The experimental results support to a large extent our conjecture that the motor program for executing a learned voluntary movement is represented by a sequence of strokes that is embedded in any execution of the action. These results have shown that such a sequence contains as many strokes as the number of target points to reach during the movement, thus providing further support to the double form of the motor program representation we have formulated in our previous work. The results also support, but to a smaller extent, our claim that the difference in the actual movements by repeated executions of the same task by the same subject are due to the feedback loop implemented by the spinal cord. Our current investigation is focused on modeling and simulating the neural network of the spinal cord to possibly uncouple its effects on the movement from movements controlled as defined in the stored motor program. As we have already noted, this cannot be done looking at the lognormal representation of the handwriting. Our ultimate aim is eventually developing algorithms, possibly working with different representation of handwriting, to extract *the* motor program from the action.

## Acknowledgment

The work reported in this paper was partially funded by the “Bando PRIN 2015 - Progetto HAND” under Grant H96J16000820001 from the Italian Ministero dell’Istruzione, dell’Università e della Ricerca.

## References

Carmona-Duarte, C., Ferrer, M. A., Parziale, A. and Marcelli, A. (2017). Time evolution in synthetic handwriting, *Pattern Recognition*, 68, pp. 233–244.

Diaz, M., Fischer, A., Ferrer, M. A. and Plamondon, R. (2018). Dynamic Signature Verification System Based on One Real Signature, *IEEE Transactions on Cybernetics*, 48(1), pp. 228–239.

Djioua, M. and Plamondon, R. (2009). Studying the variability of handwriting patterns using the Kinematic Theory, *Human Movement Science*, 28(5), pp. 588–601.

Ferrer, M. A., Diaz, M., Carmona, C. and Plamondon, R. (2018). iDeLog: Iterative Dual Spatial and Kinematic Extraction of Sigma-Lognormal Parameters. *IEEE Trans. on Pattern Analysis and Machine Intelligence (PAMI)*. 42(1): 114–125.

Frontini, S., Giordano, G., Dellavalle, F., Parziale, A. and Marcelli, A. (2017). Looking at the Ink Distribution for Assessing Writing Modalities in Forensic Handwriting Examination, *Proceedings of 18th Conference of International Graphonomics Society*, IGS, pp. 37–40.

Lashley, K. S. (1951). *Cerebral mechanisms in behavior*, eds. L. A. Jeffress, Chapter “The problem of serial order in behavior”, Wiley, New York, pp. 112–136.

Liu, M., Guo, X. and Wang, G. (2015). Stroke Parameters Identification Algorithm in Handwriting Movements Analysis by Synthesis, *IEEE Journal of Biomedical and Health Informatics*, 19(1), pp. 317–324.

Marcelli, A. and Senatore, R. (2012). A Neural Scheme for Procedural Motor Learning of Handwriting, *Proceedings of 13th International Conference on Frontiers in Handwriting Recognition*, ICFHR, pp. 659–664.

O'Reilly, C. and Plamondon, R. (2009). Development of a Sigma-Lognormal representation for on-line signatures, *Pattern Recognition*, 42(12), pp. 3324–3337.

O'Reilly, C. and Plamondon, R. (2009b). *Rehabilitation Engineering*, eds. Tan Yen Kheng, Chapter 19 "A Lognormal Framework for Human Movement Rehabilitation", INTECH Publishing, pp. 157–172.

Parisi, R., Parziale, A. and Marcelli, A. (2018). Some observations on lognormality and motor control in handwriting, *Proceedings of the 1st International Conference on Pattern Recognition and Artificial Intelligence,* ICPRAI, pp. 732–737.

Parziale, A., Santoro, A., Marcelli, A., Rizzo, A., Molinari, C., Cappuzzo, A. G. and Fontana, F. (2014). An Interactive Tool for Forensic Handwriting Examination, *Proceedings of 14th International Conference on Frontiers in Handwriting Recognition*, ICFHR, pp. 440–445.

Parziale, A. (2016). *A neurocomputational model of reaching movements*, PhD Thesis, Salerno, Italy: University of Salerno, doi: http://dx.doi.org/10.14273/unisa-756.

Parziale, A., Diaz, M., Ferrer, M. A. and Marcelli, A. (2019). SM-DTW: Stability Modulated Dynamic Time Warping for signature verification, *Pattern Recognition Letters*, 121, pp. 113–122.

Parziale, A., Della Cioppa, A., Senatore, R. and Marcelli, A. (2019) A Decision Tree for Automatic Diagnosis of Parkinson's Disease from Offline Drawing Samples: Experiments and Findings, *Proceedings of Image Analysis and Processing – ICIAP 2019*, Lecture Notes in Computer Science, 11751, pp. 196–206.

Plamondon, R. (1995). A kinematic theory of rapid human movements. Part I. Movement representation and generation, *Biological Cybernetics*, 72(4), pp. 295–307.

Plamondon, R. (1995b). A kinematic theory of rapid human movements. Part II. Movement time and control, *Biological Cybernetics*, 72(4), pp. 309–320.

Plamondon, R. and Guerfali, W. (1998). The generation of handwiritng with delta-lognormal synergies, *Biological Cybernetics*, 78, pp. 119–132.

Plamondon, R. and Djioua, M. (2006). A multi-level representation paradigm for handwriting stroke generation, *Human Movement Science*, 25(4), pp. 586–607.

Plamondon, R., O'Reilly, C., Rémi, C. and Duval, T. (2013). The lognormal handwriter: learning, performing, and declining, *Frontiers in Psychology*, 4, 945.

Raibert, M. H. (1977). *Motor control and learning by the state space model*, Cambridge: Artificial Intelligence Laboratory, MIT.

Senatore, R. and Marcelli, A. (2019). A paradigm for emulating the early learning stage of handwriting: Performance comparison between healthy controls and Parkinson's disease patients in drawing loop shapes, *Human Movement Science*, 65, pp. 89–101.

Summers, J. J. and Anson, J. G. (2009). Current status of the motor program: Revisited, *Human Movement Science*, 28(5), pp. 566–577.

Teulings, H., Contreras-Vidal, J. L., Stelmach, G. and Adler C. (2002). Adaptation of handwriting size under distorted visual feedback in patients with Parkinson's disease and elderly and young controls, *Journal of Neurology, Neurosurgery & Psychiatry*, 72, pp. 315–324.

Tucha, O., Mecklinger, L., Thome, J., Reiter, A., Alders, G. L., Sartor, H., Naumann, M. K. and Lange, K. W. (2005). Kinematic analysis of dopaminergic effects on skilled handwriting movements in Parkinson's disease, *Journal of Neural Transmission*, 113, pp. 609–623.

Ünlü, A., Brause, R. and Krakow, K. (2006). Handwriting Analysis for Diagnosis and Prognosis of Parkinson's Disease, Biological and Medical Data Analysis - ISBMDA 2006, Lecture Notes in Computer Science, 4345, Springer, Berlin, Heidelberg.

Wing, A. M. (2000). Motor control: mechanisms of motor equivalence in handwriting, *Current Biology*, 10, pp. 245–248, 2000.

https://doi.org/10.1142/9789811226830_0014

# Chapter 14

# On the Use of Interval Arithmetic for the Branch and Bound Delta-Lognormal Parameter Extraction of Rapid Human Movements

Simon Pierre Boyogueno Bidias*, Jean-Pierre David†, Yvon Savaria‡ and Réjean Plamondon§

*Laboratoire Scribens, Département de génie électrique, Polytechnique Montréal, Montréal, Canada H3C 3A7*

**simon-pierre.boyogueno-bidias@polymtl.ca, †JPDavid@polymtl.ca, ‡yvon.savaria@polymtl.ca, §rejean.plamondon@polymtl.ca*

This chapter proposes an algorithm based on interval arithmetic to construct envelope models that contain all possible parameter values of a delta-lognormal function that can match an observed human movement. In the context of the Kinematic Theory of rapid human movements, these parameters represent the basic timing properties of a neuromuscular system, as well as the control strategies used to produce a rapid movement. The basic idea behind the proposed algorithm is to exploit the isotonicity and monotonicity over an interval to sharply bound the range of the delta-lognormal function within a finite space of its parameters. The new algorithm is described, tested (under various handwritten strokes) and compared with another algorithm previously developed for the same purpose. The numerical results show that the proposed algorithm gives better results in terms of speed and accuracy and enables the creation of new tools for the basic study of single strokes used as primitives for understanding rapid human movements, according to the Kinematic Theory.

## 1. Introduction

Human movements can be very complex, and their studies in terms of motor control have been intensively used in many fields of research. From a general perspective, the complex levels of manual dexterity that humans exhibit can be attributed and demonstrated in tasks controlled by the nervous system. Specifically, handwriting gestures are part of this type of rapid movements, where we find an intervention of the motor system to achieve them. In neurosciences, handwritten strokes constitute a specific class of rapid human movements, notably used to study neurodegenerative processes such as those involved in Parkinson and Alzheimer's diseases [Schröter *et al.*, 2003; Teulings *et al.*, 1997]. These simple movements can be seen as the basic primitives [Plamondon, 1995] that are combined to produce more complex movements like handwriting signatures. They were deeply studied by psychologists, neuroscientists, physicists and computer scientists over the last century. Handwriting movements are also used in cybersecurity applications [Rusu and Govindaraju, 2005; De santis *et al.*, 2005] as well as for the rehabilitation of patients having suffered from cerebrovascular accidents [Van Gemmert and Teulings, 2006; Dipietro *et al.*, 2009; Plamondon *et al.*, 2014]. In forensic science, handwritten strokes are used to validate authorship of known suspected writers [Simmer, 2000; Fairhurst *et al.*, 2014].

In this context, the need to study the elementary properties of a single stroke for a realistic analysis becomes a key step to understand how the motor control system accomplishes complex movements. Modelling this is a powerful way to understand the basic principles of human motor control and its disorders [Henderson, 2008; Lorch, 2013]. Among the various models of rapid human movements proposed in the literature, the Kinematic Theory is commonly used because of its ability to reproduce faithfully a handwritten stroke, with a set of parameters having a physiological meaning for these movements [Plamondon, 1995, 1998]. This theory presents handwritten strokes as primitives that contain information on both the motor control and the neuromuscular system [Woch and Plamondon, 2004]. The study of such strokes relies on the extraction of parameters that characterize their velocity profile [Guerfali and Plamondon, 1995; Plamondon *et al.*, 2007].

To meet this need, three deterministic algorithms [Djioua and Plamondon, 2009; Plamondon *et al.*, 2007; Guerfali and Plamondon, 1995], have been developed in our laboratory in order to extract the lognormal parameters derived from handwriting velocity profiles. These algorithms have successfully led, for example, to applications for verification of signatures and recognition of handwriting [Pirlo *et al.*, 2015]. Although the parameter extractors developed in [Djioua and Plamondon, 2009; Plamondon *et al.*, 2007; Guerfali and Plamondon, 1995], show good data fitting, none of these prior work can guarantee the *optimality* or *uniqueness* of the solution obtained. Note that defining what constitutes an optimal solution is not so simple, but in loose terms, for medical doctor to accept extracted parameters as part of a diagnosis process, they ask for a level of certainty that there is no other *good enough* solution that departs significantly from the one provided by a parameter extraction system. Such a requirement is not always critical for some applications, since excellent curve fitting is sufficient for many pattern recognition tasks. However, globally optimal solutions are important in biomedical applications where a confidence interval on the value of an extracted parameter is necessary in developing a benchmark, or a pre-screening. To address this need, a first branch and bound algorithm was developed in [O'Reilly and Plamondon, 2012].

It is based on building a tree where each node represents a possible solution of the problem considered. The nodes are delta-lognormal envelopes that correspond to a bounding box, built by studying the variations of the delta-lognormal function, according to its seven parameters. Each subset or child node is created by subdividing the parent node into three children. The dimension according to which a parent node is subdivided is selected depending on the variables that are estimated to minimize the area of the parent delta-lognormal envelope. The so-constructed tree is explored in a way that non-promising solutions are pruned based on a monotonicity test. In summary, the branch and bound algorithms proposed in [O'Reilly and Plamondon, 2012] is based on a delta-lognormal envelope that bounds all the possible delta-lognormal parameters within a subspace, and then computes a bound on the error between the signal and the nearest envelope. This method, despite the fact that it clearly circumscribed the challenge, suffers from lower bound problems (according to [O'Reilly and Plamondon, 2012], the previously proposed procedure for computing an envelope to the velocity profile did not always hold) and each calculation of the estimated area of the envelope requires long processing time. Thus, the total processing time could be reduced significantly if these calculations were not performed so often. The aim of this chapter is to propose a new algorithm to relieve the bounding problem of the extractor proposed in [O'Reilly and Plamondon, 2012] and show its effectiveness when combined in a branch and bound scheme to extract model parameters from ideal velocity profile data. The proposed method is used to bound the delta-lognormal parameters representing synthetic velocity profile data.

The rest of the chapter is organized as follows. Section 2 reviews the delta-lognormal equation and the modelling tools that will be used to bound the objective function. Section 3 introduces some theoretical aspects of interval arithmetic and proposes an algorithm for computing enclosures of the bound of the objective function. Section 4 presents numerical results and performance comparisons with [O'Reilly and Plamondon, 2012]. Section 5 discusses the importance of the proposed algorithm, and finally Sec. 6 proposes possible future research and summarizes the main contributions of this chapter.

## 2. The Delta-lognormal Model

The Kinematic Theory of rapid human movements, from which the delta-lognormal model is derived, considers the trajectory of a pen tip as a single stroke having a delta-lognormal velocity profile.

$$\Delta\Lambda(t;p) = D_1\Lambda(t;t_0,\mu_1,\sigma_1^2) - D_2\Lambda(t;t_0,\mu_2,\sigma_2^2) \tag{1}$$

where

$$\Lambda(t;t_0,\mu,\sigma^2) = \begin{cases} \frac{1}{\sigma\sqrt{2\pi}(t-t_0)}\exp^{-\frac{(\ln(t-t_0)-\mu)^2}{2\sigma^2}}, & for\ \ t_0 < t, \\ 0, & elsewhere. \end{cases} \tag{2}$$

with $\sigma > 0$ ; $\mu, t_0 \in \mathbb{R}$ that is the lognormal impulse response function. In this model, a stroke is produced by a synergy of two neuromuscular systems: the agonist, acting in the direction of the movement and the antagonist, acting in the opposite direction. The agonist and antagonist are lognormals convolved with their corresponding input commands, represented by the subscript 1 and 2 in Eq. (1) respectively. The subtraction of the two system responses produces the velocity profile $\Delta\Lambda(t;p)$ described by the seven parameters $P$ shown in Eq. (3).

$$P = [\,t_0, D_1, \mu_1, \sigma_1, D_2, \mu_2, \sigma_2] \tag{3}$$

In (3), $t_0$ represents the system activation time, and $\mu_1$, $\sigma_1$, $\mu_2$, $\sigma_2$ correspond to the timing properties (on a logarithmic scale) of the two neuromuscular commands, in reaction to their input amplitudes $D_1 > 0$ and $D_2 > 0$ respectively.

To bound the model parameters from velocity data, we need to find the bounding box P that contains all the possible values of the seven parameters $p$ that fit the original velocity movement $v_t(t)$ (collected from synthetic human movements) most accurately. This problem can be solved by building an envelope according to the nonlinear least square function $f$ (4) that maximizes the signal to noise ratio SNR (5).

$$f = \int (v_t(t) - \Delta\Lambda(t;p))^2 dt \tag{4}$$

$$SNR = 10\log\left(\frac{\int v_t^2(t)dt}{f}\right) \tag{5}$$

## 3. Interval Arithmetic

Before formulating the proposed algorithm, some definitions and notations must be established. Consider the following problem:

Bound $f(x)\ subject\ to\ x \in \boldsymbol{X^I}$, where $f$ is an $n$ dimensional continuously differentiable function $\mathbb{R}^{\mathbb{N}} \to \mathbb{R}$ and $\mathbf{X}^{\boldsymbol{I}} \subseteq \mathbb{R}^{\mathbb{N}}$ is an $n$ dimensional interval vector.

We denote by $I = [a, b] = \{x \in \mathbb{R}: a \leq x \leq b\}$ the set of real numbers and one-dimensional closed intervals. In this chapter, an interval will be denoted with a variable in upper case and the set of values contained in the interval, in lower case. Thus $X = [\underline{x}, \overline{x}]$ is an interval variable where $\underline{x}$ and $\overline{x}$ represent the left and right end points of $X$ respectively. The width of $X$ is defined and denoted by $w(X) = \overline{x} - \underline{x}$ and the midpoint is given by:

$$m(X) = \frac{1}{2}(\overline{x} + \underline{x}).$$

We denote by $\boldsymbol{X^I} = (X_1, X_2, \dots \dots, X_n)$ the set of $n$ dimensional interval vectors or box such that:

$$\boldsymbol{X^I} = (X_1, X_2, \dots. X_n) = ([\underline{x}_1, \overline{x}_1], [\underline{x}_2, \overline{x}_2], \dots [\underline{x}_n, \overline{x}_n]).$$

The width and the midpoint of an interval vector are defined as:

$$w(\boldsymbol{X^I}) = \max_i w(\boldsymbol{X_i^I})$$

$$m(\boldsymbol{X^I}) = (m(X_1), m(X_2), \dots, m(X_n)).$$

We call $F$ an inclusion function of $f: \mathbb{R}^{\mathbb{N}} \to \mathbb{R}$, the interval extension of $f$, such that for $x \in X \Rightarrow f(x) \subseteq F(X)$. $f(X)$ is the real range of $f$ on $X$ and $f(X) \subseteq F(X)$. Now that we have these definitions and notations, it is possible to use interval arithmetic [Moore, 1966; Hansen and Walster, 2003] to bound a function defined by a mathematical expression such as the delta-lognormal function.

### 3.1. *Interval arithmetic operations*

In this section, we summarize some principles of interval arithmetic needed to derive the bounding operations of the objective function stated in (4). For more details, the reader can consult [Moore, 1966; Hansen and Walster, 2003]. The basic arithmetic operations between two intervals $X = [\underline{x}, \overline{x}]$ and $Y = [\underline{y}, \overline{y}]$ are defined in Table 1. Apart from interval arithmetic operations, interval functions are also defined as the image that corresponds to the set of interval arguments of a given function into an interval. The basic operations on usual (rational and transcendental) functions of one variable are summarized in Table 2.

Now that we have defined the basic interval operations, we extend the concept to more general functions of $n$ interval variables such as the delta-lognormal (see Eqs. (1) and (2)).

Table 1. Basic arithmetic operations between two intervals.

| Operations | Results |
|---|---|
| $X+Y$ | $[\underline{x}+\underline{y}, \overline{x}+\overline{y}]$ |
| $X-Y$ | $[\underline{x}-\overline{y}, \overline{x}-\underline{y}]$ |
| $X\times Y$ | $[\min S, \max S]$ with $S=\{\underline{x}\underline{y}, \underline{x}\overline{y}, \overline{x}\underline{y}, \overline{x}\overline{y}\}$ |
| $X\div Y$ | $X\times\left(\frac{1}{Y}\right)$, with $\left(\frac{1}{Y}\right)=\left[\frac{1}{\overline{y}}, \frac{1}{\underline{y}}\right]$ and $0\notin Y$ |

Table 2. Basic arithmetic functions.

| Functions | Results |
|---|---|
| exp $(X)$ | $[\exp(\underline{X}), \exp(\overline{X})]$ |
| ln $(X)$ | $[\ln(\underline{X}), \ln(\overline{X})]$ $for\ 0<\underline{X}$ |
| $X^{1/2}$ | $[\underline{X}^{1/2}, \overline{X}^{1/2}]$ |
| $X^n$ | $\begin{cases}[\underline{X}^n, \overline{X}^n]\ if\ \underline{X}>0\ or\ n\ is\ odd\\ [\overline{X}^n, \underline{X}^n]\ if\ \overline{X}<0\ and\ n\ is\ even\\ [0, \max\|\underline{X}^n, \overline{X}^n\|]\ if\ 0\in X and\ n\ is\ even\end{cases}$ |

**Definition 1:** the natural interval extension of a given function $f(x_1, x_2, \dots x_n)$ of n variables is given by the interval function $F(X_1, X_2, \dots X_n)$, which is obtained by replacing the real variable x with the corresponding interval variable X.

**Example 1:** The natural interval extension of

$$f(x,y)=(1-x)^2+100(y-x^2)^2 \quad (6)$$

$$g(w,z)=\exp\left(\frac{w+z}{z^2+w}\right) \quad (7)$$

Are:

$$F(X,Y)=(1-X)^2+100(Y-X^2)^2$$

$$G(W,Z)=\exp\left(\frac{W+Z}{Z^2+W}\right)$$

Where X, Y, W and Z are intervals. If we assign for example $X = [-1, 2]$ , $Y = [-1, 3]$, W= $[0, 2]$, $Z = [-1, 1]$ we have:

$$F([-1,2],[-1,3]) = (1-[-1,2])^2 + 100([-1,3]-[-1,2]^2)$$
$$= [0\ ,\ 2504] \tag{8}$$

$$G([0,2],[-1,1]) = \exp\left(\frac{[-1,3]}{[0,1]+[1,1]}\right)$$
$$= exp([-1,3]) = [\exp(-1), \exp(3)] \tag{9}$$

It is of interest that, in these examples:
in (8), if we take any interval, say $X' = [-1\ ;\ 1] \subseteq X$, the value of $F(X',Y) = [0\ ,\ 904]$ is contained in $F(X,Y)$. This important property is called inclusion isotonicity.

In the right-hand side of Eq. (9), we use the monotonicity of the exponential function to get the result. This property is also very important for bounding operations and is called inclusion monotonicity.

To this end, the following fundamental theorem of interval arithmetic can be stated:

**Fundamental Theorem. Let $F(X_1, X_2, \ldots X_n)$ be the natural interval extension of $f(x_1, x_2, \ldots x_n)$ then**

$$f(X_1, X_2, \ldots X_n) \subseteq F(X_1, X_2, \ldots X_n),$$

**and for all intervals,**

$$Y_k \subset X_k\ , \text{for } k = 1, \ldots\ldots n\ , f(Y_1, Y_2, \ldots Y_n) \subseteq F(Y_1, Y_2, \ldots Y_n),$$

**where**

$$f(Y_1, Y_2, \ldots Y_n) = \{\, f(x_1, x_2, \ldots x_n) : x_k\ \in Y_k \text{ for } k = 1, .. n\}.$$

This theorem due to Moore [Moore, 1966] was extended by Hansen [Hansen and Walster, 2003]. It states that, the evaluation of a function in any point inside a closed interval where both the function and its variables are defined is guaranteed to produce a result contained in the natural interval extension of the function. Its proof can be found in [Hansen and Walster, 2003]. This Fundamental Theorem shows how easy it is to bound the range of a function, and it makes possible the resolution of the bounding problem stated in Sec. 1. In the next sub-section, we will apply this theorem to the delta-lognormal function as a specific sequence of interval arithmetic operations to bound the range of Eq. (4) within an interval vector.

### 3.2. *Algorithm for computing the upper and lower bound of the delta-lognormal function*

To compute a bound for Eq. (4), we first define the bounding space of its variables as:

$$\boldsymbol{P^I} = \left\{\left[\underline{D}_1,\overline{D}_1\right],\left[\underline{\mu}_1,\overline{\mu}_1\right],\left[\underline{\sigma}_1,\overline{\sigma}_1\right],\left[\underline{D}_2,\overline{D}_2\right],\left[\underline{\mu}_2,\overline{\mu}_2\right],\left[\underline{\sigma}_2,\overline{\sigma}_2\right],\left[\underline{t}_0,\overline{t}_0\right]\right\} \tag{10}$$

Then, through the Fundamental Theorem, we apply the natural interval extension on *f*. This yield:

$$\begin{aligned} f(\boldsymbol{P^I}) \in F(\boldsymbol{P^I}) &= \left[\underline{F}(\boldsymbol{P^I}),\overline{F}(\boldsymbol{P^I})\right] \\ &= \int (v_t(t) - \Delta\Lambda(t;\boldsymbol{P^I}))^2 dt \end{aligned} \tag{11}$$

Equation (11) denotes the quadratic error between a handwritten velocity profile and its delta-lognormal approximation. We have chosen this error rather than other because it takes into account the fact that a large error on a small portion of the signal is considered more problematic than a small error on a large portion of the signal.

As provided by the theorem and the natural interval extension, the result of the right-hand side of Eq. (11) is an interval $\left[\underline{F}(\boldsymbol{P^I}),\overline{F}(\boldsymbol{P^I})\right]$ that bounds $f$. To compute that result, we break down the last term $\Delta\Lambda(t;\boldsymbol{P^I})$ in the integral of Eq. (11), into a unique finite sequence of interval arithmetic operations [Boyogueno Bidias *et al.*, 2018] presented below as Algorithm 1.

**Algorithm 1: Eval_Lognormal** $\left(\left[\underline{\boldsymbol{\mu}}_1,\ \overline{\boldsymbol{\mu}}_1\right],\left[\underline{\boldsymbol{\sigma}}_1,\ \overline{\boldsymbol{\sigma}}_1\right],\left[\underline{\boldsymbol{t}}_0,\ \overline{\boldsymbol{t}}_0\right],\boldsymbol{t}\right)$.

Set the constant variable const = $\sqrt{2\pi}$;
Compute Eq. (2) using the following list of expressions (';' are used as separators for brevity).

$$T_1 = \left[\underline{\sigma}_1,\ \overline{\sigma}_1\right]^2;\ T_2 = t - \left[\underline{t}_0,\ \overline{t}_0\right];$$

$$T_3 = ln(T_2);\ T_4 = T_3 - \left[\underline{\mu}_1,\ \overline{\mu}_1\right];$$

$$T_5 = {T_4}^2;\ T_6 = \frac{T_5}{T_1};\ T_7 = 0{,}5 \times T_6;$$

$$T_8 = -T_7;\ T_9 = exp(T_8);$$

$$T_{10} = const \times \left[\underline{\sigma}_1,\ \overline{\sigma}_1\right];\ T_{11} = T_2 \times T_{10};$$

$$T_{12} = \frac{T_9}{T_{11}};$$

return $T_{12}$.

This algorithm starts by expressing *f* as a code list, where each arithmetic function (±, ×, ÷, etc.) and unary elementary functions of the type (exp, ln...) are expressed with only one unique variable in their respective expressions. In this way, we ensure that the desirable properties (inclusion isotonicity, etc.) of the Fundamental Theorem are satisfied. Another advantage of using this method is the sharpness of the bound that results from this calculation process. In summary, Algorithm 1 takes as input each lognormal parameter $P_i$ and the sample time t. From steps 1 to 3, the delta-lognormal function is computed by calling Eval_Lognormal for each lognormal.

### 3.3. *Applying Algorithm 1 in a general branch and bound scheme*

The general idea behind the branch and bound technique is to bound sub-boxes of a feasible area in which we perform a search to try obtaining the sharpest one. The bounding operations can be carried out by procedures such as in Algorithm 1 and, if the bounds under consideration are not sufficiently tight, some boxes are divided into smaller one according to a splitting rule. The procedure is repeated until solutions is obtained within a specified target accuracy. Such a solution is called in this work the optimal solution. Of course, requesting a more accurate solution would normally lead to further refinement and thus a different final solution. What we aim to do in this work is prohibiting that such a different better solution would not be enclosed in some previous best solution obtained with a less stringent accuracy criterion.

In other words, in the context of this chapter, we seek to find the best solution that fits an input velocity profile inside a defined finite space. The optimal solution is a set of points that maximizes the SNR. Finding it exactly remains an elusive goal. In practice, the proposed algorithm returns a set of boxes for which the respective lowest SNR of each box is higher than a target value and the overall best solution is guaranteed to be in one of the returned boxes. The objective function to be minimized is provided by Eq. (4) and the accuracy is fixed by Eq. (5) to 100dB SNR. To facilitate the reading and comprehension of this chapter, a summary of the algorithm explored here is listed below as Algorithm 2. This algorithm is adapted from [O'Reilly and Plamondon, 2012]. A notable difference adopted for simplicity is that algorithm 2 splits boxes in 2 while the one in [O'Reilly and Plamondon, 2012] was trisecting boxes. In this summary, the various operations proposed by [O'Reilly and Plamondon, 2012] have been replaced by the operations presented in Tables 2, 3 and Algorithm1. For a more elaborate version of this description, we refer the reader to [O'Reilly and Plamondon, 2012].

**Algorithm 2**: General multidimensional interval branch and bound

1. Set the initial parameter $\boldsymbol{P}^{\boldsymbol{I}}$. Set the working list $L_1 := \{(\underline{F}(\boldsymbol{P}^{\boldsymbol{I}}), \boldsymbol{P}^{\boldsymbol{I}})\}$, the final list $L_2 := \{\}$ and the upper bound $\overline{f}^{\sim} = f(m(\boldsymbol{P}^{\boldsymbol{I}}))$
2. While $(L_1 \neq \{\})$
3. Select box $\boldsymbol{P}^{\boldsymbol{I}}$ from $L_1$
4. $L_1 -= \boldsymbol{P}^{\boldsymbol{I}}$
5. Bisect $\boldsymbol{P}^{\boldsymbol{I}}$ into $\boldsymbol{P}_i^{\boldsymbol{I}}$ sub boxes $i = 1..2$
6. For $i = 1$ *to* 2
7. Compute the bound $f(m(\boldsymbol{P}_i^{\boldsymbol{I}})), \underline{F}(\boldsymbol{P}_i^{\boldsymbol{I}})$ (Algorithm 1)
8. If *not* $(\underline{F}(\boldsymbol{P}_i^{\boldsymbol{I}}) > \overline{f})$
9. $\overline{f}^{\sim} = min\{\overline{f_i}^{\sim}, f(m(\boldsymbol{P}_i^{\boldsymbol{I}}))\}$; Update upper bound
10. If $\boldsymbol{P}_i^{\boldsymbol{I}}$ satisfy the ending criterion (Eq. (5))
11. $L_2 += \{(\underline{F}(\boldsymbol{P}_i^{\boldsymbol{I}}), \boldsymbol{P}_i^{\boldsymbol{I}})\}$;
12. Else $L_1 += \{(\underline{F}(\boldsymbol{P}_i^{\boldsymbol{I}}), \boldsymbol{P}_i^{\boldsymbol{I}})\}$;
13. Return $L_2, f^* = \overline{f}^{\sim}$

As presented above, Algorithm 2 takes as input a feasible area or box P. It uses a working list $L_1$, and produce a final list $L_2$. The objective function $f$ (see 4) and an upper bound $\overline{f}^{\sim}$ are computed by evaluating $f$ at the midpoint of the initial box. The feasible area is restricted to a search region where we know that the solution can be found. The values of the intervals defining the initial the bounding box are chosen as wide as necessary to almost certainly safely enclose the solution. The working list $L_1$, is the tree generated throughout the search by the division of boxes into sub-boxes. The final list contains solutions that meets the ending criteria presented in step 10 of the algorithm. The algorithm starts by bisecting a box and compute the bounds of each resulting sub-box using Algorithm 1. This is done in step 5 and 7 respectively. Next, the sub-boxes previously obtained are subjected to a monotonicity test (see step 8) where we validate if their lower bound is less than the current best upper bound known. If it is the case, the box is kept in the working list and the process continues. Otherwise, the box is rejected because the test has just proven that the box cannot contain a better solution. The ending criterion is evaluated in step 10, where Eq. (5) is evaluated for each box having passed the monotonicity test. We add to the list $L_2$ the boxes that have satisfied a given SNR. This is illustrated in step 11, where the subscript += and −= (step 4) in front of each list, denote a push and a pop of a box through the list. The algorithm returns the final best solution $f^*$ and the solution list $L_2$ where we can find all the possible solution satisfying the ending criterion.

## 4. Testing the Algorithm Under Ideal Conditions

The algorithm has been tested through two experiments to evaluate the ability of the resulting envelope:

- To maintain, in a defined finite space, all the enclosed spaces of this set.
- To enhance the performance of the extractor in [O'Reilly and Plamondon, 2012], when used as a bounding operation. In this latter test, we did not take into account the dependencies between the different parameters. The bounding interval of parameter space was set according to its domain.

### 4.1. *Tests conditions*

The experiments performed to validate Algorithm 1 use the database proposed in [Djioua and Plamondon, 2009] and [O'Reilly and Plamondon, 2012]. This database is composed of 7000 delta-lognormal curves, grouped into categories. These ideal curves were generated automatically using parameter sets derived from human handwriting movements, previously extracted using the IIX algorithm presented in [Djioua and Plamondon, 2009]. The choice to use synthetic data for the test is justified by the fact that the exact values of the parameters of the speed profiles to be tested are known. Therefore, the extraction algorithm can be validated based on its ability to find the same values as in the ground-truth table. We chose randomly 500 ideal delta-lognormal parameters from class Ca0, Ca2, Cb1 and Cbi (see Fig. 1). Each ideal velocity profile constructed from the ideal delta-lognormal parameters (by evaluating Eq. (1)), were sampled at 200 Hz

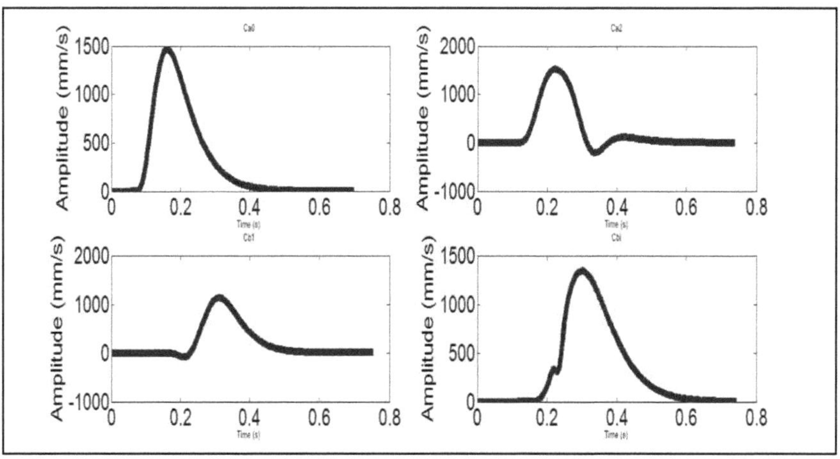


Fig. 1. Delta-lognormal velocity profile examples.

to simulate the data collected with a digitizer [Faci *et al.*, 2018]. In the second experiment, we replaced the boundary operations contained in the branch and bound algorithm proposed in [O'Reilly and Plamondon, 2012] by Algorithm 1. The created speed profiles were then fed to the modify algorithm (Algorithm 2) to try to recover the true parameters values. We carried out the tests in a 3.60 GHz Intel core (i7) with 32 GB of RAM. The implementation was done in C++ using our custom interval arithmetic library based on [Moore, 1966] and [Knüppel, 1994]. All computations were performed using rounded interval arithmetic.

### 4.2. *Results and discussion*

For the first experiment, Table 3 presents the search region, which is restricted for simplicity to a finite box where the solution can be found. The bounding box was constructed by taking the minimum of the minimums and the maximum of the maximums of each parameter value for all the classes.

Table 3. The initial box.

| Parameters | $t_0$ | $D_1$ | $\mu_1$ | $\sigma_1$ | $D_2$ | $\mu_2$ | $\sigma_2$ |
|---|---|---|---|---|---|---|---|
| Minimum | 0.0001 | 50 | -5 | 0.001 | 10 | -5 | 0.001 |
| Maximum | 0.9 | 500 | 0.005 | 1.0 | 300 | 0.005 | 0.9 |

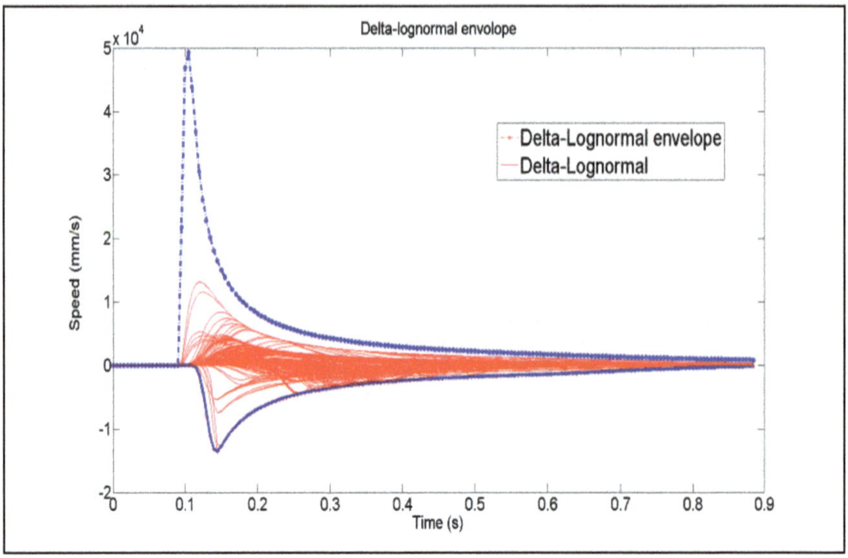


Fig. 2. Experimental testing of the delta-lognormal envelope.

As shown in Fig. 2, thick blue lines represent the envelope produced by Algorithm 1 using the natural interval extension and the bounding box values of Table 3. Inside this envelope, more than 3000 delta-lognormal curves are shown as thin red lines. The parameters of these curves have been generated by splitting the bounding box into sub-boxes such that $P_k \subset P$ for $k = 1 \ldots n$. As can be seen, the envelope bounds the possible values of Eq. (2) inside the given space quite well, since no curve extends beyond these borders.

In the second experiment, a box was considered as a solution if the upper bound of its quadratic error, or its maximum width, was less than $10^{-5}$ *mm*/*s*. Another criterion was based on Eq. (5), where the signal to noise ratio of the chosen box has to be larger than 100 dB. To this end, we computed the noise by evaluating Eq. (4) with real arithmetic using as reference, the middle of the satisfying box.

Tables 4 and 5 present the main statistics of the absolute error on each estimated parameter. As can be seen for the 500 tested curves, the errors on the estimates of $t_0$ and $D_1$ are on average $10^{-9}$ *s* and $10^{-6}$ *mm* respectively. We can also observe that the new algorithm always finds not only the true value within an accuracy of $\varepsilon = 10^{-6}$, but a confidence interval around the true value (Table 5). This interval contains all the possible solutions satisfying at least the ending criteria. Thus, if we decide to generate a solution with a smaller error (up to the limit of numerical precision), we just need to continue the search from that bounding box. This is an advantage of using interval arithmetic. All the results compiled in this table were obtained for a signal to noise ratio larger than 100 dB.

Table 6 shows a performance comparison in terms of processing time and number of boxes explored for the proposed algorithm and the one proposed in [O'Reilly and Plamondon, 2012]. In [O'Reilly and Plamondon, 2012], the algorithm was executed in parallel, on a 12-CPU node grid. As can be seen in Table 6, this algorithm requires a run time of about 50 s to perform 11 iterations

Table 4. Statistics of the absolute error on the parameters estimated from ideal delta-lognormal velocity profiles.

| Parameters | Mean error | STD | Minimum | Maximum |
|---|---|---|---|---|
| $t_0(s)$ | 1,18E-09 | 1,0E-09 | 1,6E-10 | 2,1E-09 |
| $D_1(mm)$ | 2,90E-06 | 2,2E-06 | 6,7E-07 | 5,1E-06 |
| $\mu_1\ (s)$ | 1,08E-08 | 3,5E-09 | 7,1E-09 | 1,4E-08 |
| $\sigma_1(s)$ | 4,95E-09 | 5,2E-10 | 4,4E-09 | 5,4E-09 |
| $D_2(mm)$ | 2,01E-06 | 1,2E-06 | 7,6E-07 | 3,2E-06 |
| $\mu_2\ (s)$ | 1,00E-08 | 3,1E-09 | 6,8E-09 | 1,3E-08 |
| $\sigma_2\ (s)$ | 8,58E-09 | 5,4E-09 | 3,1E-09 | 1,4E-08 |

of its main loop (i.e. passing 11 times through all the operation defined in algorithm1), while the algorithm proposed in this chapter requires only 702 ms on a single core. Of course, the processing time could be reduced with the new algorithm, if it is parallelized to exploit multiple cores. That refinement is left for future research. On the other hand, the number of boxes needed to find a solution of the desired quality could reach one million with the previous algorithm, while ours requires at most 70 thousand boxes to extract such a solution.

Table 5. Statistics of the absolute error on the parameters estimated from ideal delta-lognormal velocity profiles (the bounding box).

| | Bounding box | |
|---|---|---|
| Parameters | Minimum | Maximum |
| $t_0$ | 0,0 | 9,4E-06 |
| $D_1$ | 0,0 | 2,1E-05 |
| $\mu_1$ | 0,0 | 1,9E-06 |
| $\sigma_1$ | 0,0 | 7,2E-06 |
| $D_2$ | 0,0 | 3,3E-05 |
| $\mu_2$ | 0,0 | 3,1E-07 |
| $\sigma_2$ | 0,0 | 1,9E-07 |

Table 6. Comparison of the number of boxes explored to extract one velocity profile and the processing time required to perform 11 iterations in the main loop.

| | Methods | |
|---|---|---|
| Performance metric | (O'Reilly and Plamondon, 2012) | Proposed |
| Number of Boxes | Min 800000 | Max 70000 |
| Processing time (s) | 50 (12 cores) | 0.702 (1 core) |

### 4.3. *Discussion*

The results presented above show the efficiency of Algorithm 1 as well as the power of interval arithmetic in resolving nonlinear high-dimensional problems such as the delta-lognormal parameters extraction on ideal data (synthetic). The different rules used to reject unpromising solutions are unique in the field. They guarantee at all times through the construction of a bounding box, the unique and global character of the final solution with a reasonable computer cost. In order to validate the accuracy of Algorithm 1 and prove its efficiency, we anticipated a test on real human hand-written speed profiles of class c1. Despite the fact that this experiment is part of a subsequent study and not the subject of this chapter, we present in Fig. 3 an example in which the delta-lognormal parameters of a real human generated velocity profile of a pen-tip movement is extracted using Algorithm1 combined with [O'Reilly and Plamondon, 2012]. As can be seen in the middle trace of Fig. 3, the extracted velocity profile fits quite well (with an SNR = 30 dB) the human speed profile of the pen-tip movement shown in the top trace of Fig. 3. The bottom trace in Fig. 3 shows the extracted velocity profile decomposed into two neuromuscular systems acting in direct opposition as predicted by the delta-lognormal model. The agonist and the antagonist outputs are respectively plotted in the figure by dashed and thin lines around the continuous line movement.

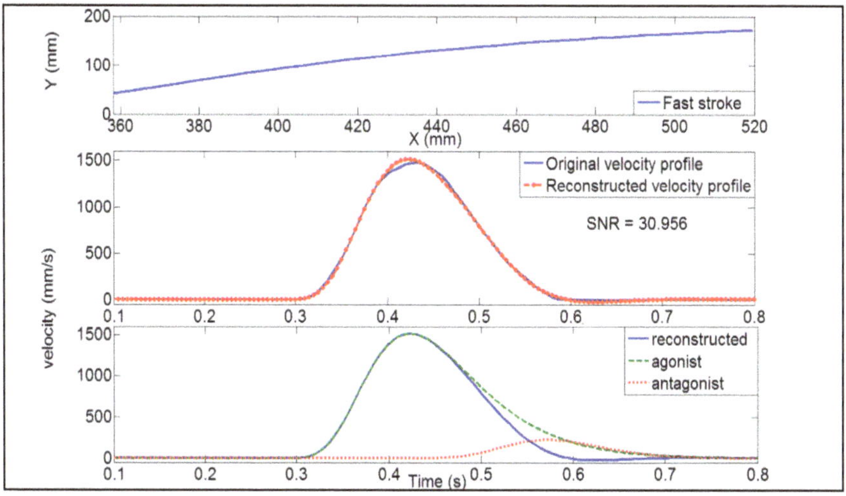


Fig. 3. Example of application of Algorithm 1 in extracting real human pen-tip movement (the top trace in blue is the fast stroke, the middle traces are respectively the original (blue solid line) and the reconstructed (red stars line) velocity profile. The bottom traces are respectively the reconstruct (blue line), the agonist (green dash line) and the antagonist (red stars line).

## 5. Conclusion

In this chapter, we have shown that, with a good bounding method applied to the delta-lognormal function, when embedded in a branch and bound algorithm, the parameters of delta-lognormal velocity profiles can be extracted with less computational costs than previous works can achieve. The effectiveness of the proposed bounding method is quite remarkable. The number of evaluations was reduced at least 71-fold to obtain results of a same target quality. This algorithm exploits the natural interval extension and the fundamental theorem of interval arithmetic to compute the bounding operations of the delta-lognormal function. Although the proposed algorithm is promising in its ability to recover a global solution in a processing time much smaller than a previously reported method, more work remains to be done. The proposed algorithm should be tested on all classes of velocity profiles that the delta-lognormal expression can model, and it should be more thoroughly tested on real human movement data. This will be the goal of a follow up study.

## References

Boyogueno Bidias, S., David, J.-P., Savaria, Y. and Plamondon, R. (May 2018). On the use of Interval Arithmetic to Bound Delta-Lognormal Rapid Human Movements Models. *Proceedings of ICPRAI 2018. First International Workshop on the Lognormality Principle and its Applications*, Montréal, Canada, pp. 738–742.

De Santis, A., Caggiano, V., Siciliano, B., Villani, L. and Boccignone, G. (2005). Anthropic inverse kinematics of robot manipulators in handwriting tasks. *In: 12th Conference of the International Graphonomics Society.*

Dipietro, L., Krebs, H. I., Fasoli, S., Volpe, B. T. and Hogan, N. (2009). Submovement changes characterize generalization of motor recovery after stroke. *Cortex*, *45*(3), 318–324.

Djioua, M. and Plamondon, R. (2009). A New Algorithm and System for the Characterization of Handwriting Strokes with Delta-Lognormal Parameters. *IEEE Transactions on Pattern Analysis and Machine Intelligence*, *31*(11), 2060–2072.

Faci, N., Boyogueno Bidias, S., Plamondon, R. and Bergeron, N. (May 2018). A new experimental Set-up to Run neuromuscular tests. *Proceedings of ICPRAI 2018. First International Workshop on the Lognormality Principle and its Applications*, Montréal, Canada, pp. 753–757.

Fairhurst, M., Erbilek, M. and Li, C. (2014. pp. 1–6). Enhancing the forensic value of handwriting using emotion prediction. *Biometrics and Forensics (IWBF), International Workshop on IEEE.*

Guerfali, W. and Plamondon, R. (1995). Signal Processing for the parameter extraction of the delta lognormal model (AA). *Research in Computer and Robot Vision*, 217.

Hansen, E. and Walster, G. W. (2003). *Global optimization using interval analysis: Revised and expanded.* CRC Press.

Henderson, V. W. (2008). Alexia and agraphia contrasting perspectives of J.-M. Charcot and J. Hughlings Jackson. *Neurology*, 70(5), 391–400.

Knüppel, O. (1994). PROFIL/BIAS—a fast interval library. *Computing*, *53*(3), 277–287.

Lorch, M. (2013). Written Language Production Disorders: Historical and Recent Perspectives. *Current Neurology and Neuroscience Reports*, *13*(8), 369.

Moore, R. E. (1966). *Interval analysis*. Englewood Cliffs: Prentice-Hall.

O'Reilly, C. and Plamondon, R. (2012). A globally optimal estimator for the delta-lognormal modeling of fast reaching movements. *IEEE Transactions on Systems, Man and Cybernetics*, *42*(5), 1428–1442.

Pirlo, G., Diaz, M., Ferrer, M. A. *et al.* (Springer, Cham, 2015, pp. 290–297). Early diagnosis of neurodegenerative diseases by handwritten signature analysis. *In: International Conference on Image Analysis and Processing.*

Plamondon, R. (1995). A kinematic theory of rapid human movements. *Biological Cybernetics*, *72*(4), 295–307. Consulté le 11 12, 2017,

Plamondon, R. (1995). A kinematic theory of rapid human movements. Part II. Movement time and control. *Biological Cybernetics*, *72*(4), 309–320.

Plamondon, R. (1998). A kinematic theory of rapid human movements: Part III. Kinetic outcomes. *Biological Cybernetics*, *78*(2), 133–145.

Plamondon, R., Li, X. and Djioua, M. (2007). Extraction of delta-lognormal parameters from handwriting strokes. *Frontiers of Computer Science in China*, *1*(1), 106–113.

Plamondon, R., O'Reilly, C. and Ouellet-plamondon, C. (2014). Strokes against stroke—strokes for strides. *Pattern Recognition*, *47*(3), 929–944.

Rusu, A. I. and Govindaraju, V. (2005). *On the challenges that handwritten text images pose to computers and new practical applications*. Consulté le 11 12, 2017, sur spiedigitallibrary.org/conference-proceedings-of-spie/5676/1/on-the-challenges-that-handwritten-text-images-pose-to-computers

Schröter, A., Mergl, R., Bürger, K., Hampel, H., Möller, H.-J. and Hegerl, U. (2003). Kinematic Analysis of Handwriting Movements in Patients with Alzheimer's Disease, Mild Cognitive Impairment, Depression and Healthy Subjects. *Dementia and Geriatric Cognitive Disorders*, *15*(3), 132–142.

Simmer, M. ( 2000). Advance in forensic document examination. *J. Forensic Document Examination*, *13*, 1–109.

Teulings, H.-l., Contreras-Vidal, J. L., Stelmach, G. E. and Adler, C. H. (1997). Parkinsonism Reduces Coordination of Fingers, Wrist, and Arm in Fine Motor Control. *Experimental Neurology*, *146*(1), 159–170.

Van Gemmert, A. W. and Teulings, H.-l. (2006). Advances in graphonomics: Studies on fine motor control, its development and disorders. *Human Movement Science*, *25*(4), 447–453.

Woch, A. and Plamondon, R. (2004). Using the framework of the kinematic theory for the definition of a movement primitive. *Motor Control*, *8*(4), 547–557.

https://doi.org/10.1142/9789811226830_0015

# Chapter 15

# Modeling 3D Movements with the Kinematic Theory of Rapid Human Movements

Andreas Fischer[⋆,†], Roman Schindler[⋆], Manuel Bouillon[⋆] and Réjean Plamondon[‡]

[⋆]*DIVA Group, University of Fribourg, Switzerland*
[†]*iCoSys Institute, HES-SO Fribourg, Switzerland*
[‡]*Laboratoire Scribens, Département de génie électrique, Polytechnique Montréal, Canada*
{*andreas.fischer,roman.schindler,manuel.bouillon*}*@unifr.ch, rejean.plamondon@polymtl.ca*

The Kinematic Theory of rapid human movements analytically describes pen tip movements as a sequence of elementary strokes with lognormal speed. The theory has been confirmed in a large number of experimental evaluations, achieving a high reconstruction quality when compared with observed trajectories and providing pertinent features for biomedical applications as well as biometric identification. So far, the Kinematic Theory has focused on one-dimensional movements with the Delta-Lognormal model and on two-dimensional movements with the Sigma-Lognormal model. In this chapter, we present a model for movements in three dimensions, which naturally extends the Sigma-Lognormal approach. We evaluate our method on two action recognition datasets and an air-writing dataset, demonstrating a high reconstruction quality for modelling rapid 3D movements in all cases.

## 1. Introduction

The Kinematic Theory of rapid human movements [Plamondon, 1995a,b, 1998, Plamondon *et al.*, 2003] is one of the most comprehensive theories on the production process of handwriting. It postulates that pen tip movements are the result of a series of elementary neuromuscular strokes, whose speed is a lognormal function. When compared with other frameworks for human movement modeling, such as coupled oscillator models [Hollerbach, 1981] or minimum jerk models [Flash and Hogans, 1985], the Kinematic Theory stands out with an excellent reconstruction quality of observed trajectories [Plamondon *et al.*, 1993], i.e. the analytical representation is highly accurate.

Moreover, an increasing number of experiments suggest that the analytical representation provided by the Kinematic Theory indeed captures neuromuscular properties of the writer, which can be exploited for the analysis of human motor control. Successful applications include learning tools for children [Djeziri *et al.*, 2002, Rémi *et al.*, 2017], biomedical applications [Plamondon *et al.*, 2013, O'Reilly *et al.*, 2014], gesture recognition [Almaksour *et al.*, 2011], handwriting recognition [Fischer *et al.*, 2014, Martin-Albo *et al.*, 2014], and signature verification [Galbally *et al.*, 2012a,b, Diaz *et al.*, 2018, Fischer and Plamondon, 2017], to name just a few. Besides the analysis of finger movements, the Kinematic Theory has also been applied to analyze wrist, arm, head, and eye movements [Plamondon, 1995a], and recently on head trunk rotations [Lebel *et al.*, 2017] as well as for speech processing [Carmona-Duarte *et al.*, 2016].

Except for a few published and unpublished exploratory studies [Leduc and Plamondon, 2001, Djioua, 2007], the Kinematic Theory has focused on movements in one and two dimensions. Rapid 1D movements are modeled by two opposed neuromuscular systems generating agonist and antagonist movements, according to the Delta-Lognormal model ($\Delta\Lambda$) [Plamondon and Guerfali, 1998]. Complex 2D movements, such as signatures, are modeled by a vectorial sum of strokes, i.e. several movements that are overlapping in time during execution, according to the Sigma-Lognormal model ($\Sigma\Lambda$) [Plamondon and Djioua, 2006].

In both cases, algorithms are needed to extract the model parameters from observed trajectories. The Robust XZERO algorithm [Djioua and Plamondon, 2008] has been proposed to estimate the parameters of the $\Delta\Lambda$ model from the velocity profile. For estimating the parameters of the $\Sigma\Lambda$ model, the Robust XZERO algorithm is complemented with an estimation of the start and end angles of the two-dimensional strokes [O'Reilly and Plamondon, 2009]. Improved variants of this approach include a breadth-first search strategy that minimizes the number of strokes [Martin-Albo *et al.*, 2015] and the recently introduced iDeLog method [Ferrer *et al.*, 2020], which estimates the lognormal parameters not only from the velocity profile but also from the trace, taking into account visual feedback.

In this chapter, we present a three-dimensional model for the Kinematic Theory. It naturally extends the $\Sigma\Lambda$ model by introducing an additional angle for the third dimension. The present chapter builds upon our recent conference paper [Schindler *et al.*, 2018], which has introduced the method. Additional contributions include a more comprehensive description of the 3D model as well as an extended experimental evaluation on two supplementary motion datasets. In our experiments, we measure the reconstruction quality of the 3D model in terms of signal-to-noise-ratio (SNR) as well as classification accuracy on two action recognition datasets, namely HDM05 [Müller *et al.*, 2007] and UTKinect [Xia *et al.*, 2012], as well as an air-writing dataset [Chen *et al.*, 2016]. While the air-writing dataset consists of finger movements, the action recognition datasets are based on wrist and ankle movements.

The remainder of this chapter is organized as follows. Section 2 reviews the standard $\Sigma\Lambda$ model of the Kinematic Theory, Sec. 3 details our proposed extension to three dimensions, and Sec. 4 reports experimental results. Finally, we draw conclusions in Sec. 5 and discuss future lines of research for modeling 3D movements with the Kinematic Theory.

## 2. Sigma-Lognormal Model

The Sigma-Lognormal model ($\Sigma\Lambda$) [Plamondon and Djioua, 2006] of the Kinematic Theory represents a two-dimensional pen tip movement as a series of elementary *strokes*.[a] An individual stroke is initiated at time $t_0$ in the central nervous system to cover distance $D$. It is then executed with lognormal speed

$$|\vec{v}(t)| = \frac{D}{\sqrt{2\pi}\cdot\sigma(t-t_0)}\exp\left(-\frac{[\ln(t-t_0)-\mu]^2}{2\sigma^2}\right) , \tag{1}$$

where $\mu$ is the log time delay and $\sigma$ is the log response time, which depend on the neuromuscular system. The distance traveled at time $t$ is

$$d(t) = \int_0^t |\vec{v}(\tau)|d\tau = \frac{D}{2}\left[1+\operatorname{erf}\left(\frac{\ln(t-t_0)-\mu}{\sigma\sqrt{2}}\right)\right] . \tag{2}$$

The $\Sigma\Lambda$ model assumes that each 2D stroke follows a circular trajectory with start angle $\theta_s$ and end angle $\theta_e$. The angular position is

$$\theta(t) = \theta_s + (\theta_e - \theta_s)\frac{d(t)}{D} . \tag{3}$$

To sum up, each lognormal stroke $s$ is characterized by a total of six parameters

$$s = (t_0, D, \mu, \sigma, \theta_s, \theta_e) . \tag{4}$$

[a]Note that the term *stroke* is sometimes used to describe the pen tip movement between pen-down and pen-up. In the present context of kinematic analysis, it refers to a movement primitive described by a lognormal velocity profile [Woch and Plamondon, 2004].

Finally, the velocity of the 2D movement is given by vectorial summation of $n$ individual strokes

$$\vec{v}(t) = \sum_{i=1}^{n} \vec{v_i}(t) \ . \tag{5}$$

The underlying assumption is that the strokes are scheduled in an action plan according to their initiation time $t_0$ and overlap in their execution.

For reconstructing the velocity and the trace of a 2D pen tip movement based on the analytical $\Sigma\Lambda$ representation, it follows that

$$v_x(t) = \sum_{i=1}^{n} |\vec{v_i}(t)| \cos(\theta_i(t)) \ , \tag{6}$$

$$v_y(t) = \sum_{i=1}^{n} |\vec{v_i}(t)| \sin(\theta_i(t)) \ , \tag{7}$$

$$x(t) = \int_0^t v_x(\tau) d\tau \ , \tag{8}$$

$$y(t) = \int_0^t v_y(\tau) d\tau \ . \tag{9}$$

Figure 1 illustrates a handwritten signature and its $\Sigma\Lambda$ model. The observed as well as the reconstructed signature is shown, together with the circular trajectories of the individual strokes. The signal-to-noise ratio (SNR) is 27dB for this example, which is highly accurate (see Sec. 2.2).

For more details, we refer to the articles that introduce the Kinematic Theory [Plamondon, 1995a,b, 1998, Plamondon *et al.*, 2003].

### 2.1. *Stroke extraction and parameter estimation*

One of the most widely used algorithms for stroke extraction and estimation of the $\Sigma\Lambda$ parameters is based on an iterative detection of strokes in the velocity profile [O'Reilly and Plamondon, 2009].

In order to improve the stability of the algorithm, certain signal preprocessing steps are recommended. First, it can be helpful to stop the pen tip artificially at the beginning and at the end of the movement during $200ms$, to ensure zero velocity. Secondly, if the acquisition device has a low or unstable sampling rate, an interpolation of the velocity profile with cubic splines and a resampling at $200Hz$ is suggested. Finally, a low-pass filter can be used to remove noise introduced by the acquisition device.

After preprocessing, strokes are extracted iteratively from the signal. The following five steps describe the extraction of a single stroke:

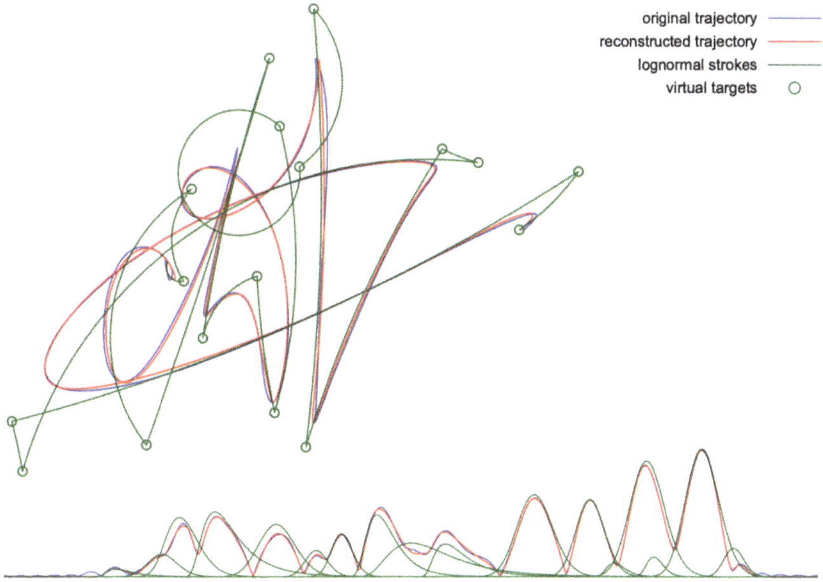


Fig. 1. Sigma-Lognormal reconstruction of the trace and the velocity profile of a handwritten signature [Fischer and Plamondon, 2017].

(1) Detect new stroke in the speed profile $|\vec{v}(t)|$.
(2) Estimate parameters $D$, $t_0$, $\mu$, and $\sigma$.
(3) Estimate angular parameters $\theta_s$ and $\theta_e$.
(4) Optimize parameter estimation.
(5) Remove stroke from the speed profile.

These steps are repeated until there are no more strokes detected in the speed profile.

Step 1 detects a local maximum in the speed profile together with neighboring infexion points and minima. The maximum speed and the area under curve have to exceed a minimum value.

Step 2 applies the Robust XZERO algorithm [Djioua and Plamondon, 2008] to estimate the speed-related parameters $D$, $t_0$, $\mu$, and $\sigma$. It is based on zero crossings of the first and second derivatives of the lognormal function.

Step 3 estimates the start angle $\theta_s$ and the end angle $\theta_e$ based on five characteristic times of the lognormal stroke illustrated in Fig. 2. The start and end are chosen as $t_1 = t_0 + \exp(\mu - 3\sigma)$ and $t_5 = t_0 + \exp(\mu + 3\sigma)$,

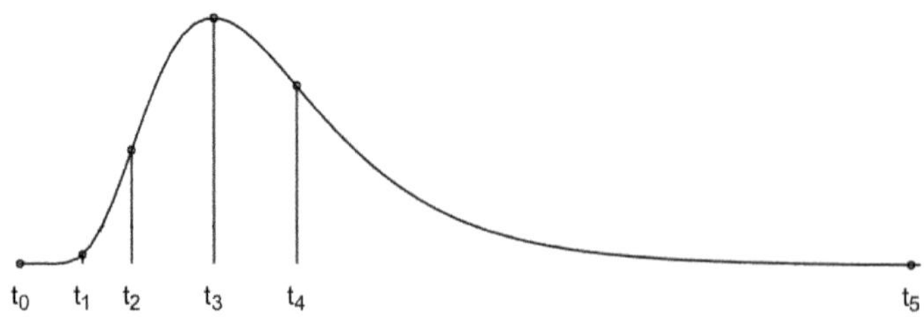


Fig. 2. Characteristic times of a lognormal stroke.

respectively. The maximum speed is reached at $t_3 = t_0 + \exp(\mu - \sigma^2)$, and the inflection points are $t_2 = t_0 + \exp(\mu - 1.5\sigma^2 - \sigma\sqrt{0.25\sigma^2 + 1})$ and $t_4 = t_0 + \exp(\mu - 1.5\sigma^2 + \sigma\sqrt{0.25\sigma^2 + 1})$, respectively.

The angular parameters are obtained by means of linear extrapolation. Considering the velocity angle

$$\theta(t) = \arctan \frac{v_y(t)}{v_x(t)} \ , \tag{10}$$

the angular derivative is calculated with respect to the distance traveled at time $t_4$ and $t_2$ (see Eq. (2))

$$\Delta\theta = \frac{\theta(t_4) - \theta(t_2)}{d(t_4) - d(t_2)} \ . \tag{11}$$

Finally, the start and end angles are estimated using

$$\hat{\theta}_s = \theta(t_3) - (d(t_3) - d(t_1))\Delta\theta \ , \tag{12}$$

$$\hat{\theta}_e = \theta(t_3) + (d(t_5) - d(t_3))\Delta\theta \ . \tag{13}$$

Step 4 optimizes the initial parameter estimation using non-linear least squares curve fitting.

Step 5 removes the new stroke from the speed profile before the next stroke is detected.

For more details on stroke extraction and parameter estimation, we refer to the original article [O'Reilly and Plamondon, 2009].

## 2.2. *Model quality*

The analytical $\Sigma\Lambda$ model allows to reconstruct the movement using Eqs. (6)–(9). The quality of the model can be evaluated by means of a signal-to-noise ratio (SNR) between the observed velocity $\vec{v}_o(t)$ and the reconstructed velocity $\vec{v}_r(t)$. With respect to the start time $t_s$ and end time

$t_e$ of the movement, the SNR is

$$SNR = 10 \cdot \log \left( \frac{\int_{t_s}^{t_e} |\vec{v}_o(\tau)|^2 d\tau}{\int_{t_s}^{t_e} |\vec{v}_o(\tau) - \vec{v}_r(\tau)|^2 d\tau} \right) . \tag{14}$$

In practice, we approximate the integral with the trapezoidal rule over all sampling points.

## 3. 3D Model Extension

The proposed 3D model naturally extends the $\Sigma\Lambda$ model by introducing an additional angle $\phi(t)$ for the third dimension as illustrated in Fig. 3, where the radius $\rho = |\vec{v}(t)|$ corresponds to the speed.

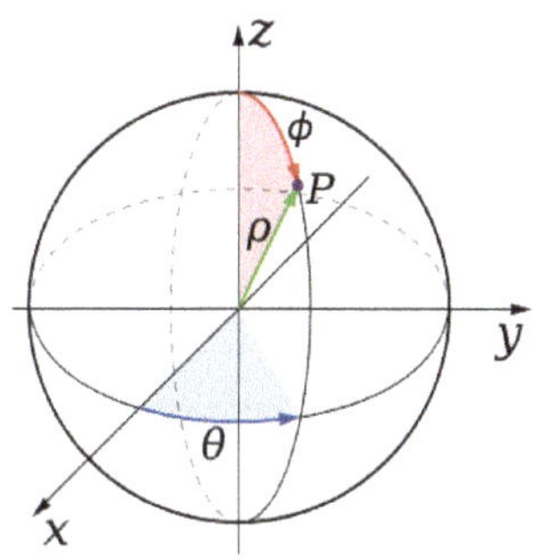


Fig. 3. Spherical coordinates of the 3D model.

Assuming that strokes act along a pivot direction, we model the new angle similar to Eq. (3) by

$$\phi(t) = \phi_s + (\phi_e - \phi_s)\frac{d(t)}{D} . \tag{15}$$

That is, the change in the angle is proportional to the distance traveled.

Two additional parameters are thus introduced to the stroke model, i.e. the start angle $\phi_s$ and the end angle $\phi_e$ in the third dimension

$$s_{3D} = (t_0, D, \mu, \sigma, \theta_s, \theta_e, \phi_s, \phi_e) . \tag{16}$$

For reconstructing the velocity based on the 3D model, Eqs. (6) and (7) have to be modified. The three velocity components are

$$v_x(t) = \sum_{i=1}^{n} |\vec{v}_i(t)| \sin(\phi_i(t)) \cos(\theta_i(t)) \quad , \tag{17}$$

$$v_y(t) = \sum_{i=1}^{n} |\vec{v}_i(t)| \sin(\phi_i(t)) \sin(\theta_i(t)) \quad , \tag{18}$$

$$v_z(t) = \sum_{i=1}^{n} |\vec{v}_i(t)| \cos(\phi_i(t)) \quad . \tag{19}$$

For reconstructing the position, Eqs. (8) and (9) remain the same. The three position components are

$$x(t) = \int_0^t v_x(\tau)d\tau \quad , \tag{20}$$

$$y(t) = \int_0^t v_y(\tau)d\tau \quad , \tag{21}$$

$$z(t) = \int_0^t v_z(\tau)d\tau \quad . \tag{22}$$

The stroke extraction and parameter estimation procedure follows the same procedure as in 2D (see Sec. 2.1). In Step 3, the additional angles $\phi_s$ and $\phi_e$ are estimated similar to $\theta_s$ and $\theta_e$.

Considering the velocity angle

$$\theta(t) = \arccos \frac{v_z(t)}{\rho} \quad , \tag{23}$$

where $\rho = |\vec{v}(t)|$ is the speed, we estimate the new angles similar to Eqs. (11)–(13):

$$\Delta\phi = \frac{\phi(t_4) - \phi(t_2)}{d(t_4) - d(t_2)} \quad , \tag{24}$$

$$\hat{\phi}_s = \phi(t_3) - (d(t_3) - d(t_1))\Delta\phi \quad , \tag{25}$$

$$\hat{\phi}_e = \phi(t_3) + (d(t_5) - d(t_3))\Delta\phi \quad . \tag{26}$$

The proposed modifications of the $\Sigma\Lambda$ model allow us to extract 3D strokes from any three-dimensional movement. By vectorial summation of the strokes, the analytical model enables us to reconstruct the movement. The model quality is measured in terms of SNR as described in Sec. 2.2.

## 4. Experimental Evaluation

The proposed 3D model has been evaluated on three publicly available motion datasets described in Sec. 4.1. Besides reporting the SNR in Sec. 4.2, we also investigate the reconstructed 3D movements in the context of a classification experiment in Sec. 4.3.

### 4.1. *Datasets*

A summary of the motion datasets is provided in Table 1, including the number of subjects who performed 3D movements, the number of classes (types of movements), the total number of samples in the dataset, the acquisition device, and the sampling rate.

Table 1. 3D motion datasets.

| | HDM05 | UTKinect | Air-Writing |
|---|---|---|---|
| Type of activities | Actions | Actions | Words |
| Number of subjects | 5 | 10 | 5 |
| Number of classes | 11 | 10 | 100 |
| Number of samples | 249 | 199 | 500 |
| Acquisition device | Vicon | Kinect | Leap |
| Sampling rate | 120 Hz | 15 Hz | 60 Hz |

Exemplary 3D trajectories together with their reconstructions are shown in Figs. 4–6 for each of the datasets.

**HDM05.** The HDM05 dataset [Müller *et al.*, 2007] contains around 1500 motion samples of 100 different action classes, performed by five subjects. The movements were recorded with an optical marker-based motion capture suit of Vicon with a sampling rate of 120 Hz.

We use a common subset of eleven classes and 249 samples. The classes are *deposit floor*, *elbow to knee*, *grab high*, *hop both legs*, *jog*, *kick forward*, *lie down floor*, *rotate both arms backward*, *sneak*, *squat* and *throw basketball*.

Four 3D trajectories are considered for our experiments, namely the wrists relative to the shoulders and the ankles relative to the hips [Boulahia *et al.*, 2016]. These trajectories are then normalized by the length of arms and legs of the subject, which supports action recognition across different persons [Kulpa *et al.*, 2005].

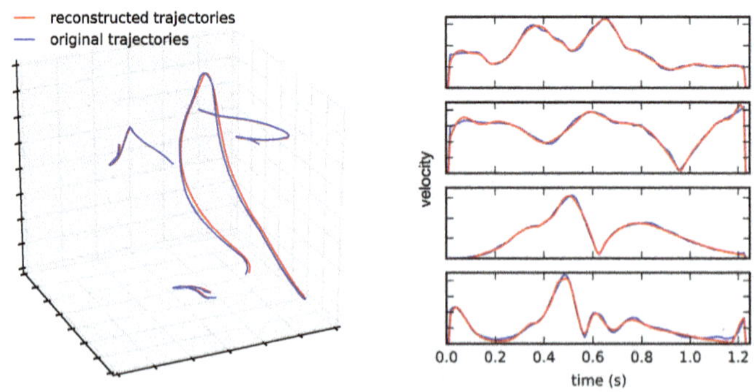


Fig. 4. Original (blue) and reconstructed (red) traces and velocity profiles of the four limbs for a kick action from the HDM05 dataset.

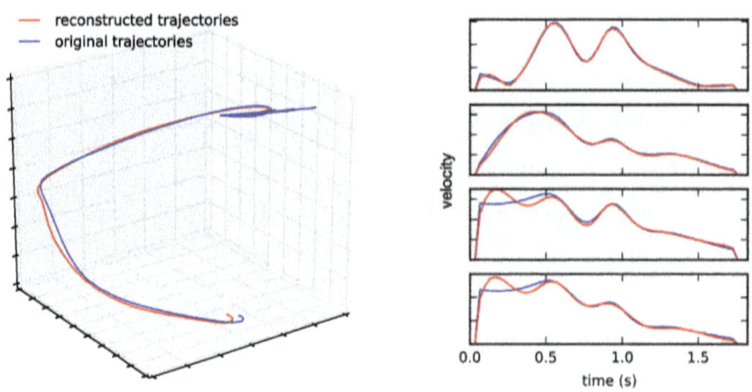


Fig. 5. Original (blue) and reconstructed (red) traces and velocity profiles of the four limbs for a throwing action from the UTKinect dataset.

**UTKinect.** The UTKinect dataset [Xia *et al.*, 2012] is composed of 199 samples of 10 actions performed by 10 subjects. The movements are recorded by a Kinect camera with a sampling rate of 30 Hz. However, frames were only recorded when the skeleton was tracked, causing missing frames and a final sampling rate of about 15 Hz.

Ten indoor activities include *walk*, *sit down*, *stand up*, *pick up*, *carry*, *throw*, *push*, *pull*, *wave* and *clap hands*.

Similar to the HDM05 dataset, we consider four 3D trajectories of normalized wrist and ankle movements.

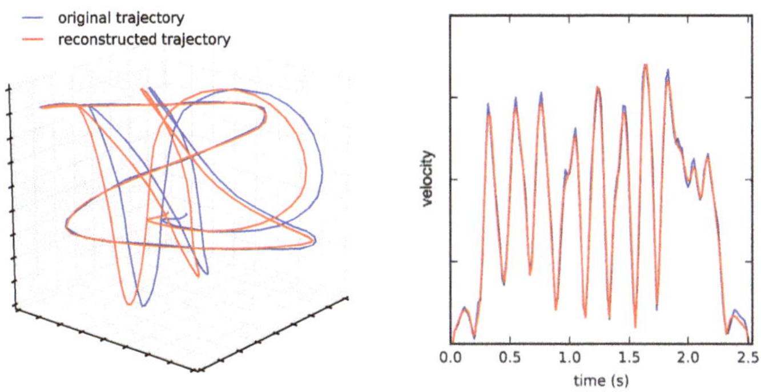


Fig. 6. Original (blue) and reconstructed (red) traces and velocity profiles of the fingertip for the word *ZIP* from the Air-Writing dataset.

**Air-Writing.** The Air-Writing dataset [Chen *et al.*, 2016] includes 150 short words written with the index finger into the air by 18 subjects. They consist of uppercase letters A-Z and were written with a specified movement order in box-writing style, i.e. on top of each other. The writing is recorded without markers or gloves using a Leap camera with a sampling rate of 60 Hz.

For our experiments, we consider 100 common words written by 5 subjects (C3, J1, M3, T3, W1), resulting in a total of 500 fingertip trajectories.

### 4.2. *SNR results*

Table 2 reports the mean SNR of our 3D model together with the standard deviation for all three datasets. It is compared with the SNR achieved with the unmodified $\Sigma\Lambda$ model when the z-direction is disregarded. Detailed bar plots are provided in Fig. 7.

Despite the increased complexity of modeling 3D movements, the proposed method achieves a high model quality that is only slightly below the 2D results. Actions recorded by the Vicon suit (HDM05) and the Kinect

Table 2. Reconstruction quality in terms of SNR.

| | HDM05 | | UTKinect | | Air-Writing | |
|---|---|---|---|---|---|---|
| | 2D | 3D | 2D | 3D | 2D | 3D |
| Mean | 19.19 | 18.52 | 21.87 | 20.21 | 12.60 | 12.52 |
| Standard deviation | 3.77 | 4.09 | 3.62 | 4.40 | 2.02 | 2.02 |

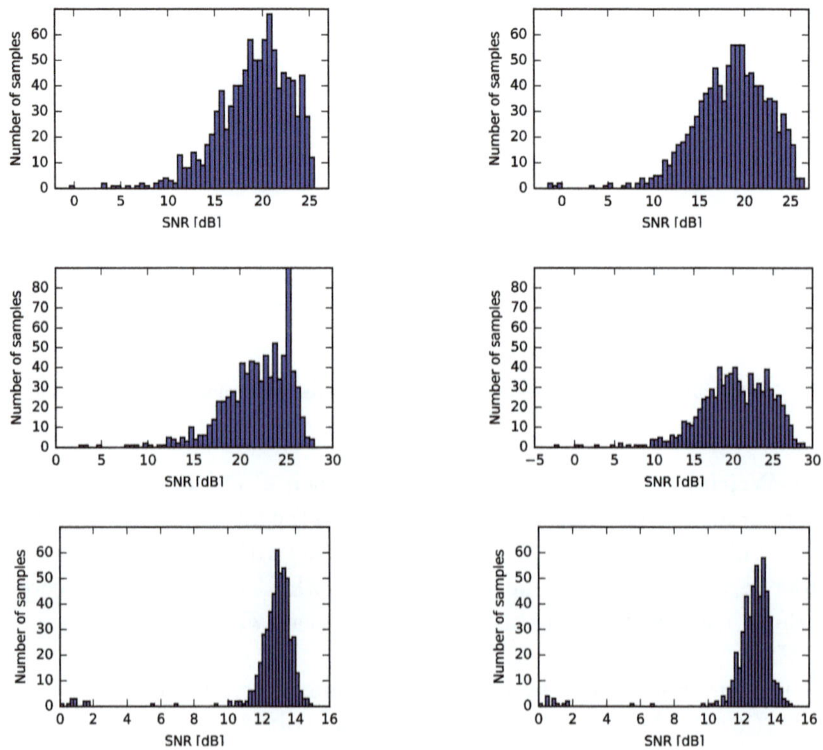


Fig. 7. SNR histograms for 2D reconstructions (left) and 3D reconstructions (right) of the HDM05, UTKinect, and Air-Writing datasets (top to bottom).

camera (UTKinect) have a significantly higher SNR than the finger movement recorded with the Leap camera (Air-Writing), which may be due to the larger scale of the whole-body actions when compared with the relatively small finger movements. In all three datasets, the movements are rather rapid with a duration of less than five seconds on average.

These results also demonstrate that the proposed algorithm can work with low frequency sampled signals as compared to most of the previous studies where the sampling frequency is above 100 Hz, generally 200 Hz.

### 4.3. *Classification results*

In a final experiment, we have classified the different types of movements using dynamic time warping (DTW) [Schindler *et al.*, 2018].

During preprocessing, we resample the original trajectories at 30 Hz for HDM05 and at 15 Hz for UTKinect and Air-Writing using cubic spline interpolation, in order to obtain a stable sampling rate and to speedup the DTW computation. The reconstructed trajectories are computed at the same sampling times. Afterwards, we use a second-order regression to compute velocity and acceleration as additional features leading to a total of nine features $(x, y, z, v_x, v_y, v_z, a_x, a_y, a_z)$ per sample point.

The classifier is optimized for each dataset separately with respect to feature selection and feature normalization. For HDM05, the best results are achieved when considering only the velocity for each of the four limbs, normalized to zero mean and unit variance for each movement. For UTKinect, only the position is considered for each of the four limbs and no normalization is applied. For Air-Writing, position and velocity are considered for the finger movement, normalized to zero mean.

For classification, the DTW distance is computed for all pairs of movements. Each movement is then assigned to the same class as the nearest neighbor in the dataset performed by other subjects. In order to speedup the distance computation and to avoid unrealistic warping paths, we use DTW with a Sakoe-Chiba band of width $|n_1 - n_2| + 10$, where $n_1$ and $n_2$ are the number of sampling points of the two movements. The resulting DTW distance is normalized with the total number of sampling points $n_1 + n_2$.

The classification results are reported in Table 3 both for original and reconstructed samples, respectively. In all cases, we observe only a slight deviation in the classification accuracy, which emphasizes the high quality of the proposed 3D model.

Table 3. Classification accuracy using original and reconstructed 3D movements.

| | Original | Reconstructed |
|---|---|---|
| HDM05 | 96.4 | 96.1 |
| UTKinect | 94.0 | 92.0 |
| Air-Writing | 99.0 | 98.2 |

## 5. Conclusions

The proposed 3D model of the Kinematic Theory of rapid human movements has demonstrated a high reconstruction quality for rapid 3D movements of ankles, wrists, and fingertips recorded with diverse equipment.

There are several promising lines of future research to pursue. First, there seems to be potential to improve the current stroke extraction and parameter estimation method, for example by minimizing the number of strokes [Martin-Albo *et al.*, 2015] or by taking into account visual feedback [Ferrer *et al.*, 2020]. Secondly, the synthesis of 3D movements may be highly rewarding. It can support classifiers with realistic training data. Thirdly, there is a wide range of new applications that may emerge from modelling 3D motions with the Kinematic Theory in biomedicine, biometrics, and robotics.

From a more fundamental perspective, our results support the universality of the lognormality principle and its potential use to study any kind of human movements.

## References

Almaksour, A., Anquetil, E., Plamondon, R. and O'Reilly, C. (2011). Synthetic handwritten gesture generation using sigma-logormal model for evolving handwriting classifier, in *Proc. 15th Int. Conf. of the International Graphonomics Society*, pp. 98–101.

Boulahia, S. Y., Anquetil, É., Kulpa, R. and Multon, F. (2016). HIF3D: handwriting-inspired features for 3d skeleton-based action recognition, in *23rd International Conference on Pattern Recognition, ICPR 2016, Cancún, Mexico, December 4–8, 2016*, pp. 985–990.

Carmona-Duarte, C., Plamondon, R., Gomez-Vilda, P., Ferrer, M., Alonso, J. and Londral, A. R. M. (2016). Application of the lognormal model to the vocal tract movement to detect neurological diseases in voice, *Innovation in Medicine and Healthcare* **60**, pp. 25–35.

Chen, M., AlRegib, G. and Juang, B.-H. (2016). Air-writing recognition – part ii: Detection and recognition of writing activity in continuous stream of motion data, *IEEE Trans. on Human-Machine Systems* **46**, 3, pp. 436–444.

Diaz, M., Fischer, A., Ferrer, M. and Plamondon, R. (2018). Dynamic signature verification system based on one real signature, *IEEE Trans. on Cybernetics* **48**, 1, pp. 228–239.

Djeziri, S., Guerfali, W., Plamondon, R. and Robert, J. (2002). Learning handwriting with pen-based systems: computational issues, *Pattern Recognition* **35**, 5, pp. 1049–1057.

Djioua, M. (2007). *Contributions à la compréhension, à la généralisation et à l'application de la Théorie Cinématique dans l'analyse et la synthèse du mouvement humain*, Ph.D. thesis, Polytechnique Montréal.

Djioua, M. and Plamondon, R. (2008). A new algorithm and system for the extraction of delta-lognormal parameters, Tech. Rep. EPM-RT-2008-04, École Polytechnique de Montréal.

Ferrer, M., Diaz, M., Carmona, C. and Plamondon, R. (2020). iDeLog: Iterative dual spatial and kinematic extraction of sigma-lognormal parameters, *IEEE Trans. PAMI* **PP**, pp. 1–12.

Fischer, A. and Plamondon, R. (2017). Signature verification based on the kinematic theory of rapid human movements, *IEEE Trans. on Human-Machine Systems* **47**, 2, pp. 169–180.

Fischer, A., Plamondon, R., O'Reilly, C. and Savaria, Y. (2014). Neuromuscular representation and synthetic generation of handwritten whiteboard notes, in *Proc. 14th Int. Conf. on Frontiers in Handwriting Recognition*, pp. 222–227.

Flash, T. and Hogans, N. (1985). The coordination of arm movements: An experimentally confirmed mathematical model, *Journal of Neuroscience* **5**, pp. 1688–1703.

Galbally, J., Plamondon, R., Fierrez, J. and Ortega-Garcia, J. (2012a). Synthetic on-line signature generation. Part I: Methodology and algorithms, *Pattern Recognition* **45**, 7, pp. 2610–2621.

Galbally, J., Plamondon, R., Fierrez, J. and Ortega-Garcia, J. (2012b). Synthetic on-line signature generation. Part II: Experimental validation, *Pattern Recognition* **45**, 7, pp. 2622–2632.

Hollerbach, J. (1981). An oscillation theory of handwriting, *Biological Cybernetics* **39**, 2, pp. 139–156.

Kulpa, R., Multon, F. and Arnaldi, B. (2005). Morphology-independent representation of motions for interactive human-like animation, in *Eurographics.*

Lebel, K., Nguyen, H., Duval, C., Plamondon, R. and Boissy, P. (2017). Capturing the cranio-caudal signature of a turn with inertial measurement systems: Methods, parameters robustness and reliability, *Frontiers in Bioengineering and Biotechnology* **5**, 51, pp. 1–13.

Leduc, N. and Plamondon, R. (2001). A new approach to study human movements: The three dimensional delta-lognormal model, in *Proc. 10th Conf. of the International Graphonomics Society*, pp. 98–102.

Martin-Albo, D., Plamondon, R. and Vidal, E. (2014). Training of on-line handwriting text recognizers with synthetic text generated using the kinematic theory of rapid human movements, in *Proc. 14th Int. Conf. on Frontiers in Handwriting Recognition*, pp. 543–548.

Martin-Albo, D., Plamondon, R. and Vidal, E. (2015). Improving sigma-lognormal parameter extraction, in *Proc. 13th Int. Conf. on Document Analysis and Recognition*, pp. 286–290.

Müller, M., Röder, T., Clausen, M., Eberhardt, B., Krüger, B. and Weber, A. (2007). Documentation mocap database HDM05, Tech. Rep. CG-2007-2, Universität Bonn.

O'Reilly, C. and Plamondon, R. (2009). Development of a sigma-lognormal representation for on-line signatures, *Pattern Recognition* **42**, 12, pp. 3324–3337.

O'Reilly, C., Plamondon, R. and Lebrun, L.-H. (2014). Linking brain stroke risk factors to human movement features for the development of preventive tools, *Frontiers in Aging Neuroscience* **6**.

Plamondon, R. (1995a). A kinematic theory of rapid human movements. I. movement representation and generation, *Biological Cybernetics* **72**, 4, pp. 295–307.

Plamondon, R. (1995b). A kinematic theory of rapid human movements. II. movement time and control, *Biological Cybernetics* **72**, 4, pp. 309–320.

Plamondon, R. (1998). A kinematic theory of rapid human movements: III. kinetic outcomes, *Biological Cybernetics* **78**, 2, pp. 133–145.

Plamondon, R., Alimi, A., Yergeau, P. and Leclerc, F. (1993). Modelling velocity profiles of rapid movements: a comparative study, *Biological Cybernetics* **69**, 2, pp. 119–128.

Plamondon, R. and Djioua, M. (2006). A multi-level representation paradigm for handwriting stroke generation, *Human Movement Science* **25**, 4–5, pp. 586–607.

Plamondon, R., Feng, C. and Woch, A. (2003). A kinematic theory of rapid human movements: IV: A formal mathematical proof and new insights, *Biological Cybernetics* **89**, 2, pp. 126–138.

Plamondon, R. and Guerfali, W. (1998). The generation of handwriting with delta–lognormal synergies, *Biological Cybernetics* **78**, pp. 119–132.

Plamondon, R., O'Reilly, C., Rémi, C. and Duval, T. (2013). The lognormal handwriter: learning, performing, and declining, *Frontiers in Psychology* **4**, 945.

Rémi, C., Nagau, J., Vaillant, J. and Plamondon, R. (2017). Preliminary study of t0, a sigma-lognormal parameter extracted from young children's controlled scribbles, in *Proc. 18th Conf. of the International Graphonomics Society*, pp. 109–113.

Schindler, R., Bouillon, M., Plamondon, R. and Fischer, A. (2018). Extending the sigma-lognormal model of the kinematic theory to three dimensions, in *Proc. 1st Int. Conf. on Pattern Recognition and Artificial Intelligence*, pp. 748–752.

Woch, A. and Plamondon, R. (2004). Using the framework of the kinematic theory for the definition of a movement primitive, *Motor Control* **8**, 4, pp. 547–557.

Xia, L., Chen, C. and Aggarwal, J. (2012). View invariant human action recognition using histograms of 3d joints, in *Proc. IEEE Conf. on Computer Vision and Pattern Recognition Workshops (CVPRW)*, pp. 20–27.

https://doi.org/10.1142/9789811226830_0016

# Conclusions

*Réjean Plamondon, Angelo Marcelli and Miguel Ferrer, Editors*

This book, we hope, is the first of a series on the Lognormality Principle. Each of the 16 chapters brought in a specific facet of the elusive diamond of this principle, shining through potential applications with its underlined open questions. Each manuscript has its own story from a computer artist seeking for the perfect graffiti to a neuropsychologist looking for objective criteria to characterize children with ADHD or teachers interested in evaluating how pupils learn how to write, from computer scientists interested in reverse engineering the brain through several lines of abstract code to the challenge of finding specific parameters to pre-screen, monitor and follow up Parkinson's disease or mild traumatic brain injuries. Some chapters point out, for example, what are the common threads that link basic command gestures and complex signatures to the need to communicate with a machine and to be identified, others focus on the parameter extraction algorithms. Data analyses are performed in 1, 2, 2.5 and 3 dimensions using digitizing tablets, drawing boards, touch screens, cameras as well as networks of inertial sensors, sometimes combining multimodal signal processing. Integrated equipment for implementing psychophysical neuromotor tests, the use of interval arithmetics to capture the global solutions and the definition of new performance criteria for handwriting learning evaluation are among the novelties presented in this book.

All together, they show that the lognormality of velocity profiles of the end-effector emerges as a *distinctive* feature of human movements, and as such it provides a powerful tool to investigate how movements are planned by the brain and executed by the body and how fluency is learned through practice. Deviations from the lognormality can then be used to infer alterations in the functioning of the brain areas involved in movement planning, learning and execution.

From a different perspective, and because of its distinctiveness, the lognormality principle may prove to be a requirement for an autonomous agent that aims at moving like a human to facilitate their interactions with humans.

From a more fundamental perspective, each paper brings us back to basics, asking very challenging questions. How come that the lognormal function seems to have won the evolutionary competition in motor control through the establishment of proportionality relationships? Which driving forces, chemical, physical or biological have been at the root of this emerging phenomenon? Is lognormality a distinctive feature of human movements, or does it characterize the movements of other living organisms? Each chapter of the book is a candlelight that can help to traverse the dark forest of the unknown where knowledge is the ultimate goal.

But this book is also the result of many matches that have been built through the friendship of science. Sharing the same hope, pursuing the same quest is ultimately the only satisfactory answer to our need to discover. And in the end what remains, above equations, algorithms, measurements is the random chance that we have had, as authors, co-authors and co-editors, to meet each other at a given time, a given place on this erratic planet in search of peace and wisdom.

The interest of the scientific community toward the Kinematic Theory has exploded in the last ten years when we have decided to share our expertise about lognormal parameter extraction with a selection of international experts wishing to explore some specific applications in their own domain of expertise. A whole research program has been built on the national and international synergies created around the official collaboration agreements between Polytechnique Montréal and 7 national and 19 international research centers from 8 countries. Various teams have played a key role, providing both basic knowledge on how to successfully use our technology and define original research directions. And what is reported in this volume is the tip of the iceberg. Indeed, many other projects are going on actually and to give a flavor of their pertinence and diversity, we have invited many of these collaborators to write down a short abstract on their proposals. Some are starting, some are going on, others are at pause waiting for new students to get involved. These are listed below and described in the following appendix. It is expected that many of these will become chapters in a second book of this series.

**Projects about neurodegenerative diseases:**

Exploring the 3D Kinematics for Brain Stroke Rehabilitation
*Alicia Fornés, Asma Bensalah, Cristina Carmona-Duarte, Miguel A. Ferrer, Andreas Fischer, Josep Lladós, Eloy Opisso, Réjean Plamondon, Josep Maria Tormos*

Rationale of Use of the Lognormal Model in the Discovery of Clinical Biomarkers in Parkinson Syndromes
*Pierre Blanchet, Karina Lebel, Cristina Carmona Duarte, Nadir Faci, Réjean Plamondon*

Cycling Training and Parkinson Disease
*Alexandra Nadeau Arnaud Boré, Ovidiu Lungu, Julien Doyon, Réjean Plamondon*

Eye Saccades and Alzheimer Disease
*Joan Carbo, Nicole Vincent, Koi Zapula, Réjean Plamondon*

The Lognormal Model for the Early Diagnosis of Neurodegenerative Diseases
*Angelo Marcelli, Antonio Parziale, Rosa Senatore, Claudio De Stefano, Francesco Fontanella and Nicole Dalia Cilia*

**Projects on learning**

Characterizing the Graphic Skills of Children with the Sigma-Lognormal Model
*Ana Rita Matias Batalha, Helena Coradinha, Filipe Melo, Nadir Faci, Réjean Pamondon*

Assessing and Comparing Handwriting Skills of Kindergarten and First Grade Pupils with Kinematic Theory and Lognormal Models.
*Denis Alamargot, Marie-France Morin, Nadir Faci and Réjean Plamondon*

**Applications to sports**

Applying the Kinematic Theory to Detect Shoulder Neuromuscular Fatigue: Case of a Horizontal Oscillatory Movement
*Anaïs Laurent, Mickael Begon, Réjean Plamondon*

The Link Between the Lognormality Principle and the Monitoring of High-Performance Paracyclists
*François Bergeron, Miriam Paquette, Nadir Faci, Réjean Plamondon*

Applying the Kinematic Theory to Track Athletes' Performances Overtime
*Nadir Faci, Baptiste Galant, Réjean Plamondon*

**Handwriting applications of lognormality**

Lognormality for Signature Biometrics Analysis:
Age Effect and Forgeries Detection
*Marta Gomez-Barrero, Javier Galbally, Réjean Plamondon*
Sigma-Lognormal Model for Synthetic Handwriting Generation
*Chetan Ramaiah, Venu Govindaraju, Réjean Plamondon*

**Projects on livestock monitoring**

Studying the Principle of Lognormality in Livestocks
*Moises Diaz, Miguel A. Ferrer, Sergio Martin, Julio Rufo, Elena Carreton, Cristina Carmona, Anselmo Gracia*

**New algorithms**

Biophysical Interpretations of the Proportionality Hypothesis
*Ben Braithwaite, Mickael Begon, Réjean Plamondon*

Applying the Delta-Lognormal Model to Simple 3D Hand movements
*Saad Chidami, Mylène Archambault-Caron, Réjean Plamondon*

3D iDeLog: Sigma-Lognormal Model for 3D Human Movements
*Miguel Ferrer Ballester, Moises Diaz, Cristina Carmona Duarte, Réjean Plamondon*

Finally, we expect that this book will attract other groups all over the world from numerous disciplines to join us and continue to work on this challenging topic with such a huge social impact.

# Appendixes

https://doi.org/10.1142/9789811226830_0017

# Appendix 1

# Exploring the 3D Kinematics for Brain Stroke Rehabilitation

Alicia Fornés[1,*], Asma Bensalah[1,†], Cristina Carmona-Duarte[2,‡], Miguel A. Ferrer[2,§], Andreas Fischer[3,ǂ], Josep Lladós[1,**], Eloy Opisso[4,††], Réjean Plamondon[5,‡‡] and Josep Maria Tormos[4,§§]
[1]*Computer Vision Center, Computer Science Department, Universitat Autònoma de Barcelona, Barcelona, Spain*
[2]*Instituto Universitario para el Desarrollo Tecnológico y la Innovación en Comunicaciones, University of Las Palmas de G.C., Las Palmas, Spain*
[3]*Institute of Complex Systems, University of Applied Sciences and Arts Western Switzerland, Fribourg, Switzerland*
[4]*Institut Guttmann, Neurorehabilitation Institute, Camí de Can s/n 08916 Badalona, Spain*
[5]*Laboratoire Scribens, Département de génie électrique, Polytechnique Montréal, Montréal, Canada*
[*]*afornes@cvc.uab.es*, [†]*abensalah@cvc.uab.es*, [‡]*ccarmona@idetic.eu*, [§]*miguelangel.ferrer@ulpgc.es*, [ǂ]*andreas.fischer@unifr.ch*, [**]*josep@cvc.uab.es*, [††]*eopisso@guttmann.com*, [‡‡]*rejean.plamondon@polymtl.ca*, [§§]*jmtormos@guttmann.com*

## 1. Abstract

A brain stroke occurs when the flow of oxygen-rich blood to a portion of the brain is obstructed or when a blood vessel leaks or bursts. It is nowadays the second leading cause of death in Europe, and experts estimate that strokes will rise dramatically in the next 20 years due to an ageing population.[a] In Canada and United-States, it is the third cause of mortality, causing direct and indirect costs of Can$ 3.6 billion (evaluated in 2000) and US$ 68.9 billion (evaluated in 2009) [Lloyd *et al.* (2009)].[b] Moreover, 60% of the survivors have different degrees of disability, with a socio-economic impact of the first magnitude for the patient,

---

[a]Stroke Alliance for Europe Report, http://strokeeurope.eu/
[b]Fondation des maladies du coeur et de l'AVC du Canada, https://www.coeuretavc.ca/articles/les-liens-en-chiffres

their environment, the health system and society in general. Indeed, for surviving patients, the consequences can be various. Among subjects of 65 years old or more of the Framingham Study cohort, six-month post-stroke, 50% were suffering from hemiparesis, 30% were unable to walk without assistance, 26% needed assistance for the daily activities and 26% were institutionalized. Also, aphasia, depression, incontinence, sensitive deficits, social integration difficulties, and vision problems were frequent symptoms (between 15 % and 40 %) [Kelly *et al.* (2003)]. Thus, it is crucial to find personalized and suitable treatments during stroke rehabilitation.

The Kinematic Theory of Rapid Human Movements [Plamondon *et al.* (2013)] provides a mathematical description of the movements made by individuals, reflecting the behavior of their neuromuscular system. It has demonstrated a great potential for monitoring neuromuscular diseases, among others, in terms of the alteration of the ideal parameters. Concerning strokes, O'Reilly *et al.* [O'Reilly *et al.* (2014)] showed that brain stroke risk factors can be associated with the deterioration of many cognitive and psychomotor characteristics. Indeed, the analysis of a computerized battery of nine neuromuscular tests administered to 120 subjects with or without stroke risk factors (obesity, diabetes, hypertension, hyper-cholesterolemia, cigarette smoking, and cardiac disease) was performed using the Kinematic Theory. Logistic regression and linear discriminant analysis with leave-one-out cross-validation revealed that the probability of presence of brain stroke risk factors could be assessed by computing area under the receiver operating characteristic curve (ROCAUC). The results supported the feasibility of the proposed methodology and its potential application for the development of brain stroke prevention tools. However, the use of the Kinematic Theory in monitoring rehabilitation processes is still a challenge: it requires to collect and to analyze the movement data using robust, efficient and task oriented lognormal parameter extraction algorithms. So far, such algorithms have focused mainly on 1D and 2D movements in a controlled scenario, e.g. pen movements on a tablet computer. This constraint makes the approach unrealistic as a worldwide tool for treating brain stroke rehabilitation in a more natural environment.

As illustrated in Fig. 1, the main goal of the project '3D Kinematics for Remote Patient Monitoring'[c] is to explore the powerfulness of the Kinematic Theory for continuous 3D movements with smartwatches to provide clinicians an objective estimator of movements and the improvement of the patients' motor abilities during brain stroke rehabilitation. Smartbands and smartwatches are

---

[c]RPM3D (ATTRACT EU): http://dag.cvc.uab.es/patientmonitoring/about/

increasingly popular and non-intrusive, which make us propose the arm movements as a source to extract the lognormal patterns. For this purpose, it is necessary to adapt the recently proposed 3D algorithms [Schindler *et al.* (2018); Fischer *et al.* (2019)] to a sequence of continuous 3D movements recorded by smartwatches in unconstrained scenarios.

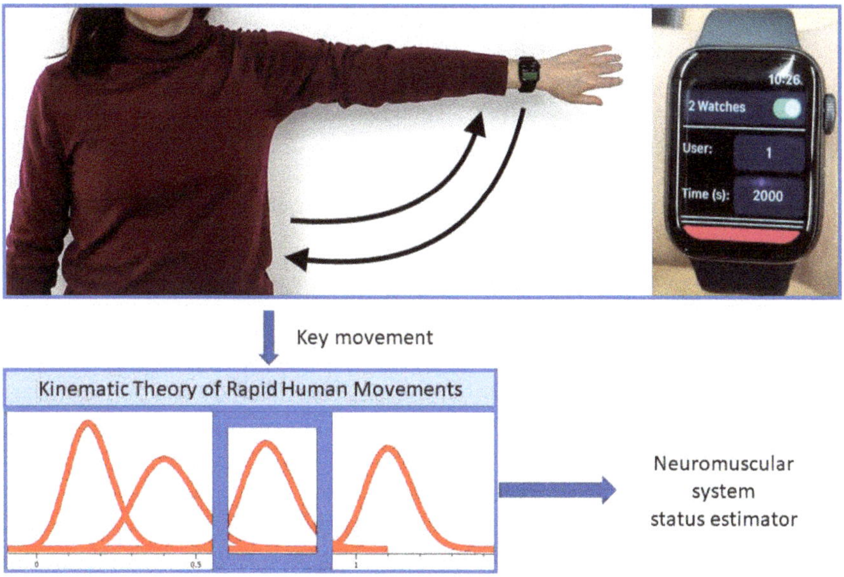


Fig. 1. A typical test.

The challenges are the following. First, the use of sensors from consumer devices instead of clinical devices can decrease the quality of the data for the application of the kinematic model. Secondly, the extraction of the model parameters from the continuous 3D movement sequences from the patient's arm movements for their posterior analysis. This involves the detection and segmentation of the meaningful 3D strokes (gestures) in uncontrolled environment conditions.

To validate our research, we will implement a proof of concept in a real case scenario for monitoring the early stages of brain stroke rehabilitation at the Guttmann Institute. Concretely, we will analyze and compare the kinematic model parameters extracted from sensor's data recorded by smartwatches from both healthy individuals and stroke patients.

In the mid-long term, our work aims to have a great impact on homecare telehealth tasks. The integration of an analytic tool in a consumer and affordable technology such as smartwatches could be used for continuous remote patient monitoring in the rehabilitation stages and during the routine daily lives of patients, improving the medical efficiency and reducing healthcare costs.

## References

A. Fischer, R. Schindler, M. Bouillon and R. Plamondon. Modeling 3D Movements with the Kinematic Theory of Rapid Human Movements (2021). In R. Plamondon, A. Marcelli and M. Ferrer (Eds.), *The Lognormality Principle: Applications for e-Security, e-Health and e-Learning*. World Scientific Publishing, pp. 327–342.

M. Kelly-Hayes, A. Beiser, C. S. Kase, A. Scaramucci, R. B. D'Agostino and P. A. Wolf. The influence of gender and age on disability following ischemic stroke: the Framingham study, *Journal of Stroke and Cerebrovascular Diseases*. 12(3), 119–126 (2003).

D. Lloyd-Jones, R. Adams, M. Carnethon, G. De Simone, T. B. Ferguson, K. Flegal, E. Ford, K. Furie, A. Go *et al.*, Heart disease and stroke statistics-2009 update: a report from the american heart association statistics committee and stroke statistics subcommittee, *Circulation*. 119(3), 480–486 (2009).

C. O'Reilly, R. Plamondon and L.-H. Lebrun, Linking brain stroke risk factors to human movement features for the development of preventive tools, *Frontiers in Aging Neuroscience*. 6, 150 (2014).

R. Plamondon, C. O'Reilly, C. Rémi and T. Duval, The lognormal handwriter: learning, performing, and declining, *Frontiers in Psychology*. 4, 945 (2013).

R. Schindler, M. Bouillon, R. Plamondon and A. Fischer. Extending the sigma-Lognormal model of the kinematic theory to three dimensions. *Proceedings of ICPRAI 2018. First International Workshop on the Lognormality Principle and its Applications*, Montréal, Canada, pp. 748–752.

https://doi.org/10.1142/9789811226830_0017

# Appendix 2

# Rationale of the Use of the Lognormal Model in the Discovery of Clinical Biomarkers in Parkinson Syndromes

Pierre J. Blanchet[1,2,*], Karina Lebel[3,†], Cristina Carmona-Duarte[4,‡] and Réjean Plamondon[5,§]
[1]*Dept. of Stomatology, Faculty of Dental Medicine, University of Montreal Montreal (QC), Canada*
[2]*Dept. of Medicine, University of Montreal Hospital Centre (CHU Montreal) Montreal (QC), Canada*
[3]*Faculty of Engineering, University of Sherbrooke, Sherbrooke (QC), Canada;*
[4]*Instituto Universitario para el Desarrollo Tecnológico y la Innovación en Comunicaciones, University of Las Palmas de G.C., Las Palmas de G.C., Spain*
[5]*Laboratoire Scribens, Département de génie électrique, Polytechnique Montréal, Montréal (QC), Canada*
[*]*pierre.j.blanchet@umontreal.ca,* [†]*Karina.Lebel@USherbrooke.ca,*
[‡]*criscarmonad@gmail.com,* [§]*rejean.plamondon@polymtl.ca*

This appendix introduces how the lognormal model can be applied in the discovery of clinical biomarkers in Parkinson syndromes.

## 1. Introduction

A biomarker is defined as "any substance, structure, or process that can be measured in the body or its products and influence or predict the incidence of outcome or disease" [WHO, 2001]. In spite of world-wide sustained investments made in the last 20 years in the exploration of a variety of indicators, advances in the field of Parkinson's disease (PD) remain impeded by the lack of a reliable, widely available and relevant biomarker for its early diagnosis, capable of discriminating clinically similar conditions (phenocopies) that adversely affect the basal ganglia pathways intrinsic to motor function. Thus, up to 20% of patients recruited in clinical studies are found not to display the typical presynaptic nigrostriatal binding loss on neuroimaging studies of the dopamine transporter density. Amongst different types of biomarkers, much effort focused on biomarkers for early diagnosis (state markers), even during the premotor stage of the illness, in order to promote early neuroprotective interventions. Studies involved clinical parameters, neuroimaging ligands, key patho-physiologic mechanisms, as well as -omic (genomic, proteomic, metabolomic) and epigenetic (microRNA) approaches in various biospecimens. Thus far, none of these single modalities yielded a positive biomarker, justifying current cost-ineffective strategies combining approaches [Südmeyer *et al.*, 2011]. The validation process in biomarker research remains difficult due to lack of pathological confirmation.

Following the lognormal primitive concept [Woch *et al.*, 2011], it is now possible to analyze complex motor output through the summation of lognormal vectors necessary over time to complete a given task. The basic mathematical convergence toward lognormality would then interpret handwriting learning as a migration toward an ideal representation of perfectly mastered movements, whilst aging and health problems constitute a progressive divergence from lognormality. Very early in PD and even in its premotor stage, subtle changes in handwriting and graphical tasks, fine motor control, balance and gait, and vocal control, difficult to appreciate clinically, may be captured with instrumental neuromotor assessment methods. Several studies used a graphic tablet and digitizer pen to assess bradykinesia (movement time, velocity), micrographia (writing size), and tremor, yielding significant changes in PD relative to healthy controls [Van Gemmert *et al.*, 2003; Smits *et al.*, 2017]. The complementary motor tasks examined are based on specific aspects of human motor control, including motor sequencing, visuomotor integration, agonist-antagonist muscle coupling, and biomechanic oscillatory coupling. The lognormal velocity patterns generated, which are the results of an asymptotic convergence, can be interpreted as the ideal behavior of subjects in perfect control of their movements, which reflect the integrity of the entire neuromuscular circuitry recruited for the performance of the task. This circuitry is distinctly altered in each disease process, thereby resulting in a potentially identifiable motor output signature.

## 2. Work Description

The lognormal principle has found applications in the kinematic analysis of distal upper limb, axial, or vocal tasks. For instance, in graphical tasks with a digitizer pen (series of interconnected loops or letter "l"), one can automatically segment the output and generate average lognormal parameters such as stroke length and duration, normalized jerk, maximal linear velocity, number of lognormal primitives required to reconstruct the entire motor trajectory, signal-noise ratio, and irregularities, as dependent variables. Thus, the less smooth and regular performance is readily apparent in PD patients, who typically require a greater number of lognormal primitives to complete each task relative to controls [Van Gemmert *et al.*, 2013]. The lognormal approach has also been used to assess the change in relative head-to-trunk angle and relative head-to-trunk angular velocity profile achieved during the turn in the 10-m Timed-Up-and-Go task [Lebel *et al.*, 2017, 2018]. In that task, the head and trunk are viewed as two distinct neuromuscular systems having their own lognormal impulse response, and the model allowed to accurately capture the cranio-caudal signature of the turn and assessed its efficiency.

Another application of great interest is the use of novel lognormal features to discriminate parameters of normal and pathological voices [Carmona-Duarte et al., 2017]. Akin to graphical tasks, the phonatory database consists of different tasks under distinct motor control, generated in separate blocks, requiring subjects to naturally speak in a microphone connected to a tablet. For example, participants sustain one of three vowels in succession (e.g., \aaaaa\iiiii\ooooo\) for 4 s each, with a brief pause between each vowel. They then generate a repetitive 5-s vocal sequence rapidly alternating (oscillating) between two vowels (e.g., \a\i\a\i\a\i\), followed by another repetitive 5-s triangular sequence between three vowels (e.g., \a\i\o\a\i\o\a\i\o\). Measured and estimated lognormal curves fit perfectly and distinguish the performance of neurologically intact volunteers and Parkinson's disease patients. It is unknown at present whether this analysis will generate an early state biomarker with sufficient discriminative power to tease out disease phenocopies.

## 3. Discussion

Much remains to be done before the implementation of the lognormal model in PD research. The cohort size must be sufficient for a given lognormal parameter in order to make a sound analysis of the performance of each task in the early discrimination of PD from its phenocopies. As in other biomarker research efforts, instrumental tasks involving different body segments and different neuromuscular responses should be tested in combination to determine whether axial vs. appendicular, or distal vs. proximal, motor performance is best predictive of the diagnosis.

Ultimately, two or more lognormal parameters used in conjunction may be necessary to achieve sufficient power. The validation process ideally requires to conduct such experiments in two separate patient cohorts. Short of pathological confirmation, the clinical diagnosis should be based on as many tools as possible, ideally including cardiac or neuroimaging modalities, and confirmed by the test of time, for instance 5 years after the onset of symptoms. Since many other approaches have failed thus far to provide a promising biomarker in PD, expectations are high that the lognormal model will fill this gap and propose a powerful, widely available, costefficient method to assist the research and development in PD and related syndromes.

## References

Carmona-Duarte C., Gomez-Vilda P., Plamondon R., Rodellar-Biarge V., Palacio-Alonso D., Ferrer-Ballester M.A., Alonso-Hernandez J.B. Phonatory and articulatory correlates in kinematic neuromotor degeneration. *Converging Clinical and Engineering Research on Neurorehabilitation 2017*;15:203–208.

Lebel K., Nguyen H., Duval C., Plamondon, R., Boissy, P. Capturing the cranio-caudal signature of a turn with inertial measurement systems: methods, parameters robustness and reliability. *Front Bioeng. Biotechnol.* 2017; 5:51. doi 10:3389/fbioe.2017.00051.

Lebel K., Duval C., Nguyen H.P., Plamondon, R., Boissy, P. Cranio-caudal kinematic turn signature assessed with inertial systems as a marker of mobility deficits in Parkinson's disease. *Front Neurol.* 2018 Jan 29; 9:22. doi: 10.3389/fneur.2018.00022. eCollection 2018.

Simits E.J., Tolonen A.J., Cluitmans L., van Gils, M., Conway, B.A., Zietsma, R.C. Graphical tasks to measure upper limb function in patients with Parkinson's disease: validity and response to dopaminergic medication. *IEEE J. Biomed. Health Informatics* 2007; 21:283–289.

Südmeyer M., Antke C., Zizek T., Beu, M., Nikolaus, S., Wojtecki, L., Schnitzler, A., Müller H-W. Diagnostic accuracy of combined FP-CIT, IBZM, and MIBG scintigraphy in the differential diagnosis of degenerative parkinsonism: a multidimensional statistical approach. *J. Nucl. Med.* 2011; 52:733–40.

Van Gemmert A.W.A., Adler C.H., Stelmach G.E. Parkinson's disease patients undershoot target size in handwriting and similar tasks. *J. Neurol. Neurosurg. Psychiat.* 2013; 74:1502–1508.

Van Gemmert A., Plamondon R., O'Reilly C. Using the Sigma-lognormal model to investigate handwriting of individuals with Parkinson's disease. *Proc. 16th Biennial Conf. on the International Graphonomics Society*, Nara, Japan, June 2013, pp. 119–122.

WHO International Programme on Chemical Safety Biomarkers in Risk Assessment: Validity and Validation. 2001. Retrieved from http://www.inchem.org/documents/ehc/ehc/ehc222.htm.

Woch A., Plamondon R., O'Reilly C. Kinematic characteristics of bidirectional delta-lognormal primitives in young and older subjects. *Human Mov. Sci.* 2011; 30:1–17.

https://doi.org/10.1142/9789811226830_0017

# Appendix 3

## Cycling Training and Parkinson Disease

Alexandra Nadeau[1,*], Arnaud Boré[1,†], Ovidiu Lungu[1,‡], Julien Doyon[1,§] and Réjean Plamondon[2,ǂ]
[1]*Centre de recherche de l'Institut universitaire de gériatrie de Montréal, Montréal, Canada*
[2]*Laboratoire Scribens, Département de Génie électrique, Polytechnique Montréal, Montréal, Canada*
**alexandra.nadeau.3@umontreal.ca; †arnaud.bore@gmail.com; ‡o.lungu@umontreal.ca; §julien.doyon@mcgill.ca; ǂrejean.plamondon@polymtl.ca*

### 1. Introduction

A few studies have suggested that aerobic exercise, like stationary cycling, could be an effective mean by which to improve specific upper limb functions in Parkinson's disease patients [Ridgel *et al.*, 2012]. This assumption has not been tested directly yet and we thus investigated whether the Kinematic Theory [Plamondon, 1995a, 1995b, 1998; Plamondon *et al.*, 2003] could help in assessing the efficiency of exercise in improving upper limb functioning in PD [Nadeau *et al.*, 2018]. Eighteen PD patients and twenty healthy adults (HA), aged between 40 and 80 years old, took part in the study. They were right-handed, sedentary, neurologically intact. HA participants were matched with PD patients at the group level with respect to sex distribution, age, number of years of education as well as cognitive and fitness levels.

### 2. Method

The duration and frequency of aerobic exercise training (AET) sessions were set at 12 weeks, 3 times per week. Time of each exercising began at 20 minutes and 60% of intensity and was then increased by steps of 5 minutes and 5% of intensity every week, until participants reached 40 minutes of training at 80% intensity. The kinematic properties of the upper limb movement were assessed with a target directed fast simple reaction time task using the Delta-Lognormal model [Djioua and Plamondon, 2009]. This task employs a digitizer and an electronic display to present visual stimuli, participants being asked to draw straight lines on the graphic tablet, using a non-inking pen. The task consisted in executing simple arm movements in response to a visual stimulus, each participant using his preferred hand in his preferred direction. Figure 1 depicts the test sheet used. It was placed under a transparent plastic fold on the tablet to reduce friction and indicate the starting dot and the grey target zone to subjects.

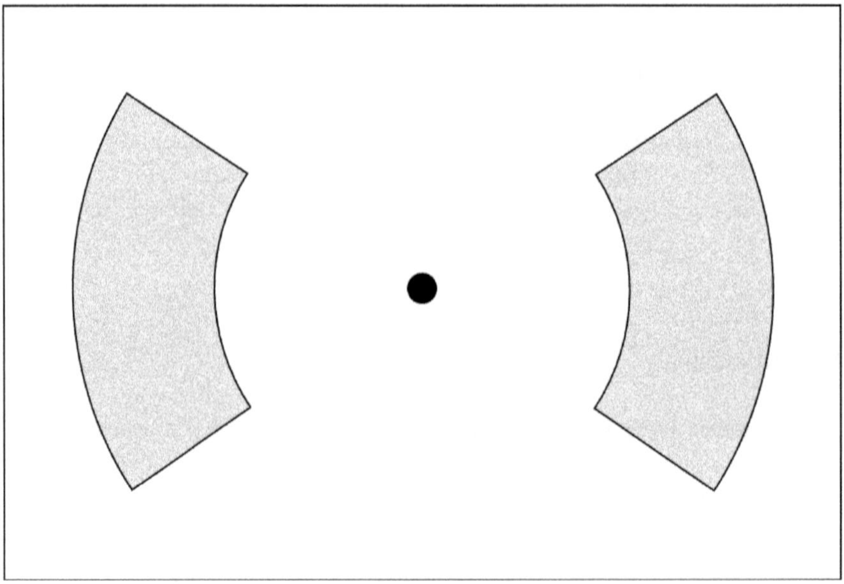

Fig. 1. The guide sheet used during this test. The black dot indicates the starting point and the grey areas are the target zone.

Participants' trials were used to extract several kinematic parameters of the movement based on the Kinematic Theory and its Delta-Lognormal model. A repeated model ANOVA (with pre- and post-AET assessments as repeated measures) was used to test the effect of cycling on primary and secondary outcomes in PD participants. In addition, measures at the baseline that showed significant inter-group differences (i.e. depression as measured with BDI) as well as age were used as covariates in a mixed-model ANCOVA; the latter allowing to test for group differences across assessments. We also tested for the effect of training within each group and for group differences, separately at the baseline and after AET, for all kinematic movement parameters. In order to account for the effect of multiple comparisons, statistical significance was adjusted using the Bonferroni method. Paired t-tests were used to evaluate AET-related changes in UPDRS sub-scores in PD participants. Analyses were conducted using SPSS 21.0 (IBM, Armonk, NY: IBM Corp.). The level of statistical significance for all tests was set at $p < 0.05$.

## 3. Preliminary Results

One main advantage of the Delta-Lognormal approach is that the reconstruction of the velocity profile gives access to the agonist and antagonist component of the synergy involved in the stroke generation. Figure 2 illustrates typical mean changes observed in PD subjects after exercise.

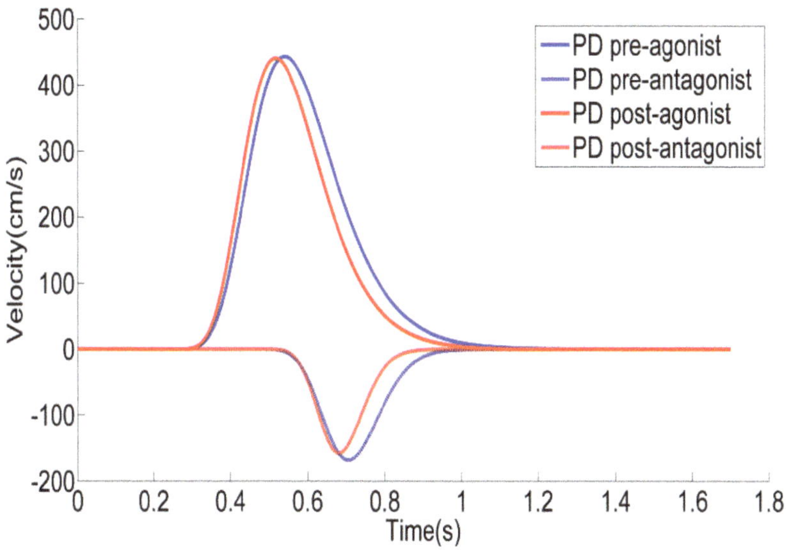


Fig. 2. The typical mean changes observed on PD subject after exercise.

In brief, significant difference between PD and HA groups were observed at the baseline in regards to three variables: D2, the amplitude of the antagonist input command ($p < 0.05$), ET, the elapsed time that is the delay between the moment where the visual stimulus was sent until the immobilization of the stylus on the digitizer ($p < 0.05$) and MT, the delay between the moment where the movement was initiated until the immobilization of the stylus on the tablet ($p < 0.01$). Generally speaking, our results suggest that PD patients exhibit a larger antagonist response, a longer elapsed time between the visual stimulus and the end of the movement, and a longer time of the stylus displacement.

Following a 12-week aerobic exercise training regimen, repeated measures ANOVA indicated that PD participants showed significant decreases on the D1 ($F1, 17 = 8.916$, $p < 0.01$), D2 ($F1, 17 = 5.039$, $p < 0.05$) and $\sigma 2$ ($F1, 17 = 6.553$, $p < 0.05$), those are the agonist command, the antagonist command and its response spread, respectively. The mixed ANCOVA model revealed a significant group*session interaction on D1 ($F1, 34 = 8.679$, partial $R2 = 0.203$, $p < 0.01$) and $\sigma 2$ ($F1, 34 = 5.359$, partial $R2 = 0.136$, $p < 0.05$), indicating a significant change for these two parameters only in PD patients after cycling training. While differences were observed at the baseline for D2, ET and MT, the groups did not differ significantly in post-AET comparisons for D2 and ET. These results indicate that an increase in response spread of the antagonist component or in movement time are associated with a decrease in inhibition effect. The original contribution of the current study is the use of a relatively simple kinematic task, which can be easily performed by the PD patients and, most importantly, could offer an objective clinical measure of the state of the global neurophysiological mechanisms involved in the motor control of the upper limb.

## References

Djioua, M., Plamondon, R. (2009). A New Algorithm and System for the Characterization of Handwriting Strokes with Delta-Lognormal Parameter, *IEEE Transactions on Pattern Analysis and Machine Intelligence*, *31*(11), 2060–2072. Consulté le 11 12, 2017, sur https://ncbi.nlm.nih.gov/pubmed/19762931

Nadeau, A., Lungu, O., Duchesne, C., Robillard, M-È., Boré, A., Bobeuf, F., Plamondon, R., Lafontaine, A-L., Gheysen, F., Bherer, L., Doyon. J. (2018). A 12-week cycling training regimen improves upper limb functions in people with Parkinson's disease. *Frontiers in human neuroscience*, *12*, 351.

Plamondon, R. (1995a). A kinematic theory of rapid human movements: Part I: Movement representation and generation. *Biological Cybernetics*, 72, 295–307.

Plamondon, R. (1995b). A kinematic theory of rapid human movements: Part II. Movement time and control. *Biological Cybernetics*, 72, 309–320.

Plamondon, R. (1998). A kinematic theory of rapid human movements: Part III. Kinetic outcomes. *Biological Cybernetics*, 78(2), 133–145.

Plamondon, R., Feng, C., Woch, A. (2003). A kinematic theory of rapid human movement. Part IV: a formal mathematical proof and new insights. *Biological Cybernetics*, 89(2), 126–138.

Ridgel, A. L., Peacock, C. A., Fickes, E. J., Kim, C. H. (2012). Active-assisted cycling improves tremor and bradykinesia in Parkinson's disease. *Archives of physical medicine and rehabilitation*, *93*(11), 2049–2054.

https://doi.org/10.1142/9789811226830_0017

# Appendix 4

# Eye Saccades and Alzheimer's Disease

Joan Carbo[1,*], Nicole Vincent[2,†], Koi Zapula[3,‡] and Réjean Plamondon[1,§]
[1]*Laboratoire Scribens, Département de Génie électrique, Polytechnique Montréal, Montréal, Canada*
[2]*LIPADE - Université de Paris, F75006 Paris, France*
[3]*Laboratoire IRIS, Physiopathologie de la Vision et Motricité Binoculaire, CNRS, Université Paris Descartes, France*
[*]*johan.carbo@gmail.com;* [†]*Nicole.Vincent@u-paris.fr;*
[‡]*zoi.kapoula@gmail.com;* [§]*rejean.plamondon@polymtl.ca*

## 1. Abstract

Many studies have investigated and reported alterations to oculomotor function and viewing behaviour as a result of Alzheimer's disease (AD) [Molitor *et al.*, 2015]. In this project, we investigate if the Kinematic Theory could be of any use to model and study eye saccades. For many years, it has been known that the delta-lognormal model could be used to model single eye saccade [Plamondon, 1995]. In this study, we want to first check if the model could be used to reconstruct log sequences of saccades as currently used in some Alzheimer disease tests, as illustrated in Fig. 1. These data were captured with an eye movement tracker (the 2008 version of the gaze-driven camera EyeSeeCam.

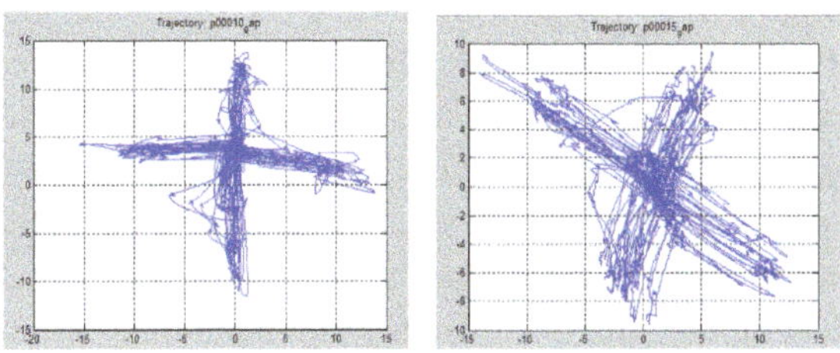

Figure 1.

To do so the patients have to follow some protocol. The task is to follow with the eye, a moving target on a computer display. First the target appears in the center of the screen during a delay of 1 500 ms and disappears during a random time lap and comes back on a peripheral point on the screen during a 1 000 ms period of time. The location can take four points, either on the left, on the right, on the top or the bottom of the screen, making a cross pattern.

The hypothesis is that the structure of the oculomotor trajectories is linked to the presence or not of some neurodegenerative dementia and even the type of dementia such as Alzheimer or dementia with Lewy bodies. Indeed, studies show the reaction time as well as the variability in the ocular saccades are varying with respect to pathology or non-pathology and with the type of pathology [Kapoula *et al.*, (2010)].

In this study, the data are captured from 12 patients that have been labeled by the experts in three groups according to clinical information (severe, mild, sane) as well as some quantitative information as age and MMSE (a neuro test). The file to be analyzed contains the sequences of geometrical coordinate of the eye trajectory. Some noise has to be removed to eliminate some aberrant values and this must be done in a way as coherent as possible with the expert interpretation.

In this study, the delta-lognormal model, did not fulfill our expectations in order to have a well reconstructed velocity profile. The principal problem which caused the non-success of the delta-lognormal model was due to the difficulty of isolating individual eye movements, as we tried at the beginning of this project. We thought that we would deal with straight and simple movements, which the delta-lognormal could be capable of fitting and representing. After looking the real complexity of the movement, we decided to go ahead with the analysis but now, using the sigma-lognormal model. Using the Script Studio executable software, provided by the Scribens lab, we proceed to do the analysis of the eye movements files.

The experimental task was evaluated using the axiom that the error between the reconstructed specimen and the original one should be as small as possible, and also the extracted solution should be somehow consistent. The error is easily evaluated using the SNR. The larger the SNR, the better is the reconstruction. The conclusion we reached is that the sigma-lognormal model complies with our expectations and it is able to reconstruct the original signal with a SNR over 25 dB, as opposed to delta-lognormal. Another important thing to take into account is the number of lognormals extracted using sigma-lognormal. Once we have a good SNR, we have to focus on the number of lognormals, since a smaller number is generally better. This is a topic we have to work on, and see if we can get some information about the saccade analyzed knowing the number of lognormals extracted. Figure 2 shows a typical result.

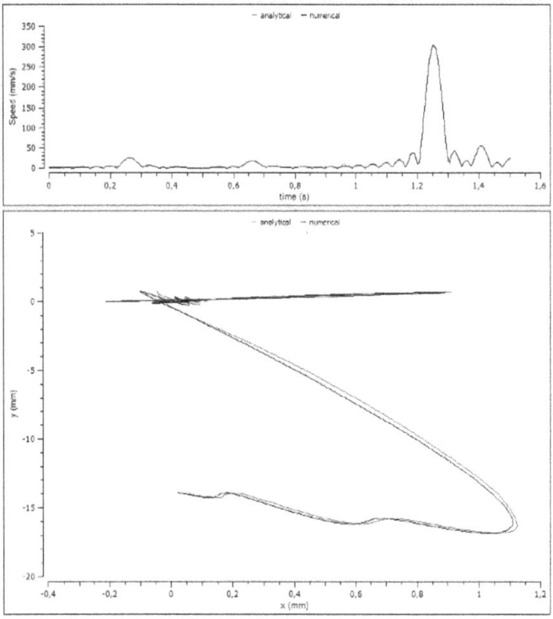


Figure 2.

Moreover, we tried our executable with greater and smaller signals than the one presented above. The aim of analyzing smaller signals is to study the behaviour of the extractor with only a unique saccade being this the predominant one of the signals.

In this experiment, we found that the extractor needs more information to reproduce the saccade. For example, for a typical signal, which gives us 27.88 dB of SNR with a 0.5 seconds duration, the same signal, now of 0.17 seconds of duration and with only the predominant lognormal lead to a 17 dB reconstruction with 2 lognormals. On the other hand, if we increase the window and we analyze the same signal but now with 1.5 seconds we get 34.54 dB of SNR and 39 lognormals. We introduce the maximum of signals that the executable able us (100 seconds from a total signal of 217 seconds). It took 4876 seconds to finalize the parameter extraction, and the results were so good. We reach a 28.64 dB of SNR with 1831 lognormals. This confirms that we can use the sigma-lognormal model to study the effect of Alzheimer's disease on eye saccades, as illustrated in Fig. 3, where in black is the original signal and the blue one, the reconstructed signal.

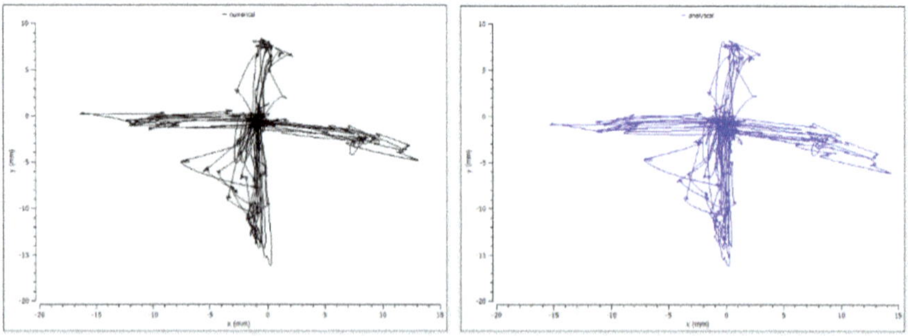

Figure 3.

## References

Kapoula, Z., Yang, Q., Vernet, M., Dieudonné, B., Greffard, S., Verny, M., (2010) Spread deficits in initiation, speed an accuracy of horizontal and vertical automatic saccades in dementia with Lewy bodies, *Frontiers in Neurology*, 1, 22.

Molitor, R.J., Ko, P.C., Ally, B.A. (2015) Eye movements in Alzheimer's disease. *J. Alzheimer's Dis.* 44(1):1–12. doi: 10.3233/JAD-141173.

Plamondon, R. (1995). Theory of Rapid Human Movements: Part 1: Movement Representation and Generation. *Biological Cybernetics 72*(4): 295–307.

https://doi.org/10.1142/9789811226830_0017

# Appendix 5

# The Lognormal Model for the Early Diagnosis of Neurodegenerative Diseases

Angelo Marcelli[*,‡], Antonio Parziale[*], Rosa Senatore[*], Claudio De Stefano[†], Francesco Fontanella[†] and Nicole Dalia Cilia[†]
[*]*DIEM – Natural Computation Lab, University of Salerno*
*Via Giovanni Paolo II, 132, 84084 Fisciano (SA), Italy*
[†]*DIEI, University of Cassino and Southern Lazio,*
*Via Di Biasio 43, Cassino, (FR) 03043, Italy*
[‡]*amarcelli@unisa.it*

## 1. Introduction

The quality of movements aimed at executing a task depends on the training time spent by the person in executing that task. During the initial learning stage force regulation requires more attention, performed movements are slower, characterized by reduced spatial precision and fluency. As learning proceeds motor performance increases, and the motor task is performed more efficiently. Being a fine motor task, also handwriting learning can be divided into two phases [Senatore and Marcelli, 2012], with motor performance depending on the learning phase [Graham and Weintraub, 1996; Rosenblum *et al.*, 2003; Smits-Engelsman *et al.*, 2001; Wann, 1987]. In particular, point-to-point movements performed by beginner writers become "embedded" into a single sequence drawn without any feedback by skilled writers that, being aware of time needed for drawing a stroke, can initiate the next stroke before the current one is completed [Plamondon and Maarse, 1989].

On the other hand, it has been observed that Parkinson's diseases (PD) patients show deficits in executing motor behaviors acquired before the onset of the disease, as well as in learning novel motor behaviors [Krebs *et al.*, 2001; Roncacci *et al.*, 1996; Worringham and Stelmach, 1990]. Similarities in the deficits observed in handwriting production between PD patients and beginner writers suggest that the fine tuning of the motor plan parameters involved in the handwriting production is impaired in PD.

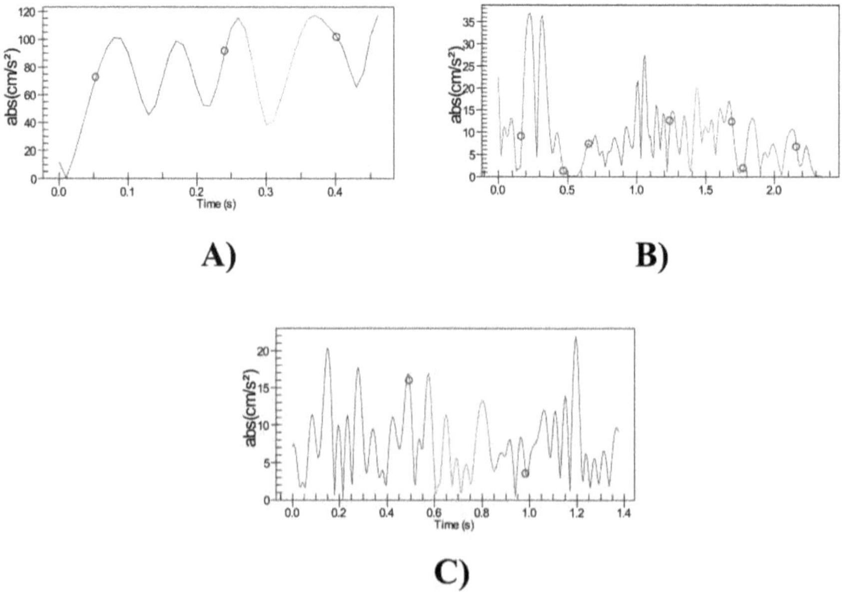


Fig. 1. Acceleration profile of the pen-tip movements. A-B) l-loops performed by a healthy subject using the habitual (A) and novel (B) motor plan; (C) l-loop performed by a PD patient using the habitual motor plan. Red circles represent the segmentation points.

To investigate this hypothesis, in a previous work [Senatore and Marcelli, 2019] we have designed an experiment in which we attempted to emulate the early learning stage of a stroke sequence: we asked to proficient healthy writers to produce a sequence of strokes representing a familiar shape, but using a motor plan that is different from that they are used to. In particular, we asked healthy controls to write a sequence of cursive 'l-loops' in two different ways: 1) drawing the loop of the letter in counterclockwise order (as they are used to); 2) drawing the loop of the letter in clockwise order (modifying the way they are used to). For the execution of the novel motor task the writers had to deal with a novel motor plan, of which they don't know the fine-tuned parameters for modulating the force of the fingers and the wrist in order to obtain the same shape.

By comparing the kinematic features of the handwriting samples produced by healthy subjects and PD patients, we showed that healthy controls outperform PD patients in producing familiar motor plans (in accordance with other studies). Indeed, acceleration profiles of samples drawn by healthy subjects using the habitual motor plan are smoother than those observed in samples drawn by PD, and mean acceleration is higher for healthy subjects than PD patients (Figs. 1(A) and 1(C)).

Furthermore, we identified performance similarities between controls and patients when healthy controls had to deal with a novel motor plan. Indeed, looking at the acceleration profiles associated to the novel l-loop written by the healthy groups we observed that they were characterized by multiple accelerations and decelerations, as observed in the acceleration profile of the handwriting performed by PD patients (Figs. 1(B) and 1(C)).

Based on those findings, this project aims at exploiting the lognormality principle to detect changes in the motor program that can be related to neurodegenerative diseases. For the purpose, we extract the motor program of a healthy subject by using the MPE algorithm [Parziale *et al.*, 2021] and perturb the parameters of the lognormal decomposition in such a way that the reconstructed velocity profile resembles that of a PD patient at different stages. Then, a machine learning technique based on grammar evolution [Fenton *et al.*, 2017; O'Neil *et al.*, 2001] is applied to infer a correspondence between the progression of the diseases and the changes in the parameter of the lognormal decomposition, with the ultimate goal of developing a predictive model able to detect the insurgence of the diseases as early as possible.

## References

Fenton, M., McDermott, J., Fagan, D., Forstenlechner, S., Hemberg, E. and O'Neill, M. (2017). Ponyge2: Grammatical evolution in python, *Proceedings of the Genetic and Evolutionary Computation Conference Companion*, pp. 1194–1201.

Graham, S. and Weintraub, N. (1996). A review of handwriting research: Progress and prospects from 1980 to 1994, *Educational Psychology Review*, 8(1), pp. 7–87.

Krebs, H., Hogan, N., Hening, W., Adamovich, S. and Poizner, H. (2001). Procedural motor learning in Parkinson's disease. *Experimental Brain Research*, 141(4), pp. 425–437.

O'Neill, M. and Ryan, C. (2001). Grammatical evolution. *IEEE Transactions on Evolutionary Computation*, 5(4), pp. 349–358.

Parziale, A., Parisi, R. and Marcelli, A. (2021). Extracting the motor program of handwriting from its lognormal representation, *The Lognormality Principle and Its Applications in e-Security, e-Learning and e-Health,* World Scientific, pp. 289–308.

Plamondon, R. and Maarse, F. J. (1989). An Evaluation of Motor Models of Handwriting, *IEEE Transactions on Systems, Man and Cybernetics,* 19(5), pp. 1060–1072.

Roncacci, S., Troisi, E., Carlesimo, G. A., Nocentini, U. and Caltagirone, C. (1996). Implicit memory in parkinsonian patients: evidence for deficient skill learning, *European Neurology*, 36(3), 154–159.

Rosenblum, S., Weiss, P. L. and Parush, S. (2003). Product and process evaluation of handwriting difficulties. *Educational Psychology Review*, 15(1), pp. 41–81.

Senatore, R. and Marcelli, A. (2012). A neural scheme for procedural motor learning of handwriting, *Proceedings of International Workshop on Frontiers in Handwriting Recognition,* IWFHR, pp. 659–664.

Senatore, R. and Marcelli, A. (2019). A paradigm for emulating the early learning stage of handwriting: Performance comparison between healthy controls and Parkinson's disease patients in drawing loop shapes, *Human Movement Science*, 65, pp. 89–101.

Smits-Engelsman, B. C. M., Niemeijer, A. S. and Van Galen, G. P. (2001). Fine motor deficiencies in children diagnosed as DCD based on poor grapho-motor ability, *Human Movement Science*, 20(1–2), pp. 161–182.

Wann, J. P (1987). Trends in the refinement and optimization of fine-motor trajectories: Observations from an analysis of the handwriting of primary school children, *Journal of Motor Behavior*, 19(1), pp. 13–37.

Worringham, C. J. and Stelmach, G. E. (1990). Practice effects on the preprogramming of discrete movements in Parkinson's disease. Journal of Neurology, *Neurosurgery & Psychiatry*, 53(8), 702–704.

https://doi.org/10.1142/9789811226830_0017

# Appendix 6

# Assessing and Comparing Handwriting Skills of Kindergarten and First Grade Pupils with Kinematic Theory and Lognormal Models*

Denis Alamargot[1,**], Marie-France Morin[2,†], Nadir Faci[3,‡] and Réjean Plamondon[3,§]
[1]*Laboratoire Cognition Humaine et Artificielle, Université Paris-Est Créteil Val de Marne, France*
[2]*Département d'études sur l'adaptation scolaire et sociale, Faculté d'éducation, Université de Sherbrooke, Canada*
[3]*Laboratoire Scribens, Département de génie électrique, Polytechnique Montréal, Canada*
[**]*Denis.Alamargot@upec.fr*, [†]*Marie-France.Morin@USherbrooke.ca*, [‡]*nadir.faci@polymtl.ca*, [§]*rejean.plamondon@polymtl.ca*

## 1. Introduction

From a cognitive perspective, handwriting requires the representation (allograph) of each letter to be retrieved from memory. The corresponding motor program must then be configured (shape, size and speed), and executed by arranging the symbols on the page [Alamargot and Morin, 2015]. Strongly dependent on lessons in school, learning to write is also constrained by gross and fine motor maturation [Blöte and Hamstra-Bletz, 1991].

One way of assessing young writers' level of motor maturation is to model the kinematics of their handwriting, based on the Kinematic Theory of rapid human movements [Plamondon *et al.*, 2003]. This modeling, confirmed both experimentally and physiologically, is the ideal mathematical description of the impulse responses of a neuromuscular system. Thus, it has been shown that the

*This study was conducted as part of the research project ANR-16-FRQSC – 0006-01 France-Quebec - NumEc entitled "Contribution of digital tools to handwriting research and learning to write in France and Quebec" (coordinators: Denis Alamargot & Marie-France Morin).

lognormality of pen speed reflects the behavior of writers who are in complete control of their movements [Plamondon *et al.*, 2014].

As a corollary, the migration toward lognormality in young writers can be studied by analyzing their hand trajectories in different situations of learning to write [Duval *et al.*, 2015]. In this context, the objective of this chapter is to probe the relevance of kinematic theory and its lognormal models for assessing the development of handwriting skills in young children between kindergarten and the first grade of elementary school.

## 2. Methodology

A database was created as part of the ANR-FRQSC France Quebec NumEc project. A total of 94 preschoolers and first graders (47 from France and 47 from Quebec) were asked to produce 30 pen strokes as fast as possible on a digitizing tablet (Wacom Cintiq 13HD). Script Studio, a software program developed in house [O'Reilly and Plamondon, 2009], extracted the lognormals contained in each stroke, and yielded two global parameters for each stroke (i.e. number of lognormal functions required to reconstruct the signal - nbLog; signal-to-noise ratio between the original speed profile and the reconstructed speed profile - SNR). Six other intrinsic parameters for each lognormal were extracted in order to describe the internal functioning of the neuromotor system (i.e. time required for the brain to produce a motor command - $t_0$; amplitude of the movement associated with each motor command - D; time taken to cover half the movement distance on a logarithmic scale - $\mu$; time taken by the neuromuscular system to respond to a command on a logarithmic scale - $\sigma$; starting angular direction of each stroke - $\theta_s$; ending angular direction of each stroke - $\theta_e$).

## 3. Results

We ran independent-samples t tests to compare the different parameters between the two groups of children (i.e. preschoolers vs. first graders). We applied Bonferroni corrections for multiple comparisons. Results show a significant difference in the two global parameters (nlLog et SNR) between the preschoolers and first graders, with lower values in first graders. It can be argued that children improve their motor control with age, and develop their calibration and control of handwriting in line with the principle of lognormality. In addition, there are significant differences in the neuromotor parameters ($t_0$, D, $\mu$, $\sigma$) reflecting better neuromotor control at both central and peripheral levels. These results highlight several significant differences in children's motor skills and changes in these skills with learning, in line with a developmental migration toward lognormality from preschool to first grade.

## 4. Conclusion

As can be seen in this study, and consistent with Duval *et al.* [2015], kinematic theory and its lognormal models appear to be entirely appropriate for studying written movements produced by kindergarten and first grade pupils. These encouraging initial results justify the use of analyses based on lognormal models to document in greater detail the effects of classroom teaching, and more particularly of different school programs, on the development of handwriting skills in elementary-school pupils [Morin and Alamargot, 2019].

## References

Alamargot, D., Morin, M.-F. (2015). Does handwriting on a tablet screen affect students' graphomotor execution? A comparison between Grades 2 and 9. *Human Movement Science, 44*, 32–41.

Blöte, A. W., Hamstra-Bletz, L. (1991). A longitudinal study on the structure of handwriting. *Perceptual and Motor Skills, 72*(3), 983–994.

Duval, T., Rémi, C., Plamondon, R., Vaillant, J., O'Reilly, C. (2015). Combining sigma-lognormal modeling and classical features for analyzing graphomotor performances in kindergarten children. *Human Movement Science*, 43, 183–200.

Morin, M.-F., Alamargot, D. (2019). Les entraînements graphomoteurs: Quelles pratiques, quels effets? *Approche Neuropsychologique des Acquisitions de l'Enfant (ANAE), 163*, 730–738.

O'Reilly, C., Plamondon, R. (2009). Development of a Sigma–Lognormal representation for on-line signatures. *Pattern Recognition, 42*(12), 3324–3337.

Plamondon, R., Feng, C., Woch, A. (2003). A kinematic theory of rapid human movement. Part IV: A formal mathematical proof and new insights. *Biological Cybernetics, 89*(2), 126–138.

Plamondon, R., O'Reilly, C., Galbally, J., Almaksour, A., Anquetil, E. (2014). Recent developments in the study of rapid human movements with the kinematic theory:Applications to handwriting and signature synthesis. *Pattern Recognition Letters, 35*(1), 225–235.

https://doi.org/10.1142/9789811226830_0017

# Appendix 7

# Characterizing the Graphic Skills of Children with the Sigma-Lognormal Model

Ana Rita Matias[1,*], Helena Coradinho[1,†], Filipe Melo[2,‡], Nadir Faci[3,§] and Réjean Plamondon[3,ǂ]
[1]*Department of Sport and Health, University of Évora*
[2]*Faculty of Human Kinetics, University of Lisbon*
[3]*Laboratoire Scribens, Département de génie électrique, Polytechnique Montréal, Montréal, Canada*
[*]*aritamatias@uevora.pt,* [†]*lena.coradinho@gmail.com,* [‡]*fmelo@fmh.ulisboa.pt,* [§]*nadir.faci@polymtl.ca;* [ǂ]*rejean.plamondon@polymtl.ca*

## 1. Abstract

One of the most discussed issues in handwriting is the question of when young children are (or not) ready to begin handwriting instruction. Several studies highlight the importance of early detection of graphomotor difficulties to better assist and remediate them in the first years of formal school. Also, it is necessary to understand how children control handwriting movements and its learning strategies.

Using sigma lognormal approach, in this study we aim to: i) characterize the variability of the graphic skills of children attending the last year of pre-school and the handwriting of children in the 2nd year of school; ii) analyze the impact of a graphomotor intervention program, in the graphic skills evolution.

Beyond perceptual readiness, linguistic readiness and maturity of pencils control, Beery (1989) argued that young children are not ready to develop handwriting until they correctly copy the first nine forms of the Beery-Buktenica Developmental Test of Visual-Motor Integration (6$^{th}$ edition) (VMI). Considering this assumption, sixty-three children attending the last year of pre-school (32 from experimental group; 31 from control group) will perform VMI on a digitizing tablet.

To address the issue related with handwriting, forty-seven second graders (24 from the experimental group and 23 from the control group) will perform The

Concise Assessment method for Children's Handwriting (BHK), the same conditions as mentioned above. A follow-up assessment will be performed six months after the end of graphomotor intervention program.

The handwriting learning process can be influenced by fine motor activities. To better understand this influence both experimental groups will participate in 16 sessions (twice a week) of a graphomotor intervention program, divided into small groups (6–8 children/group). Each session will last for 30 minutes.

All participants will be blind and randomly distributed in two groups (control and experimental group) considering the following exclusion criteria: children with suspected of neurological deficits, psychiatric and/or behavioral disorders; with uncorrected vision and hearing problems; referenced by the Special Education/National Early Intervention Service in Childhood; with one or more school retention; whose native language is not Portuguese; submitted to intervention in graphomotor skills or proposed to intervention less than1 year ago; whose participation in the intervention program is less than 80%; submitted to drug therapy that interferes with the study variables.

https://doi.org/10.1142/9789811226830_0017

# Appendix 8

# Applying the Kinematic Theory to Detect Shoulder Neuromuscular Fatigue: Case of a Horizontal Oscillatory Movement

Anaïs Laurent[1,*], Mickaël Begon[2,†] and Réjean Plamondon[1,‡]
[1]*Laboratoire Scribens, Département de génie électrique, Polytechnique Montréal, Montréal, Canada*
[2]*École de kinésiologie et des sciences de l'activité physique, Faculté de médecine, Université de Montréal*
[*]*anais.laurent@polymtl.ca*, [†]*mickael.begon@umontreal.ca*, [‡]*rejean.plamondon@polymtl.ca*

## 1. Introduction

Neuromuscular fatigue is an important risk factor for injuries. The ones affecting shoulders are recurrent among the whole population and are incapacitating, whether at work or in the daily life [Initiative, 2014]. Detection of fatigue at an early stage could prevent injuries. Existing methods for detecting fatigue are either subjective (*e.g.* questionnaires) or expensive (*e.g.* EMGs) and cannot discriminate the different types of fatigue (central *versus* peripheral *i.e.* from the central nervous system *versus* from the muscles) [Al-Mulla, Sepulveda and Colley, 2011]. The Kinematic Theory of Rapid Human Movement accurately describes someone's motor control from his end effector and a modification from the baseline reflects disorders [Laniel *et al.*, 2019]. The aim of this analysis was to apply the Kinematic Theory to study shoulder neuromuscular fatigue [Laurent *et al.*, 2020].

## 2. Methods

Twenty healthy participants performed, in a standing position, oscillatory movements between two 5×2 cm rectangles spaced 5 cm apart. The oscillations lasted 10 s and were performed at maximal speed. Data were captured on a Wacom Cintiq 13HD tablet positioned at shoulder's height. Participants had then to execute a shoulder external rotation fatigue on an isokinetic dynamometer. Fatigue was assessed when participants were not capable to reach 50% of their maximal force anymore. Another set of 10 s oscillations was realized after shoulder fatigue. Data were reconstructed using the Sigma-Lognormal model.

Parameters extracted were classified into four different categories: parameters related to the central system ($\Delta(t_0)$, $D$), parameters related to the peripheral system ($\mu$, $\sigma$), parameters related to both systems (*mode*, *median*, *time delay*, *response time*) and parameters describing the global state of the neuromotor system (*Nblog*, *SNR*, *SNR/Nblog*).

The effects of fatigue on all participants were assessed through a paired U-Mann-Whitney test. The statistical significance was $p < 0.0045$ (0.05/11), after Bonferroni corrections.

## 3. Results

Table 1. Evolution of parameters extracted on the oscillations pre and post fatigue.

| Parameters | Pre fatigue | Post fatigue | Significance | Effect size |
|---|---|---|---|---|
| **$\Delta(t_0)$** | 0.08±0.01 | 0.09±0.01 | 0.0004* | 0.72[a] |
| **D** | 125±23.7 | 126±26.4 | 0.0001* | 0.11 |
| **$\mu$** | -0.81±0.07 | -0.75±0.10 | 0.0001* | 0.83[b] |
| **$\sigma$** | 0.06±0.00 | 0.06±0.00 | 0.6666 | - |
| **Mode** | 0.53±0.04 | 0.56±0.06 | 0.0001* | 0.79[a] |
| **Median** | 0.53±0.04 | 0.57±0.06 | 0.0001* | 0.79[a] |
| **Time delay** | 0.53±0.04 | 0.57±0.06 | 0.0001* | 0.79[a] |
| **Response time** | 0.03±0.00 | 0.03±0.00 | 0.0001* | 0.81[b] |
| **Nblog** | 94.7±6.78 | 89.1±8.79 | 0.0002* | 0.80[b] |
| **SNR** | 28.4±1.40 | 27.6±1.26 | 0.0868 | - |
| **SNR/Nblog** | 0.21±0.02 | 0.23±0.03 | 0.0068 | - |

*Significant results ($p < 0.0045$). [a,b]Respectively, moderate and large effect size.

The results showed group effects of fatigue: $\Delta(t_0)$ was significantly higher after fatigue, so as $\mu$ with an effect size of 0.83. The *mode*, *median*, *time delay* and *response time* were significantly higher with a $p < 0.0001$. The *Nblog* was significantly lower after fatigue.

## 4. Discussion

Kinematic parameters extracted from horizontal oscillations tend to evolve similarly in all participants after fatigue. A general pattern is observable, which is encouraging for using the tablet as a clinical device for fatigue detection in the future. At least one parameter from each system was impacted by fatigue. This method is able to detect and to distinguish the different types of fatigue, whether central or peripheral. For the parameters describing the motor program execution, all parameters significantly increased with a moderate to large effect size. As those parameters have a more global impact on the motor control assessment, they seem to be more sensitive to detect neuromuscular fatigue than the other parameters. The number of Lognormals is lower after fatigue, meaning a general decline in the performance velocity and in the motor quality. This is in accordance with the literature, as the motor control and smoothness of a movement are modified with fatigue [Cortes *et al.*, 2014]. All these results tend to show that the Kinematic Theory offers interesting clinical possibilities in assessing shoulder neuromuscular fatigue.

## References

Al-Mulla, M. R., Sepulveda, F. and Colley, M. (2011). A Review of Non-Invasive Techniques to Detect and Predict Localised Muscle Fatigue. *Sensors*, *11*(4), 3545–3594. doi:10.3390/s110403545.

Cortes, N., Onate, J. and Morrison, S. (2014). Differential effects of fatigue on movement variability. *Gait & Posture*, *39*(3), 888–893. doi:https://doi.org/10.1016/j.gaitpost.2013.11.020.

Initiative, U. S. B. a. J. (2014). *The Burden of Musculoskeletal Diseases in the United States (BMUS)* (Third Edition ed.). Rosemont, IL.

Laniel, P., Faci, N., Plamondon, R., Beauchamp, M. H. and Gauthier, B. (2019). Kinematic analysis of fast pen strokes in children with ADHD. *Applied Neuropsychology. Child*, 1–16. doi:10.1080/21622965.2018.1550402.

Laurent, A., Plamondon, R., Begon, M. (2020). A proof of concept for pre-screening centraland peripheral shoulder fatigue using the Sigma-Lognormal model. *Frontiers in human neuroscience*. Special issue on Graphonomics and Your Brain on Art, Creativity, and Innovation. https://doi.org/10.3389/fnhum.2020.00171.

https://doi.org/10.1142/9789811226830_0017

# Appendix 9

# The Link Between the Lognormality Principle and the Monitoring of High-Performance Paracyclists

François Raymond[1,*], Nadir Faci[2,†] and Réjean Plamondon[2,‡]
[1]*Institut National du Sport du Québec, Complexe sportif, Stade Olympique, Montréal*
[2]*Laboratoire Scribens, Département de génie électrique, Polytechnique Montréal*
[*]*fraymond@insquebec.org*, [†]*nadir.faci@polymtl.ca*, [‡]*rejean.plamondon@polymtl.ca*

With growing support from the scientific community around the benefits of neuromuscular (NM) factors in endurance sports performance, the interest for NM monitoring is soaring. With this rising interest comes the realization that fatigue is a complex, elusive and multi-faceted mechanism and that current monitoring data streams are not always targeting neuromuscular markers *per se* [Stirling, 2012]. While research is almost unanimous regarding the positive role of strength training in optimizing endurance performance [Berryman, 2019], there is a growing body of evidence showing that benefits of strength training can be sub-optimal. Indeed when strength training is performed maximally, or not appropriately planned within the week, it can interfere with the development of other key physiological qualities [Mujika, 2016]. Therefore, there is a need to find simple, reliable measures of NM fatigue, which can be performed by all paracyclists, to maximize the effects of strength training without impairing aerobic training quality.

Traditionally, NM monitoring consists of dynamic tasks that involve maximal ballistic efforts or stretch-shortening mechanisms [Cormack, 2008; Gathercole, 2015]. A practitioner can pick from a host of possibilities: different styles of jumps or hops, short and maximal sprinting tasks, Olympic weightlifting movements or derivatives, traditional strength exercises, etc. While some of these possibilities are valid and reliable, they do not always match the reality of the cycling or paracycling world. Not only can the athletes be unfamiliar with some of these tasks, thus limiting the magnitude of insights, but most of the para-athletes are constrained by functional limitations that negate the possibility of accomplishing classical NM tests. Some of these tests also require the use of specialized and expensive equipment, which limit their utility in the decentralized reality of paracycling in Canada. Recent evidence shows that simple point-and-reach tasks could inform on central and peripheral fatigue [O'Reilly, Plamondon and Lebrun, 2014]. We hope to pursue this avenue in order to overcome some of the existing challenges, while also developing other novel NM markers of our own.

In order to achieve its goal of developing an accessible and reliable monitoring system, Cycling Canada is partnering up with Scribens laboratory to investigate the neuromuscular effects of training and cycling competition. The Scribens laboratory has been able to show, through the development of proprietary software and exciting studies, how the lognormality principle applies to a host of situations. Specifically, simple pencil strokes are acquired with a system that allows researchers to precisely establish the neuromotor patterns of the user. The process is based on the following rationale:

Writing is a complex operation from a neuromotor point of view. It requires the chain of different cognitive processes in order to produce the form that will be traced on a sheet [Plamondon, 1995a, 1995b; Plamondon, O'Reilly, Rémi and Duval, 2013]. It consists of a succession of pencil lines. For each movement, the agonist and antagonist muscles will act in parallel, grouped in neuromuscular networks. According to the Kinematic Theory of human movements, the impulse response of a network can be modeled by a lognormal law, which makes it possible to describe the speed of the drawing of a line [Plamondon, 1995a, 1995b]. According to the theory, the lognormal of the velocity profiles reflects the state of an individual in perfect control of his movements. From this hypothesis, the idea is to try to understand if the state of the brain and the arm influences the parameters of the models of these strokes.

Scribens laboratory created a simple system by using a Wacom CINTIQ*13HD graphic tablet sampling at 200Hz [Faci *et al.*, 2018a]. The testing protocol consists of 9 tests, each contributing to establish an individualized profile of the user [O'Reilly, Plamondon and Lebrun, 2014]. Lognormal parameters are extracted by iterative and robust algorithms [O'Reilly and Plamondon, 2009]. Visual or auditory cues prompt the subject to start a testing sequence, thus highlighting different areas of the brain.

It is our hope, based on Scribens' previous work, to show that the lognormality principle can apply to high-performance sports in a sound and reliable monitoring process. Indeed, their technology was able to, just to name a few accomplishments, discriminate between healthy patients and patients suffering from neuromotor deficiencies, highlight differences between ADHD children and a control group [Laniel, Faci, Plamondon, Beauchamp and Gauthier, 2018], notice a difference in signal quality between children who suffered head trauma and those that did not [Faci *et al.*, 2018b], show muscular fatigue after fatiguing tasks and suggest concussive damage in football players (yet to publish data).

With such a track record, we believe that we will be able to answer some crucial questions. For example, how tired is an athlete? Can we cross-reference data extracted during lognormality-constrained activities to other valid neuromuscular monitoring tools or results? Can we discriminate between peripheral and central fatigue? Do para-athletes respond to fatiguing tasks differently than able-bodied athletes?

Put simply, the goal of this project is to acquire and analyze data from high-performance para-cyclists in a variety of contexts. The first step will be to establish an individualized lognormal profile for each athlete, allowing the researchers to investigate the subtleties of this specific population displaying varied neuromotor characteristics. Then, we will longitudinally follow the evolution of their neuromotor profile through a competitive season, including training sessions, training camps and international travel. There is a scientific interest in both acute and chronic responses to training. Over time, monitoring the neuromotor response of athletes will allow us to establish a standard and contribute to the monitoring effort of tracking training loads and neuromuscular fatigue.

We believe that the development and implementation of reliable and simple NM monitoring for all our athletes will translate into improvements in both training quality and competition performance.

Improvement in training quality will come from the possibility to adopt a fluid approach to training and programming. Such an approach would lead to a paradigm shift in the way coaching is usually conducted in cycling. In effect, it would require the coach and the athlete to establish heuristics on how to adjust the training contents based on daily fluctuations of the athlete's NM state rather than following a weekly or monthly plan that typically consists of a mix of volume and intervals training. A simple example of what fluid periodization would look like in the cycling realm would be to accomplish difficult intervals during "good days" while easy volume trainings can be undertaken on suboptimal days. By aligning the training stimuli with athletes' daily NM state, we can expect better training adaptations that culminate in a summation of marginal gains. Over the course of a season, this strategy can create exponential gains given the fact that each training block rests on the previous one.

Performance in competition will be increased through the optimization of training and recovery as described above, but also by an optimization of tapering strategies. In fact, by tracking the time-course of the fitness-fatigue interplay, a coach could develop a systematic and refined plan to have an athlete peak on race day. Knowing the small margins between podium and average performances, such a tapering plan could increase the odds of an optimal neuromuscular state during crucial competitions and represent an elusive edge we are all chasing in high-performance sport.

## References

Berryman, N., Mujika, I., & Bosquet, L. (2019). Concurrent Training for Sports Performance: The 2 Sides of the Medal. *International Journal of Sports Physiology & Performance*, 14(3).

Cormack, S. J., Newton, R. U., & McGuigan, M. R. (2008). Neuromuscular and endocrine responses of elite players to an Australian rules football match. *International Journal of Sports Physiology and Performance*, 3(3), 359–374.

Faci, N., Boyogueno-Bidias, S.P., Plamondon, R., Bergeron, N. (2018). A New Experimental Set-up to Run Neuromuscular Tests. *Proceedings of ICPRAI 2018. First International Workshop on the Lognormality Principle and its Applications*, Montréal, Canada, pp. 753–757.

Faci N., Désiré N., Beauchamp, M.H., Gagnon, I., Plamondon, R. (2018). Lognormality in Children With Mild Traumatic Brain Injury: A Preliminary Pilot Study. *Proceedings of ICPRAI 2018. First International Workshop on the Lognormality Principle and its Applications*, Montréal, Canada, pp. 790–794.

Gathercole, R., Sporer, B., Stellingwerff, T., & Sleivert, G. (2015). Alternative countermovement-jump analysis to quantify acute neuromuscular fatigue. *International Journal of Sports Physiology and Performance*, 10(1), 84–92.

Laniel, Patricia, Faci, Nadir, Plamondon, Réjean, Beauchamp, Miriam H., Gauthier, Bruno. (2019). Kinematic analysis of fast pen strokes in children with ADHD. *Applied Neuropsychology: Child*: 16 pages. http://dx.doi.org/10.1080/21622965.2018.1550402.

Mujika, I., Rønnestad, B. R., & Martin, D. T. (2016). Effects of increased muscle strength and muscle mass on endurance-cycling performance. *International Journal of Sports Physiology and Performance*, 11(3), 283–289.

O'Reilly, C., Plamondon, R. (2009). Development of a Sigma-Lognormal Representation for On-Line Signatures. *Pattern Recognition, Special Issue on Frontiers in Handwriting Recognition*. 42(12): 3324–3337.

O'Reilly C., Plamondon, R., Lebrun, L.-H. (2014). Linking Brain Stroke Risk Factors to Human Movement Features for the Development of Preventive Tools. *Frontiers in Aging Neuroscience* doi: 10.3389/fnagi.2014.00150. 2014: 1–13. http://dx.doi.org/10.3389/fnagi.2014.00150.

Plamondon, R. (1995a). A kinematic theory of rapid human movements: Part I: Movement representation and generation. *Biological Cybernetics*, 72, 295–307.

Plamondon, R. (1995b). A kinematic theory of rapid human movements. Part II. Movement time and control. *Biological Cybernetics*, 72(4), 309–320.

Plamondon, R., O'Reilly, C., Rémi C., Duval, T. (2013). The lognormal handwriter: learning, performing and declining. *Frontiers in Psychology: Cognitive Science*. 2013: 1–14. http://dx.doi.org/10.3389/fpsyg.2013.00945.

Stirling, L. M., Von Tscharner, V., Fletcher, J. R., & Nigg, B. M. (2012). Quantification of the manifestations of fatigue during treadmill running. *European Journal of Sport Science*, 12(5), 418–424.

https://doi.org/10.1142/9789811226830_0017

# Appendix 10

# Applying the Kinematic Theory to Track Athletes' Performances Overtime

Nadir Faci[1,*], Baptiste Galant[1,†], Patrick Gendron[2,‡], France Brunet[2,§] and Réjean Plamondon[1,ǂ]

[1]*Laboratoire Scribens, Département de génie électrique, Polytechnique Montréal*

[2]*Clinique de physiothérapie et de médecine du sport, Université de Montréal*

**nadir.faci@polymtl.ca, †baptiste.galant@polymtl.ca,*

*‡patrick.gendron.1@umontreal.ca, §f.brunet@umontreal.ca,*

*ǂrejean.plamondon@polymtl.ca*

## 1. Introduction

Performance in sports is an important challenge and athletes are willing to use different tools to improve their performance. However, some tools are expensive, bulky and requires expertise to use it. Moreover, there is a need for tools that can evaluate the states of the athletes after sustaining injuries, whether it is concussions or upper body related. For this purpose, tools need to be developed to track the performance of the athletes and to detect injuries and fatigues. In this project, we want to evaluate the impact of practices using the Kinematic Theory of Rapid Human Movement [Plamondon, 1995a, 1995b, 1998]. This theory can be helpful to track the evolution of fatigue and upper-body injury in athletes. That is the target of this on-going study.

## 2. Methods

Seventy-five athletes from an USports American football team were recruited to participate in the study. Those athletes had to perform seven tests of drawing strokes [O'Reilly *et al.*, 2014]. The first test consisted of drawing a fast stroke to an ending zone after seeing a visual stimulus (a). We also have guide sheet for left-handed participants (a1). The four following tests were used with an audio stimulus and have a restricted target zone (b-c-d-e). This area is narrowed from one test to the next. Moreover, the distance to cover was also shrunk. The sixth test consisted of drawing a triangle after hearing the audio stimulus (f). The last test consisted of drawing oscillatory strokes between two areas as fast as possible (g). The participants needed to execute each test 30 times, except for the oscillation one. However, the oscillation lasted 10 seconds and has to be executed once. The following figure presents the guide sheet for the tests.

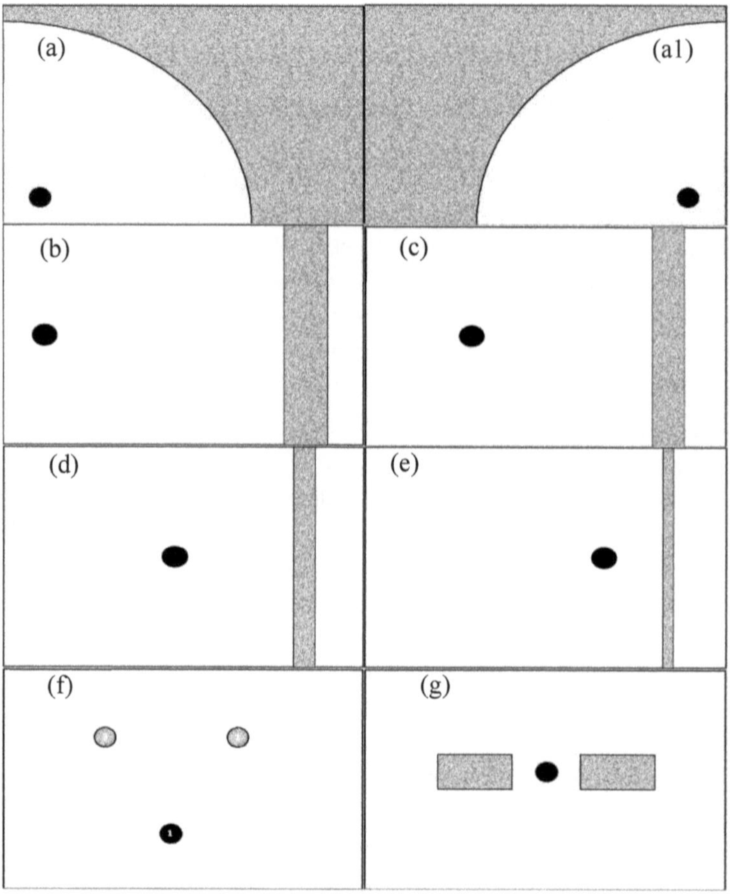


Fig. 1. The Guide sheets used for each test. The black dot is the starting point.

All the players executed the baseline tests, but only 20 players were kept for the rest of the following data collection. For the next step of the data collection, those 20 players were only executing the first test and the oscillation one. The tests were executed before and after their practices and before and after their games. The data were collected using the Lognometer, an in-house software using a Wacom Cintiq 13HD [Faci *et al.*, 2018].

## 3. Future Steps

Future steps will consist of reconstructing the data using the Sigma-Lognormal model. This model reconstructs the data using six parameters: two parameters describing the global state of the neuromotor system (SNR, NbLog) [Plamondon, *et al.*, 2013], two parameters related to the central system ($t_0$, D), two parameters related to the peripheral system ($\mu$, $\sigma$) [O'Reilly, *et al.*, 2013]. We will seek to see if the practice and the game affect the results and if there are stress-related impacts.

## References

Faci, N., Boyogueno Bidias, S., Plamondon, R. and Bergeron, N. (2018). A New Experimental Set-up To Run Neuromuscular Tests. *International Conference on Pattern Recognition and Artificial Intelligence (ICPRAI 2018)*, pp. 753–757. Montréal.

O'Reilly, C., Plamondon, R. and Lebrun, L.-H. (2014). Linking brain stroke risk factors to human movement features for the development of preventive tools. *Frontiers in Aging Neuroscience*, 6(3), 150. doi:10.3389/fnagi.2014.00150.

O'Reilly, C., Plamondon, R., Landou, M. K. and Stemmer, B. (2013). Using kinematic analysis of movement to predict the time occurrence of an evoked potential associated with a motor command. *European Journal of Neuroscience*, 32(2), 173–180.

Plamondon, R. (1995a). A kinematic theory of rapid human movements: Part I: Movement representation and generation. *Biological Cybernetics*, 72, 295–307.

Plamondon, R. (1995b). A kinematic theory of rapid human movements. Part II. Movement time and control. *Biological Cybernetics*, 72(4), 309–320. Retrieved 11 12, 2017, from https://ncbi.nlm.nih.gov/pubmed/7748960.

Plamondon, R. (1998). A kinematic theory of rapid human movements: Part III. Kinetic outcomes. *Biological Cybernetics*, 78(2), 133–145.

Plamondon, R., O'Reilly, C., Rémi, C. and Duval, T. (2013). The lognormal handwriter: learning, performing, and declining. *Frontiers in Psychology*, 4, 945.

https://doi.org/10.1142/9789811226830_0017

# Appendix 11

# Lognormality for Signature Biometrics Analysis: Age Effect and Forgeries Detection

Marta Gomez-Barrero[1,*], Javier Galbally[2,†] and Réjean Plamondon[3,‡]
[1]*Hochschule Ansbach, Germany*
[2]*European Commission - DG-Joint Research Centre, Italy*
[3]*Laboratoire Scribens, Département de génie électrique, Polytechnique Montréal, Canada*
[*]*marta.gomez-barrero@hs-ansbach.de*, [†]*javier.galbally@ec.europa.eu*, [‡]*rejean.plamondon@polymtl.ca*

## 1. Abstract

The Sigma-Lognormal model can help us detect anomalies in signatures due to different factors. Among them, we have studied two different aspects: age [Gomez-Barrero *et al.* (2013)] (i.e., how the subjects' signatures change as we get older, which in turn can lead to a decrease of the systems' recognition performance) and signature forgeries [Gomez-Barrero *et al.* (2015)] (i.e., presentation attacks in which a subject may be impersonated by an attacker who tries to imitate his/her signature).

### 1.1. *Sigmalognormality to analyse the age effect in signature biometrics*

In order to study the effects of age in signatures, we analysed how the Sigma-Lognormal model parameters evolved with time. We compared two population groups: younger subjects aged between 18 and 21, and older subjects aged between 59 and 72. As it can be observed in Fig. 1, the distribution of each parameter showed clear differences between both groups. Whereas the shape of the strokes (i.e., $\delta_d$ and $\delta_f$ parameters) remains almost the same in order to preserve our signatures, the number of strokes $N$ is higher for the elder group. As a consequence, since we tend to preserve

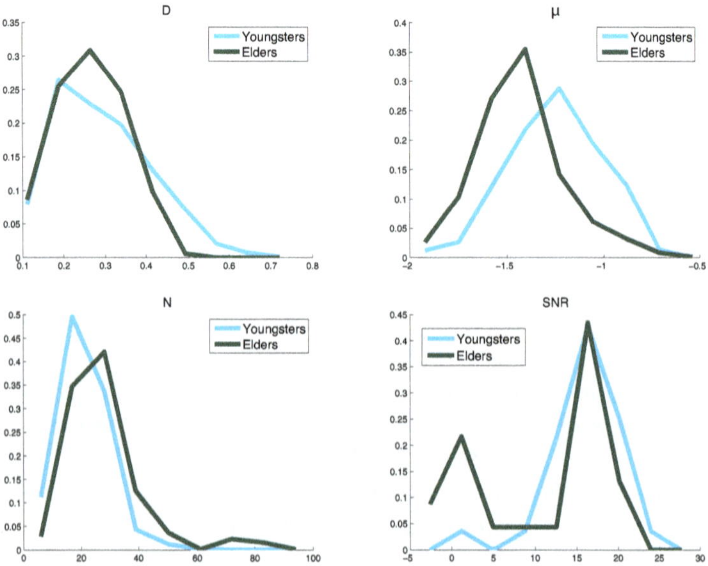


Fig. 1. Distributions of the age groups (young and elder) for selected parameters ($D$, $\mu$), for the number of strokes ($N$) and for the Signal to Noise Ratio (SNR).

the overall shape of our signature, the length $D$ of such strokes tends to be smaller for the elders, also requiring a lower logtime delay $\mu$, associated with the plannning time. This analysis also showed another issue associated with ageing: the SNR decreases with age, moving us away from the ideal lognormality of healthy young subjects, which support once again the lognormality principle [Plamondon *et al.* (2013)].

## 1.2. *Sigmalognormality for forgeries detection in signature biometrics*

The accurate detection of presentation attacks (PA) is nowadays a hot topic within the biometric community, and still remains a challenge for most biometric characteristics. This is specially the case for signature based systems, since a skilled impostor can easily forge our signature, thereby degrading the security of the system and its recognition performance.

Whereas a shallow analysis of offline signatures (i.e., only the static image) might be insufficient to detect skilled signature forgeries, a deep analysis of online (i.e., dynamic coordinate signals of the signing process) signatures can yield a good presentation attack detection (PAD) module.

To that end, the Sigma-Lognormal model can provide the necessary tools to detect the anomalies present in these PAs. We have therefore analysed how to combine the sigma lognormal parameters extracted from the signatures with state of the art online signature recognition systems to achieve accurate and secure biometrics recognition systems. In particular, given a probe input signature, four Sigma-Lognormal parameters ($N$, SNR, $\mu$, and $\sigma$) were extracted and a linear classifier based on Linear Discriminant Analysis (LDA) was used to output a PAD score (i.e., the score indicates whether the signature is a bona fide or a PA / forgery). This score was subsequently fused with the recognition score obtained with a Dynamic Time Warping (DTW) algorithm with a weighted sum. The global performance recognition analysis showed how the error rates were reduced up to 18% when skilled forgeries were present in the test set.

### 1.3. *Open challenges*

Whereas the aforementioned works have shed some light into two critical issues in signature based biometrics, there is still work to be done in these areas. First, a deeper analysis of age needs to be conducted on larger datasets, containing samples from more subjects and spanning across longer periods, ideally up to some years. Then, countermeasures and template update strategies need to be developed, in order to palliate the impact of ageing in the recognition performance.

Regarding the systems' security and their robustness to presentation attacks, more in depth analysis of the predicting capabilities of Sigma-Lognormal features will further help developing more accurate PAD modules. Using all the available parameters, or combinations thereof, with more sophisticated classifiers (e.g., random forests or the new deep learning paradigms) can help to solve this open challenge.

## References

M. Gomez-Barrero, J. Galbally, R. Plamondon, J. Fierrez and J. Ortega-Garcia. Variations of handwritten signatures with time: A Sigma-Lognormal analysis. In *Proc. Int. Conf. on Biometrics (ICB)* (06, 2013).

M. Gomez-Barrero, J. Galbally, J. Fierrez, J. Ortega-Garcia and R. Plamondon. Enhanced on-line signature verification based on skilled forgery detection using Sigma-LogNormal features. In *Proc. Int. Conf. on Biometrics (ICB)* (05, 2015).

Plamondon R., O'Reilly, C., Rémi, C., Duval, T. (2013). The lognormal handwriter: learning, performing and declining. *Frontiers in Psychology: Cognitive Science*. 2013: 1–14. http://dx.doi.org/10.3389/fpsyg.2013.00945.

https://doi.org/10.1142/9789811226830_0017

# Appendix 12

# Sigma-lognormal Model for Synthetic Handwriting Generation

Chetan Ramaiah[1,*], Venu Govindaraju[†] and Réjean Plamondon[‡]
[1]*Center for Unified Biometrics and Sensors, Department of Computer Science and Engineering, University at Buffalo*
[2]*Laboratoire Scribens, Département de génie électrique, Polytechnique Montréal*
[*]*cramaiah@salesforce.com*, [†]*govind@buffalo.edu*,
[‡]*rejean.plamondon@polymtl.ca*

## 1. Abstract

In this work, we want to develop a process for the synthetic generation of handwriting samples. Supervised machine learning for handwriting processing requires large amounts of annotated handwritten samples, and while there is no shortage of unlabeled handwriting samples, the annotations necessary for supervised learning are rare. A synthetic handwriting generation pipeline can be a cheaper alternative to manual annotation of handwriting samples, and the required annotations can be generated concurrently with the handwriting sample. However, synthetic generation of legible handwriting samples is a challenging problem. The generation process needs to satisfy certain criteria: (i) The samples must be legible, and (ii) The samples must resemble handwriting.

The sigma-lognormal model has been shown to represent complex human movement in the form of lognormal impulse responses. The model is represented as a synergy emerging from the interaction and coupling of many neuromuscular systems, each of which is believed to have a lognormal shape. In essence, the Sigma-Lognormal model can be used to obtain a parametric representation of a handwriting sample. Figure 1 shows handwriting samples and their sigma-lognormal representations.

Let us take a brief look at the process: First, an online handwriting sample, represented by a sequence of points on a two-dimensional plane and a timestamp,

[1]Work done while at University at Buffalo.

is obtained. Next, the sigma-lognormal representation of the handwriting is computed. Third, minor perturbations are added to the sigma-lognormal representation in order to generate samples with variations in handwriting style. The scale of the variations can control legibility, small changes will be very similar to the original sample, whereas very large variations can make the generated sample illegible. Adding controlled perturbations within a well-defined range has also been shown to be an effective handwriting CAPTCHA system. Figure 2 shows the effect of increasing perturbations to a handwriting sample.

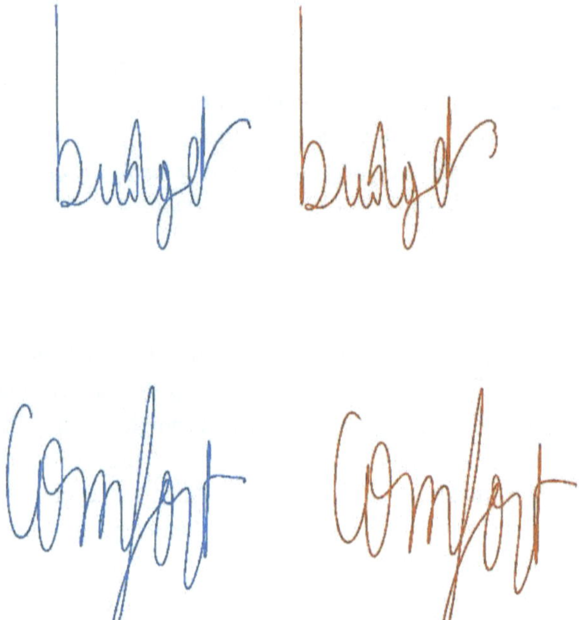

Fig. 1. Handwriting samples and their sigma-lognormal representations. The images on the left (in blue) are original handwriting samples, and the samples on the right (in orange) are rendered from their sigma-lognormal representation.

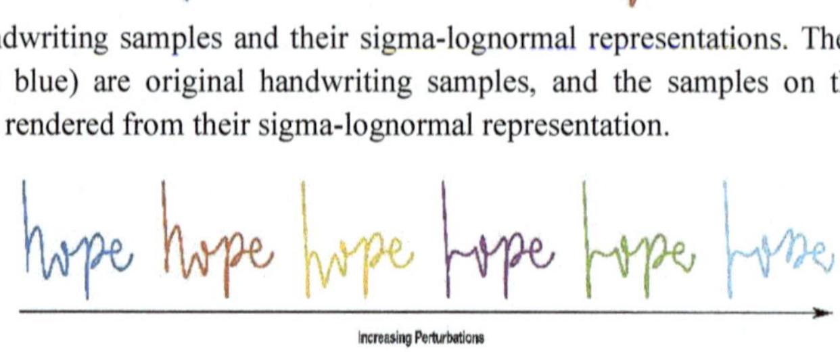


Fig. 2. The effect of adding perturbations to the sigma-lognormal parameters of a handwriting samples.

Let us now revisit the criteria for a synthetic handwriting generation process. The sigma-lognormal model is based on the Kinematic Theory of rapid human movements and is designed to generate parametric representations for complex human movements. Handwriting is a natural candidate for the model and several studies have shown the efficacy of the model to capture handwriting. Figure 1 shows the similarity in a handwriting sample and its sigma-lognormal representation. Adding perturbations in a carefully tuned range ensures legibility, and the sample still retains its handwriting like quality. in addition, the perturbations provide variability in samples.

The applications of this approach are numerous. Studies demonstrate that a carefully tuned range of perturbations make very effective handwritten CAPTCHAs. These are samples that are legible to humans, but not to automated recognition technology. A smaller range of perturbations can be applied to generate an unlimited amount of labelled data for training robust recognition systems. In addition, data generated from this technique can be used to in an adversarial training paradigm, making the handwriting processing system robust to adversarial attacks.

## References

O'Reilly, C. and Plamondon, R. (2009). Development of a sigma–lognormal representation for on-line signatures, *Pattern Recognition* **42**, 12, pp. 3324–3337.

Plamondon, R. and Djioua, M. (2006). A multi-level representation paradigm for handwriting stroke generation, *Human Movement Science* **25**, 4–5, pp. 586–607.

Ramaiah, C., Plamondon, R. and Govindaraju, V. (2014). A sigma-lognormal model for handwritten text captcha generation, in *2014 22nd International Conference on Pattern Recognition*, pp. 250–255.

Ramaiah, C., Plamondon, R. and Govindaraju, V. (2015). A sigma-lognormal model for character level captcha generation, in *2015 13th International Conference on Document Analysis and Recognition (ICDAR)*, pp. 966–970.

https://doi.org/10.1142/9789811226830_0017

# Appendix 13

# Studying the Principle of Lognormality in Livestock

Moises Diaz[1,*], Miguel A. Ferrer[2], Sergio Martin[2], Julio Rufo[1], Elena Carreton[2], Cristina Carmona[2] and Anselmo Gracia[2]
[1]*Universidad del Atlántico Medio, Las Palmas de Gran Canaria, 35017, Spain*
[2]*Universidad de Las Palmas de Gran Canaria, Spain*
[*]*moises.diaz@atlanticomedio.es*

The project covered in this appendix aims to verify the lognormal movements in pets, such as dogs or cats, and farms animals. Since both humans and animals share a similar neuromotor movement generation, our hypothesis relies on the idea that the Kinematic Theory of rapid movements could be used to describe the animal movements as well.

## 1. Introduction

In both humans and animals, the movement starts in the cognitive part of the brain. When a movement is initiated, neurons are activated and deliver an electrical impulse that spreads through the neuromotor system. Examples of these movements are walking, running, tail wags or head swinging. The Kinematic Theory of rapid movements models analytically these long and complex movements through an ordered and timely overlapped sum of lognormal functions as ultimate impulse response of a sequence of coupled subsystems [1]. Recent advances in these directions have improved the computational implementation of the theory and its associated Sigma-Lognormal model. One of the recent implementations is the iDeLog model [2], which is freely distributed for research purposes, under collaborative research agreement.

## 2. Preliminary Results

The objective of this research program is to verify the lognormality principle in animals' movements, mainly in ruminants which are the basis of the livestock farming industry. As preliminary proof of concept, we have analyzed the feasibility of this idea with pets. Basically, we have used a neuronMOCAP system, which houses an Inertial Measurement Unit. A neuron of this sensor has been introduced inside the harness attached to several dogs while they walk with their owner or just freely in a controlled space. Next, the neuronMOCAP processed internally the inertial data giving as output a processed position of the neuron in the three-axis. An orthogonal conversion of these data with principal component analysis was applied to, finally, work out the velocity in 2D. These velocity profiles were analyzed with the iDeLog model [2] obtaining promising performances in terms of SNR: over 16dB and 25dB in both velocity and trajectory. Figure 1 describes one of our preliminary tests.

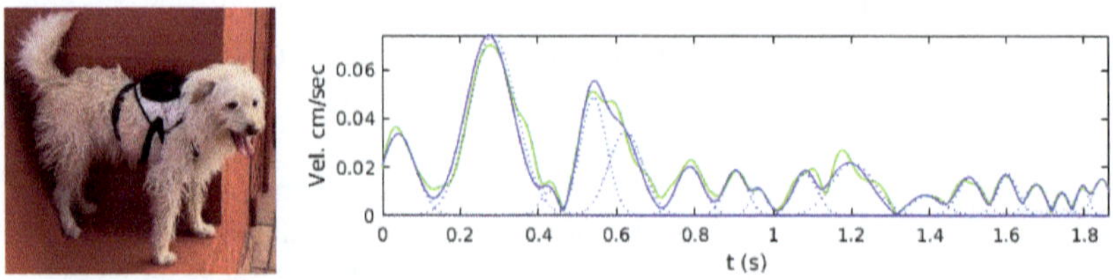


Fig. 1. Example of a proof of concept: dog with the inertial sensor inside the harness, *on the left,* and example of the original and reconstructed velocity profiles, *on the right*.

## 3. Discussion

The preliminary experiments in dogs suggest that the rapid animal movement can be modeled through the Kinematic Theory of rapid human movements. It is worth pointing out that our tests were carried out in healthy dogs. In this ongoing project, we aim to analyze animals' movement, both without clinical signs and animals with pain. We also plan to model their movements in three dimensions to explore additional information about their motor state.

## References

[1] R. Plamondon, "A kinematic theory of rapid human movements. Part I. Movement representation and generation," *Biol. Cybern.*, vol. 72, no. 4, pp. 309–320, 1995.

[2] M. A. Ferrer, M. Diaz, C. Carmona-Duarte, and R. Plamondon, "iDeLog: Iterative Dual Spatial and Kinematic Extraction of Sigma-Lognormal Parameters," *IEEE Trans. Pattern Anal. Mach. Intell.*, vol. 42, no. 1, pp. 114–125, Jan. 2020.

https://doi.org/10.1142/9789811226830_0017

# Appendix 14

# Biophysical Interpretations of the Proportionality Hypothesis

Ben Braithwaite[1,*], Réjean Plamondon[1,†] and Mickael Begon[2,‡]
[1]*Laboratoire Scribens, Département de génie électrique, Polytechnique Montréal, Montréal, Canada*
[2]*École de kinésiologie et des sciences de l'activité physique, Faculté de médecine, Université de Montréal*
*[*]ben.braithwaite@polymtl.ca, [†]rejean.plamondon@polymtl.ca,*
*[‡]mickael.begon@umontreal.ca*

## 1. Abstract

Grounding lognormal convergence in biological processes is useful for understanding why the Kinematic Theory works so well in describing human movements. That is why this study focuses on one of the main hypotheses of the kinematic theory: the proportionality of cumulative time delays. The logarithm of the cumulative time delay of the system converges towards the 1st moment of the impulse response, μ, which describes the time taken by the whole system to reach the median of its associated displacement. Finding physiologically plausible processes to provide an origin for this hypothesis is important, not only to further strengthen the lognormal model but also to create bridges between different types of muscular analyses.

The proportionality phenomenon, in its discrete form, appears in the series of the proportionality coefficients [Plamondon *et al.*, 2003]. A 1st order approximation is used in the proof when defining the convergence of the time delays and the simplified proportionality equation is completely analogous to a 1st ordinary differential equation (ODE): the problem and the solution structure are identical. This analogy is sound as discretization of the 1st order equation directly gives a proportionality relationship. This allows for the use of continuous phenomena for interpretation. We adopt 2 different approaches when it comes to interpretation: we can solve the equations for different potential known sequences of coefficients and interpret the behavior of the result and we can start from a well-known phenomenon in muscular biology and mathematize it in order to propose a relevant sequence of coefficients. So far, we have treated four types of proportionality coefficient sequences (constant coefficient, inverse, inverse power, inverse hexagonal numbers) and one muscle activation model [Millard *et al.*, 2013].

It is worth noting that the sequences will provide the cumulative time delay as a function of the number of sub-systems activated (*i.e.* muscle activation) whereas the muscle activation model provides activation as a function of time. Series of constant coefficients or inverse sequences both diverge when the number of sub-systems tends towards infinity. However, as the number of physiological sub-systems is finite, it is useful to study such series. Constant coefficients provide good results for a small number of sub-systems [Djioua and Plamondon, 2004] and the inverse sequence produces a phenomenon which propagates at a constant velocity, an adequate simplification for certain models. An inverse power sequence can be interpreted as the activation of fibres within a motor unit for instance and the exponent used can be understood as the fractal dimension of the motor unit. This is physiologically sound because larger motor units would have higher fractal dimensions. Furthermore, the convergence speed of the sequence is greater for higher powers. This means larger motor units are faster, which fits with the Henneman principle [Henneman, 1985]. The inverse hexagonal numbers series describes what would happen if the muscle fibres were distributed on a hexagonal matrix: the activation signal would be distributed to larger and larger hexagons. Interestingly, this sequence provides insight into structural organisation, in a similar way to the power sequence. Lastly, we chose to analyse the Millard muscle model as it uses a 1$^{st}$ order non-linear equation to describe the muscle activation dynamics separately from a biomechanical model which simulates force production. The model structure is similar to that of the Kinematic Theory insofar as it distinguishes between activation dynamics and mechanical properties. The analytical solution to the activation equation is a type of Lambert's W-function parameterized by motoneuronal stimulation. Its solution can be analytically approximated, in line with the Millard's model mean absolute errors ranging from 8% to 20.9%, depending on the biomechanical model used. We decided to produce an analytical solution with a maximum relative error of 20%. To determine the most relevant models, we will compare the relative errors of the five different models and then compare them to experimental data.

We have so far shown that as long as a 1$^{st}$ order process governs the activation properties of a complex system, then the system will converge towards a lognormal response. This allows for different interpretations of lognormal processes with continuous evolution models and not just discrete incremental growth. Future studies should focus on re-writing a series of 1st order muscle activation models based on the sequences that we proposed combined with an approximation of the Millard's model. This comparison will allow us to propose our most coherent interpretation of the proportionality hypothesis.

## References

Djioua, M., Plamondon, R. (2004). The Generation of Velocity Profiles With an Artificial Simulator. *IJPRAI*. 18. 1207–1219. 10.1142/S0218001404003721.

Henneman, E. (1985) The Size-Principle: A Deterministic Output Emerges From A Set Of Probabilistic Connections. *J. Exp. Biol.* 115, 105–112.

Millard, M., Uchida, T., Seth, A., Delp, S.L. (2013). Flexing Computational Muscle: Modeling and Simulation of Musculotendon Dynamics. *Journal of Biomechanical Engineering*. 135. 021005. 10.1115/1.4023390.

Plamondon, R., Feng, C. and Woch, A. *Biol. Cybern.* (2003) 89: 126. https://doi.org/10.1007/s00422-003-0407-9.

https://doi.org/10.1142/9789811226830_0017

# Appendix 15

# Applying the Delta-Lognormal Model to Simple 3D Hand Movements

Saad Chidami*, Mylène Archambault-Caron† and Réjean Plamondon‡
*Laboratoire Scribens, Département de génie électrique,*
*Polytechnique Montréal, Montréal, Canada*
**saad.chidami@polymtl,* †*mylene.archambault-caron@polymtl.ca,*
‡*rejean.plamondon@polymtl.ca*

## 1. Abstract

In this study, we are focusing on the analysis of the human upper limbs and record trajectories of short-hand movements using a target whose grip handling is similar to a pencil. As we are investigating the Delta-Lognormal model in 3D, all the recorded movements are similar to rapid pencil strokes in 3D space.

A Type II Chebyshev Filter of a 11th-order with 16 Hz cut-off frequency was applied to the x, y and z coordinates extracted from the recording system. The tangential velocity is calculated after preprocessing data with a derivative filter. Although the trajectory of the movement is slightly curved, we assume that this curvature is an intrinsic property of the human neuro-motor response. We also presume the origin of this curvature as the consequence of the wrist rotation or the elbow rotation for short and rapid hand movement. From this last hypothesis, we assume that every 3D short and fast hand stroke is in fact a limb restricted 2D movement on a 3D coordinates plane. To find the least errors plane approximation, we use a Principal Component Analysis (PCA), which allows us to extract the best projection plan for each hand movement trajectory.

The rotations around the z axis and the x axis of the plane make it possible to obtain a new trajectory parallel to the xy plane. The new coordinates of the trajectory can be considered as 2D since we can neglect the z coordinates because they are constant. The new transformed 2D coordinates are used with the starting and ending correction angles to reconstruct the signal. To do so, the starting angle is extracted from the first half sampling points of the trajectory (removing a first portion that could correspond to a break time). A linear regression is applied on this portion of trajectory and the start angle is calculated as the angle between the regression line and the axis. The ending angle is measured from the second half of the trajectory, the final portion is removed and is considered as the resting time. As for the starting angle, a linear regression is applied on this portion of the 2D trajectory and the ending angle is calculated as the angle between the regression line and the u axis. The resulting signal is reconstructed, using a 3D Delta-Lognormal model and the IIX algorithm [Djioua and Plamondon, 2009].

The following figures present a typical reconstruction result.

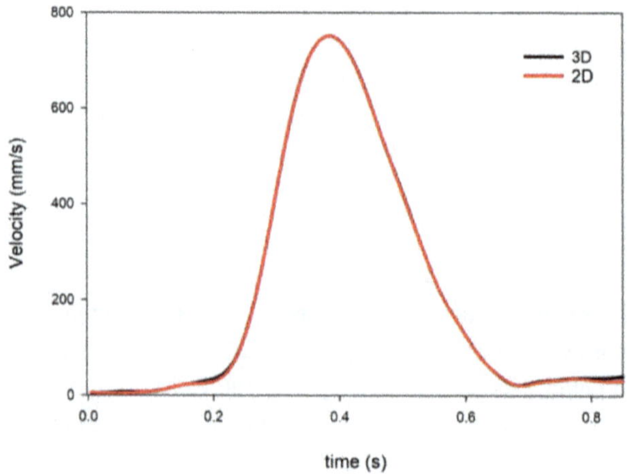


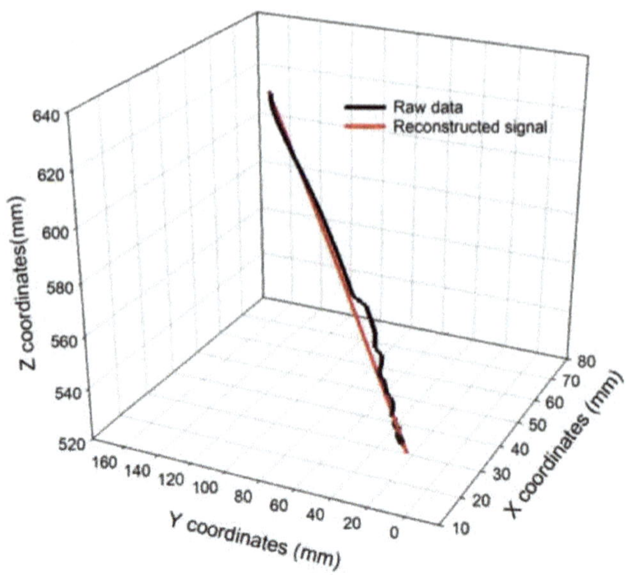

Our preliminary tests show that the Delta-Lognormal model can be applied to simple 3D hand movement. It is also concluded that the transformation of 3D to 2D by using a PCA to project the trajectory on a plane does not alter significantly the tangential velocity. The SNR values (ranging from 26 to 28 dB) obtained from this 2.5D method are equivalent to standard 2D stroke analysis. Moreover, there are no statistical differences between the extraction of Delta-Lognormal parameters from the 3D data or the 2D transformed data. However, it was observed, when analyzing highly curved rapid hand stroke that the parameter extraction from the tangential velocity was giving a high SNR, but the reconstruction was incorrect. A full 3D reconstruction will be needed to process these general cases.

## References

Chidami, S., Archambault-Caron, M., Plamondon, R. (2018). The Delta- Lognormal model in 2.5D. *Proc. ICPRAI 2018, First International Workshop on the Lognormality Principle and its Applications*, Montréal, Canada (743–747).

M. Djioua, R. Plamondon, (2009). A new algorithm and system for the characterization of handwriting strokes with delta-lognormal parameter, *IEEE Trans. on PAMI*, vol. 31, no. 11, pp. 2060–2072.

https://doi.org/10.1142/9789811226830_0017

# Appendix 16

# 3D iDeLog: Sigma-Lognormal Model for 3D Human Movements

Miguel A. Ferrer[1,*], Moises Diaz[2,†], Cristina Carmona[1,‡], Jose J. Quintana[1,§] and Réjean Plamondon[3,**]

[1]*Universidad de Las Palmas de Gran Canaria, Spain*

[2]*Universidad del Atlántico Medio, Las Palmas de Gran Canaria, 35017, Spain*

[3]*Laboratoire Scribens, Département de génie électrique, Polytechnique Montréal, Montréal PQ, Canada*

[*]*miguelangel.ferrer@ulpgc.es*, [†]*moisdc@gmail.com*, [‡]*criscarmonad@gmail.com*, [§]*jose.quintana@ulpgc.es*, [**]*rejean.plamondon@polymtl.ca*

This appendix introduces an extension of the 2D iDeLog model to approach 3D trajectories. The iDeLog model is based on the Kinematic Theory of rapid movements and its associate Sigma-Lognormal model. Preliminary results with public databases of 3D movements encourage the ongoing project.

## 1. Introduction

The Kinematic Theory of Rapid Movements and its associate Sigma-Lognormal model have been traditionally applied to model 2D trajectories. The advent of high-precision and low-cost inertial measurement units integrated into personal devices make feasible remote monitoring of three-dimensional movements. It opens the opportunity for more in-depth analysis and understanding of human movement. As such, we have developed 3D iDeLog, which is an extension of the iDeLog model, to decomposes three-dimensional spatiotemporal trajectories [1].

## 2. Work Description

3D iDeLog model is based on the Kinematic Theory of Rapid Movements and its Lognormality Principle. It refers that a stroke links two virtual target points with a spatial circumference drawn from its pivot in a lognormal temporal scale. Then, the Sigma-Lognormal rule of the temporal overlapping of strokes is used to build a complex trajectory [1].

In this way, the 3D iDeLog model three-dimensional spatio-temporal sequences as an overlapping of circumferences. As a circumference is a planar trajectory, the 3D iDeLog model generates 3D sequences by overlapping circumferences in different planes. In each plane, the lognormal parameters are obtained similarly to the 2D method. The virtual target points and start and end angles are reversely projected from the circumference plane into 3D as three-dimensional virtual target points coordinates and starting and ending polar and azimuth angles: $\theta_s$, $\theta_e$, $\varphi_s$ and $\varphi_e$.

## 3. Discussion

We followed the evaluation proposed in [2] to evaluate the 3D iDeLog ability of modelling three-dimensional sequences. Table 1 shows preliminary results in terms of SNR for both trajectory and velocity in several databases. As can be seen, all performances are over 15 dB, giving promising expectations about the 3D iDeLog model. Moreover, an example of a reconstructed trajectory and velocity is displayed in Fig. 1.

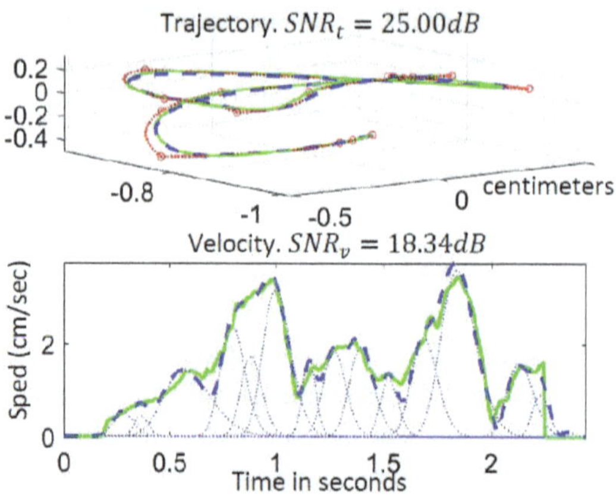


Fig. 1. Original (green continuous line) and 3D iDeLog (dashed blue line) reconstructed trajectory and velocity for a sample of Hdm05.

Table 1. 3D iDeLog Performance with HDm05, UTKinect and Airwriting databases.

| Database | $SNR_t$(dB) | $SNR_t/N$ | $SNR_v$dB | $SNR_v/N$ |
|---|---|---|---|---|
| Hdm05 | 24.25 | 1.892 | 18.22 | 1.547 |
| UTKinect | 22.05 | 4.083 | 17.68 | 3.367 |
| Airwriting | 23.66 | 0.296 | 15.42 | 0.199 |

## References

[1] M. A. Ferrer, M. Diaz, C. Carmona-Duarte, and R. Plamondon, "iDeLog: Iterative Dual Spatial and Kinematic Extraction of Sigma-Lognormal Parameters," *IEEE Trans. Pattern Anal. Mach. Intell.*, vol. 42, no. 1, pp. 114–125, Jan. 2020.

[2] R. Schindler, M. Bouillon, R. Plamondon and A. Fischer, "Extending the Sigma-Lognormal Model of the Kinematic Theory to Three Dimensions," in International Conference on Pattern Recognition and Artificial Intelligence, *Proceedings of ICPRAI 2018. First International Workshop on the Lognormality Principle and its Applications*, Montréal, Canada, pp. 748–752.

https://doi.org/10.1142/9789811226830_bmatter

# Index

Lightning Source UK Ltd.
Milton Keynes UK
UKHW021016181120
373615UK00001B/51

9 789811 226823